genetics

genetics

VOLUME **3**
K–P

Richard Robinson

**MACMILLAN
REFERENCE
USA™**

THOMSON
★
GALE

New York • Detroit • San Diego • San Francisco • Cleveland • New Haven, Conn. • Waterville, Maine • London • Munich

Genetics
Richard Robinson

Volume ISBN Numbers
0-02-865607-5 (Volume 1)
0-02-865608-3 (Volume 2)
0-02-865609-1 (Volume 3)
0-02-865610-5 (Volume 4)

LIBRARY OF CONGRESS CATALOGING- IN-PUBLICATION DATA

Genetics / Richard Robinson, editor in chief.
     p. ; cm.
Includes bibliographical references and index.
   ISBN 0-02-865606-7 (set : hd.)
   1. Genetics—Encyclopedias.
   [DNLM: 1. Genetics—Encyclopedias—English. 2. Genetic Diseases, Inborn—Encyclopedias—English. 3. Genetic Techniques—Encyclopedias—English. 4. Molecular Biology—Encyclopedias—English. QH 427 G328 2003] I. Robinson, Richard, 1956–
   QH427 .G46 2003
   576'.03—dc21

          2002003560

Printed in Canada
10 9 8 7 6 5 4 3 2 1

For Your Reference

The following section provides a group of diagrams and illustrations applicable to many entries in this encyclopedia. The molecular structures of DNA and RNA are provided in detail in several different formats, to help the student understand the structures and visualize how these molecules combine and interact. The full set of human chromosomes are presented diagrammatically, each of which is shown with a representative few of the hundreds or thousands of genes it carries.

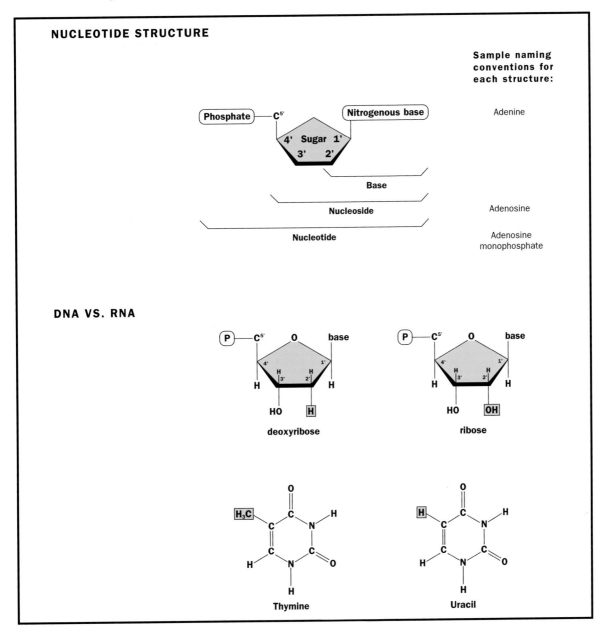

NUCLEOTIDE STRUCTURE

DNA VS. RNA

NUCLEOTIDE STRUCTURES

Purine-containing DNA nucleotides

Adenine

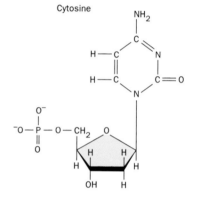

Pyrimidine-containing DNA nucleotides

Thymine

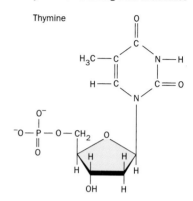

Guanine

Cytosine

CANONICAL B-DNA DOUBLE HELIX

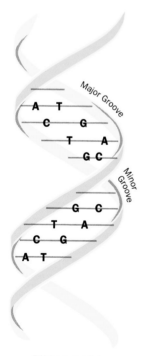

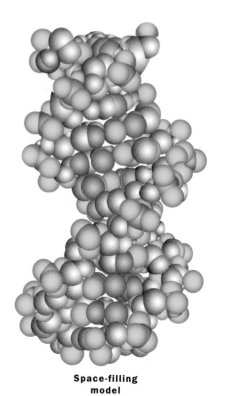

Ribbon model

Ball-and-stick model

Space-filling model

DNA NUCLEOTIDES PAIR UP ACROSS THE DOUBLE HELIX; THE TWO STRANDS RUN ANTI-PARALLEL

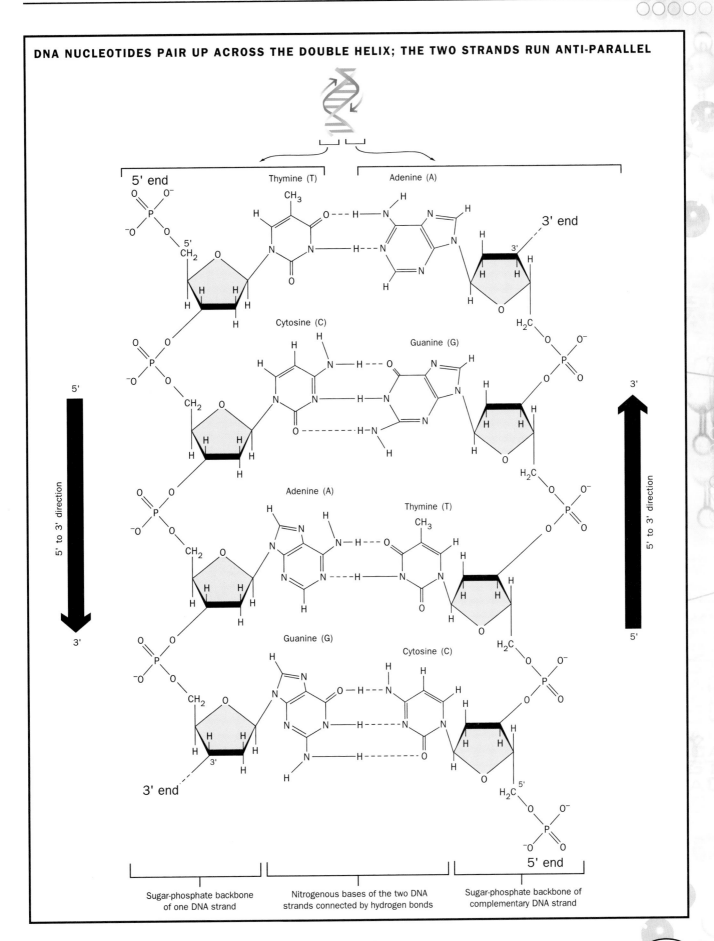

Sugar-phosphate backbone of one DNA strand

Nitrogenous bases of the two DNA strands connected by hydrogen bonds

Sugar-phosphate backbone of complementary DNA strand

SELECTED LANDMARKS OF THE HUMAN GENOME

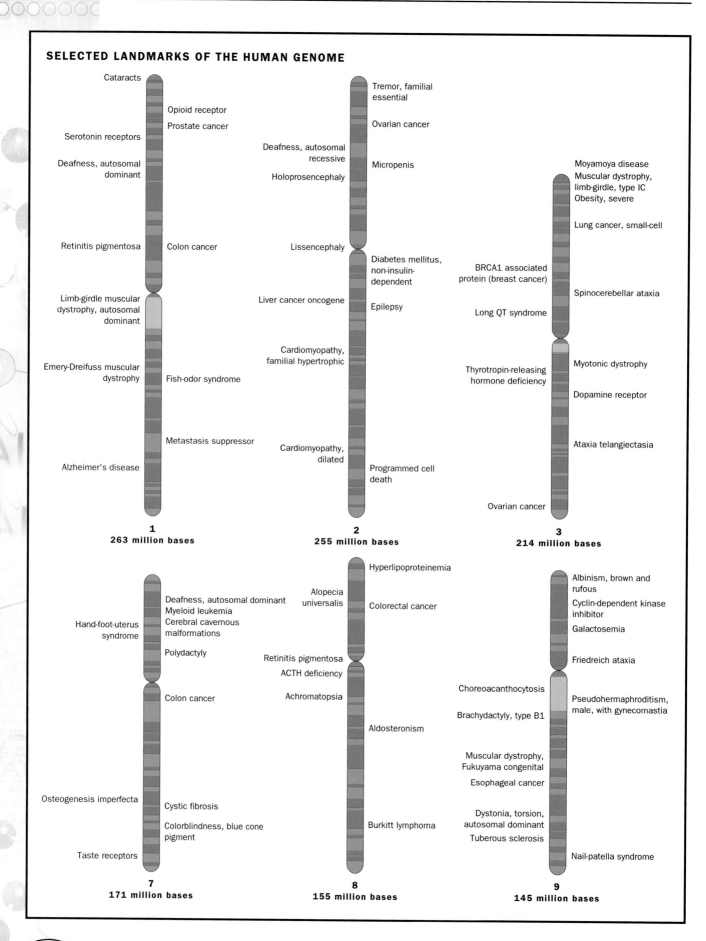

Cataracts
Opioid receptor
Prostate cancer
Serotonin receptors
Deafness, autosomal dominant
Retinitis pigmentosa
Colon cancer
Limb-girdle muscular dystrophy, autosomal dominant
Emery-Dreifuss muscular dystrophy
Fish-odor syndrome
Metastasis suppressor
Alzheimer's disease

1
263 million bases

Tremor, familial essential
Ovarian cancer
Deafness, autosomal recessive
Micropenis
Holoprosencephaly
Lissencephaly
Diabetes mellitus, non-insulin-dependent
Liver cancer oncogene
Epilepsy
Cardiomyopathy, familial hypertrophic
Cardiomyopathy, dilated
Programmed cell death

2
255 million bases

Moyamoya disease
Muscular dystrophy, limb-girdle, type IC
Obesity, severe
Lung cancer, small-cell
BRCA1 associated protein (breast cancer)
Spinocerebellar ataxia
Long QT syndrome
Thyrotropin-releasing hormone deficiency
Myotonic dystrophy
Dopamine receptor
Ataxia telangiectasia
Ovarian cancer

3
214 million bases

Deafness, autosomal dominant
Myeloid leukemia
Cerebral cavernous malformations
Hand-foot-uterus syndrome
Polydactyly
Colon cancer
Osteogenesis imperfecta
Cystic fibrosis
Colorblindness, blue cone pigment
Taste receptors

7
171 million bases

Hyperlipoproteinemia
Alopecia universalis
Colorectal cancer
Retinitis pigmentosa
ACTH deficiency
Achromatopsia
Aldosteronism
Burkitt lymphoma

8
155 million bases

Albinism, brown and rufous
Cyclin-dependent kinase inhibitor
Galactosemia
Friedreich ataxia
Choreoacanthocytosis
Pseudohermaphroditism, male, with gynecomastia
Brachydactyly, type B1
Muscular dystrophy, Fukuyama congenital
Esophageal cancer
Dystonia, torsion, autosomal dominant
Tuberous sclerosis
Nail-patella syndrome

9
145 million bases

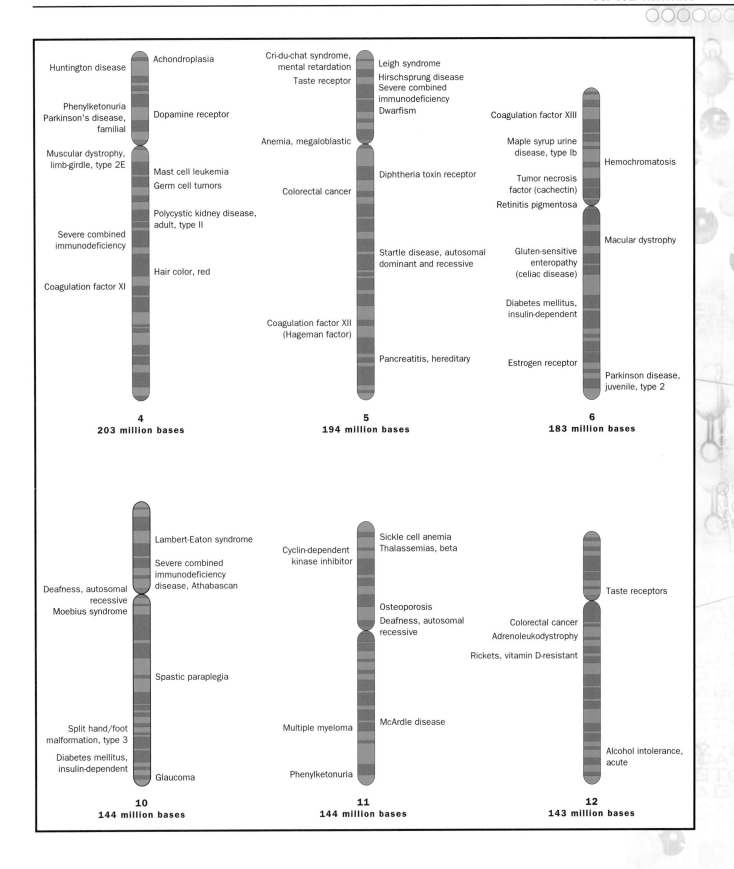

Huntington disease

Achondroplasia

Phenylketonuria
Parkinson's disease, familial

Dopamine receptor

Muscular dystrophy, limb-girdle, type 2E

Mast cell leukemia
Germ cell tumors

Polycystic kidney disease, adult, type II

Severe combined immunodeficiency

Hair color, red

Coagulation factor XI

4
203 million bases

Cri-du-chat syndrome, mental retardation
Taste receptor

Leigh syndrome
Hirschsprung disease
Severe combined immunodeficiency
Dwarfism

Anemia, megaloblastic

Diphtheria toxin receptor

Colorectal cancer

Startle disease, autosomal dominant and recessive

Coagulation factor XII (Hageman factor)

Pancreatitis, hereditary

5
194 million bases

Coagulation factor XIII

Maple syrup urine disease, type Ib

Tumor necrosis factor (cachectin)

Retinitis pigmentosa

Hemochromatosis

Macular dystrophy

Gluten-sensitive enteropathy (celiac disease)

Diabetes mellitus, insulin-dependent

Estrogen receptor

Parkinson disease, juvenile, type 2

6
183 million bases

Lambert-Eaton syndrome

Severe combined immunodeficiency disease, Athabascan

Deafness, autosomal recessive
Moebius syndrome

Spastic paraplegia

Split hand/foot malformation, type 3
Diabetes mellitus, insulin-dependent

Glaucoma

10
144 million bases

Cyclin-dependent kinase inhibitor

Sickle cell anemia
Thalassemias, beta

Osteoporosis
Deafness, autosomal recessive

Multiple myeloma

McArdle disease

Phenylketonuria

11
144 million bases

Taste receptors

Colorectal cancer
Adrenoleukodystrophy
Rickets, vitamin D-resistant

Alcohol intolerance, acute

12
143 million bases

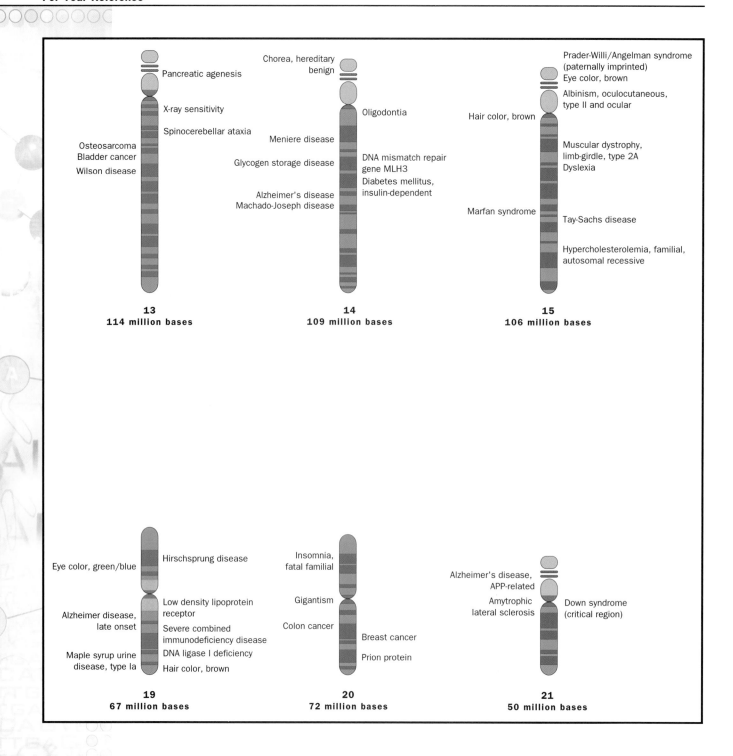

13
114 million bases

- Pancreatic agenesis
- X-ray sensitivity
- Spinocerebellar ataxia
- Osteosarcoma
- Bladder cancer
- Wilson disease

14
109 million bases

- Chorea, hereditary benign
- Oligodontia
- Meniere disease
- Glycogen storage disease
- DNA mismatch repair gene MLH3
- Diabetes mellitus, insulin-dependent
- Alzheimer's disease
- Machado-Joseph disease

15
106 million bases

- Prader-Willi/Angelman syndrome (paternally imprinted)
- Eye color, brown
- Albinism, oculocutaneous, type II and ocular
- Hair color, brown
- Muscular dystrophy, limb-girdle, type 2A
- Dyslexia
- Marfan syndrome
- Tay-Sachs disease
- Hypercholesterolemia, familial, autosomal recessive

19
67 million bases

- Eye color, green/blue
- Hirschsprung disease
- Low density lipoprotein receptor
- Alzheimer disease, late onset
- Severe combined immunodeficiency disease
- DNA ligase I deficiency
- Maple syrup urine disease, type Ia
- Hair color, brown

20
72 million bases

- Insomnia, fatal familial
- Gigantism
- Colon cancer
- Breast cancer
- Prion protein

21
50 million bases

- Alzheimer's disease, APP-related
- Amytrophic lateral sclerosis
- Down syndrome (critical region)

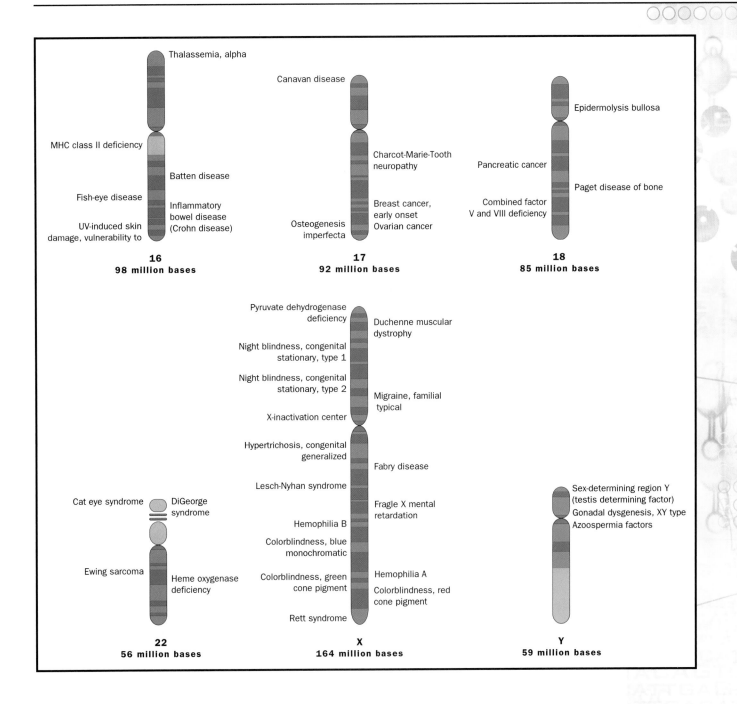

Thalassemia, alpha

MHC class II deficiency

Batten disease

Fish-eye disease

Inflammatory
bowel disease
(Crohn disease)

UV-induced skin
damage, vulnerability to

16
98 million bases

Canavan disease

Charcot-Marie-Tooth
neuropathy

Breast cancer,
early onset
Ovarian cancer

Osteogenesis
imperfecta

17
92 million bases

Epidermolysis bullosa

Pancreatic cancer

Paget disease of bone

Combined factor
V and VIII deficiency

18
85 million bases

Pyruvate dehydrogenase
deficiency

Duchenne muscular
dystrophy

Night blindness, congenital
stationary, type 1

Night blindness, congenital
stationary, type 2

Migraine, familial
typical

X-inactivation center

Hypertrichosis, congenital
generalized

Fabry disease

Lesch-Nyhan syndrome

Fragle X mental
retardation

Hemophilia B

Colorblindness, blue
monochromatic

Colorblindness, green
cone pigment

Hemophilia A

Colorblindness, red
cone pigment

Rett syndrome

X
164 million bases

Cat eye syndrome

DiGeorge
syndrome

Ewing sarcoma

Heme oxygenase
deficiency

22
56 million bases

Sex-determining region Y
(testis determining factor)
Gonadal dysgenesis, XY type
Azoospermia factors

Y
59 million bases

Contributors

Eric Aamodt
Louisiana State University Health Sciences Center, Shreveport
Gene Expression: Overview of Control

Maria Cristina Abilock
Applied Biosystems
Automated Sequencer
Cycle Sequencing
Protein Sequencing
Sequencing DNA

Ruth Abramson
University of South Carolina School of Medicine
Intelligence
Psychiatric Disorders
Sexual Orientation

Stanley Ambrose
University of Illinois
Population Bottleneck

Allison Ashley-Koch
Duke Center for Human Genetics
Disease, Genetics of
Fragile X Syndrome
Geneticist

David T. Auble
University of Virginia Health System
Transcription

Bruce Barshop
University of California, San Diego
Metabolic Disease

Mark A. Batzer
Louisiana State University
Pseudogenes
Repetitive DNA Elements
Transposable Genetic Elements

Robert C. Baumiller
Xavier University
Reproductive Technology
Reproductive Technology: Ethical Issues

Mary Beckman
Idaho Falls, Idaho
DNA Profiling
HIV

Samuel E. Bennett
Oregon State University Department of Genetics
DNA Repair
Laboratory Technician
Molecular Biologist

Andrea Bernasconi
Cambridge University, U.K.
Multiple Alleles
Nondisjunction

C. William Birky, Jr.
University of Arizona
Inheritance, Extranuclear

Joanna Bloom
New York University Medical Center
Cell Cycle

Deborah Blum
University of Wisconsin, Madison
Science Writer

Bruce Blumberg
University of California, Irvine
Hormonal Regulation

Suzanne Bradshaw
University of Cincinnati
Transgenic Animals
Yeast

Carolyn J. Brown
University of British Columbia
Mosaicism

Michael J. Bumbulis
Baldwin-Wallace College
Blotting

Michael Buratovich
Spring Arbor College
Operon

Elof Carlson
The State Universtiy of New York, Stony Brook
Chromosomal Theory of Inheritance, History
Gene
Muller, Hermann
Polyploidy
Selection

Regina Carney
Duke University
College Professor

Shu G. Chen
Case Western Reserve University
Prion

Gwen V. Childs
University of Arkansas for Medical Sciences
In situ Hybridization

Cindy T. Christen
Iowa State University
Technical Writer

Patricia L. Clark
University of Notre Dame
Chaperones

Steven S. Clark
University of Wisconsin
Oncogenes

Nathaniel Comfort
George Washington University
McClintock, Barbara

P. Michael Conneally
Indiana University School of Medicine
Blood Type
Epistasis
Heterozygote Advantage

Howard Cooke
Western General Hospital: MRC Human Genetics Unit
Chromosomes, Artificial

Denise E. Costich
Boyce Thompson Institute
Maize

Terri Creeden
March of Dimes
Birth Defects

Kenneth W. Culver
Novartis Pharmaceuticals Corporation
Genomics
Genomics Industry
Pharmaceutical Scientist

Mary B. Daly
Fox Chase Cancer Center
Breast Cancer

Pieter de Haseth
Case Western Reserve University
Transcription

Rob DeSalle
American Museum of Natural History
Conservation Geneticist
Conservation Biology: Genetic Approaches

Elizabeth A. De Stasio
Lawerence University
Cloning Organisms

Danielle M. Dick
Indiana University
Behavior

Michael Dietrich
Dartmouth College
Nature of the Gene, History

Christine M. Disteche
University of Washington
X Chromosome

Gregory Michael Dorr
University of Alabama
Eugenics

Jennie Dusheck
Santa Cruz, California
Population Genetics

Susanne D. Dyby
U.S. Department of Agriculture: Center for Medical, Agricultural, and Veterinary Entomology
Classical Hybrid Genetics
Mendelian Genetics
Pleiotropy

Barbara Emberson Soots
Folsom, California
Agricultural Biotechnology

Susan E. Estabrooks
Duke Center for Human Genetics
Fertilization
Genetic Counselor
Genetic Testing

Stephen V. Faraone
Harvard Medical School
Attention Deficit Hyperactivity Disorder

Gerald L. Feldman
Wayne State University Center for Molecular Medicine and Genetics
Down Syndrome

Linnea Fletcher
Bio-Link South Central Regional Coordinater, Austin Community College
Educator
Gel Electrophoresis
Marker Systems
Plasmid

Michael Fossel
Executive Director, American Aging Association
Accelerated Aging: Progeria

Carol L. Freund
National Institute of Health: Warren G. Magnuson Clinical Center
Genetic Testing: Ethical Issues

Joseph G. Gall
Carnegie Institution
Centromere

Darrell R. Galloway
The Ohio State University
DNA Vaccines

Pierluigi Gambetti
Case Western Reserve University
Prion

Robert F. Garry
Tulane University School of Medicine
Retrovirus
Virus

Perry Craig Gaskell, Jr.
Duke Center for Human Genetics
Alzheimer's Disease

Theresa Geiman
National Institute of Health: Laboratory of Receptor Biology and Gene Expression
Methylation

Seth G. N. Grant
University of Edinburgh
Embryonic Stem Cells
Gene Targeting
Rodent Models

Roy A. Gravel
University of Calgary
Tay-Sachs Disease

Nancy S. Green
March of Dimes
Birth Defects

Wayne W. Grody
UCLA School of Medicine
Cystic Fibrosis

Charles J. Grossman
Xavier University
Reproductive Technology
Reproductive Technology: Ethical Issues

Cynthia Guidi
University of Massachusetts Medical School
Chromosome, Eukaryotic

Patrick G. Guilfoile
Bemidji State University
DNA Footprinting
Microbiologist
Recombinant DNA
Restriction Enzymes

Richard Haas
University of California Medical Center
Mitochondrial Diseases

William J. Hagan
College of St. Rose
Evolution, Molecular

Jonathan L. Haines
Vanderbilt University Medical Center
Complex Traits
Human Disease Genes, Identification of

Mapping
McKusick, Victor

Michael A. Hauser
Duke Center for Human Genetics
DNA Microarrays
Gene Therapy

Leonard Hayflick
University of California
Telomere

Shaun Heaphy
University of Leicester, U.K.
Viroids and Virusoids

John Heddle
York University
Mutagenesis
Mutation
Mutation Rate

William Horton
Shriners Hospital for Children
Growth Disorders

Brian Hoyle
Square Rainbow Limited
Overlapping Genes

Anthony N. Imbalzano
University of Massachusetts Medical School
Chromosome, Eukaryotic

Nandita Jha
University of California, Los Angeles
Triplet Repeat Disease

John R. Jungck
Beloit College
Gene Families

Richard Karp
Department of Biological Sciences, University of Cincinnati
Transplantation

David H. Kass
Eastern Michigan University
Pseudogenes
Transposable Genetic Elements

Michael L. Kochman
University of Pennsylvania Cancer Center
Colon Cancer

Bill Kraus
Duke University Medical Center
Cardiovascular Disease

Steven Krawiec
Lehigh University
Genome

Mark A. Labow
Novartis Pharmaceuticals Corporation
Genomics
Genomics Industry
Pharmaceutical Scientist

Ricki Lewis
McGraw-Hill Higher Education; The Scientist
Bioremediation
Biotechnology: Ethical Issues
Cloning: Ethical Issues

Genetically Modified Foods
Plant Genetic Engineer
Prenatal Diagnosis
Transgenic Organisms: Ethical
Issues

Lasse Lindahl
University of Maryland, Baltimore
Ribozyme
RNA

David E. Loren
*University of Pennsylvania School of
Medicine*
Colon Cancer

Dennis N. Luck
Oberlin College
Biotechnology

Jeanne M. Lusher
*Wayne State University School of
Medicine; Children's Hospital of
Michigan*
Hemophilia

Kamrin T. MacKnight
*Medlen, Carroll, LLP: Patent,
Trademark and Copyright Attorneys*
Attorney
Legal Issues
Patenting Genes
Privacy

Jarema Malicki
Harvard Medical School
Zebrafish

Eden R. Martin
Duke Center for Human Genetics
Founder Effect
Inbreeding

William Mattox
*University of Texas/Anderson
Cancer Center*
Sex Determination

Brent McCown
University of Wisconsin
Transgenic Plants

Elizabeth C. Melvin
Duke Center for Human Genetics
Gene Therapy: Ethical Issues
Pedigree

Ralph R. Meyer
University of Cincinnati
Biotechnology and Genetic Engi-
neering, History of
Chromosome, Eukaryotic
Genetic Code
Human Genome Project

Kenneth V. Mills
College of the Holy Cross
Post-translational Control

Jason H. Moore
Vanderbilt University Medical School
Quantitative Traits
Statistical Geneticist
Statistics

Dale Mosbaugh
*Oregon State University: Center for
Gene Research and Biotechnology*

DNA Repair
Laboratory Technician
Molecular Biologist

Paul J. Muhlrad
University of Arizona
Alternative Splicing
Apoptosis
Arabidopsis thaliana
Cloning Genes
Combinatorial Chemistry
Fruit Fly: *Drosophila*
Internet
Model Organisms
Pharmacogenetics and Pharma-
cogenomics
Polymerase Chain Reaction

Cynthia A. Needham
*Boston University School of
Medicine*
Archaea
Conjugation
Transgenic Microorganisms

R. John Nelson
University of Victoria
Balanced Polymorphism
Gene Flow
Genetic Drift
Polymorphisms
Speciation

Carol S. Newlon
*University of Medicine and
Dentistry of New Jersey*
Replication

Sophia A. Oliveria
*Duke University Center for Human
Genetics*
Gene Discovery

Richard A. Padgett
Lerner Research Institute
RNA Processing

Michele Pagano
*New York University Medical
Center*
Cell Cycle

Rebecca Pearlman
Johns Hopkins University
Probability

Fred W. Perrino
*Wake Forest University School of
Medicine*
DNA Polymerases
Nucleases
Nucleotide

David Pimentel
*Cornell University: College of
Agriculture and Life Sciences*
Biopesticides

Toni I. Pollin
*University of Maryland School of
Medicine*
Diabetes

Sandra G. Porter
Geospiza, Inc.
Homology

Eric A. Postel
Duke University Medical Center
Color Vision
Eye Color

Prema Rapuri
Creighton University
HPLC: High-Performance Liq-
uid Chromatography

Anthony J. Recupero
Gene Logic
Bioinformatics
Biotechnology Entrepreneur
Proteomics

Diane C. Rein
BioComm Consultants
Clinical Geneticist
Nucleus
Roundworm: *Caenorhabditis ele-
gans*
Severe Combined Immune Defi-
ciency

Jacqueline Bebout Rimmler
Duke Center for Human Genetics
Chromosomal Aberrations

Keith Robertson
*Epigenetic Gene Regulation and
Cancer Institute*
Methylation

Richard Robinson
Tucson, Arizona
Androgen Insensitivity Syndrome
Antisense Nucleotides
Cell, Eukaryotic
Crick, Francis
Delbrück, Max
Development, Genetic Control of
DNA Structure and Function,
History
Eubacteria
Evolution of Genes
Hardy-Weinberg Equilibrium
High-Throughput Screening
Immune System Genetics
Imprinting
Inheritance Patterns
Mass Spectrometry
Mendel, Gregor
Molecular Anthropology
Morgan, Thomas Hunt
Mutagen
Purification of DNA
RNA Interference
RNA Polymerases
Transcription Factors
Twins
Watson, James

Richard J. Rose
Indiana University
Behavior

Howard C. Rosenbaum
*Science Resource Center, Wildlife
Conservation Society*
Conservation Geneticist
Conservation Biology: Genetic
Approaches

Astrid M. Roy-Engel
Tulane University Health Sciences Center
Repetitive DNA Elements

Joellen M. Schildkraut
Duke University Medical Center
Public Health, Genetic Techniques in

Silke Schmidt
Duke Center for Human Genetics
Meiosis
Mitosis

David A. Scicchitano
New York University
Ames Test
Carcinogens

William K. Scott
Duke Center for Human Genetics
Aging and Life Span
Epidemiologist
Gene and Environment

Gerry Shaw
MacKnight Brain Institute of the University of Flordia
Signal Transduction

Alan R. Shuldiner
University of Maryland School of Medicine
Diabetes

Richard R. Sinden
Institute for Biosciences and Technology: Center for Genome Research
DNA

Paul K. Small
Eureka College
Antibiotic Resistance
Proteins
Reading Frame

Marcy C. Speer
Duke Center for Human Genetics
Crossing Over
Founder Effect
Inbreeding
Individual Genetic Variation
Linkage and Recombination

Jeffrey M. Stajich
Duke Center for Human Genetics
Muscular Dystrophy

Judith E. Stenger
Duke Center for Human Genetics
Computational Biologist
Information Systems Manager

Frank H. Stephenson
Applied Biosystems
Automated Sequencer
Cycle Sequencing
Protein Sequencing
Sequencing DNA

Gregory Stewart
State University of West Georgia
Transduction
Transformation

Douglas J. C. Strathdee
University of Edinburgh
Embryonic Stem Cells
Gene Targeting
Rodent Models

Jeremy Sugarman
Duke University Department of Medicine
Genetic Testing: Ethical Issues

Caroline M. Tanner
Parkinson's Institute
Twins

Alice Telesnitsky
University of Michigan
Reverse Transcriptase

Daniel J. Tomso
National Institute of Environmental Health Sciences
DNA Libraries
Escherichia coli
Genetics

Angela Trepanier
Wayne State University Genetic Counseling Graduate Program
Down Syndrome

Peter A. Underhill
Stanford University
Y Chromosome

Joelle van der Walt
Duke University Center for Human Genetics
Genotype and Phenotype

Jeffery M. Vance
Duke University Center for Human Genetics

Gene Discovery
Genomic Medicine
Genotype and Phenotype
Sanger, Fred

Gail Vance
Indiana University
Chromosomal Banding

Jeffrey T. Villinski
University of Texas/MD Anderson Cancer Center
Sex Determination

Sue Wallace
Santa Rosa, California
Hemoglobinopathies

Giles Watts
Children's Hospital Boston
Cancer
Tumor Suppressor Genes

Kirk Wilhelmsen
Ernest Gallo Clinic & Research Center
Addiction

Michelle P. Winn
Duke University Medical Center
Physician Scientist

Chantelle Wolpert
Duke University Center for Human Genetics
Genetic Counseling
Genetic Discrimination
Nomenclature
Population Screening

Harry H. Wright
University of South Carolina School of Medicine
Intelligence
Psychiatric Disorders
Sexual Orientation

Janice Zengel
University of Maryland, Baltimore
Ribosome
Translation

Stephan Zweifel
Carleton College
Mitochondrial Genome

Table of Contents

VOLUME 1

PREFACE v

FOR YOUR REFERENCE ix

LIST OF CONTRIBUTORS xvii

A

Accelerated Aging: Progeria 1

Addiction 4

Aging and Life Span 6

Agricultural Biotechnology 9

Alternative Splicing 11

Alzheimer's Disease 14

Ames Test 19

Androgen Insensitivity Syndrome 21

Antibiotic Resistance 26

Antisense Nucleotides 29

Apoptosis 31

Arabidopsis thaliana 33

Archaea 36

Attention Deficit Hyperactivity Disorder   39

Attorney 42

Automated Sequencer 43

B

Balanced Polymorphism 45

Behavior 46

Bioinformatics 52

Biopesticides 57

Bioremediation 59

Biotechnology 62

Biotechnology Entrepreneur 65

Biotechnology: Ethical Issues 66

Biotechnology and Genetic Engineering,
    History 70

Birth Defects 74

Blood Type 82

Blotting 86

Breast Cancer 89

C

Cancer 92

Carcinogens 97

Cardiovascular Disease 101

Cell Cycle 103

Cell, Eukaryotic 108

Centromere 114

Chaperones 116

Chromosomal Aberrations 119

Chromosomal Banding 125

Chromosomal Theory of Inheritance,
    History 129

Chromosome, Eukaryotic 132

Chromosome, Prokaryotic 139

Chromosomes, Artificial 144

Classical Hybrid Genetics 146

Clinical Geneticist 149

Cloning Genes 152

Cloning: Ethical Issues 158

Cloning Organisms 161

College Professor 165

Colon Cancer 166

Color Vision 170

Combinatorial Chemistry 173

Complex Traits 177

Computational Biologist 181

Conjugation 182

Conservation Biology: Genetic
    Approaches 186

Conservation Geneticist 190

Crick, Francis 192

Crossing Over 194

Cycle Sequencing 198
Cystic Fibrosis 199

D

Delbrück, Max 203
Development, Genetic Control of 204
Diabetes 209
Disease, Genetics of 213
DNA 215
DNA Footprinting 220
DNA Libraries 222
DNA Microarrays 225
DNA Polymerases 230
DNA Profiling 233
DNA Repair 239
DNA Structure and Function, History . 248
DNA Vaccines 253
Down Syndrome 256

PHOTO CREDITS 259
GLOSSARY 263
TOPICAL OUTLINE 281
INDEX 287

VOLUME 2

FOR YOUR REFERENCE v
LIST OF CONTRIBUTORS xiii

E

Educator 1
Embryonic Stem Cells 3
Epidemiologist 6
Epistasis 7
Escherichia coli (*E. coli* bacterium) 9
Eubacteria 11
Eugenics 16
Evolution, Molecular 21
Evolution of Genes 26
Eye Color 31

F

Fertilization 33
Founder Effect 36
Fragile X Syndrome 39
Fruit Fly: *Drosophila* 42

G

Gel Electrophoresis 45

Gene 50
Gene and Environment 54
Gene Discovery 57
Gene Expression: Overview of Control .. 61
Gene Families 67
Gene Flow 70
Gene Targeting 71
Gene Therapy 74
Gene Therapy: Ethical Issues 80
Genetic Code 83
Genetic Counseling 87
Genetic Counselor 91
Genetic Discrimination 92
Genetic Drift 94
Genetic Testing 96
Genetic Testing: Ethical Issues 101
Genetically Modified Foods 106
Geneticist 110
Genetics 111
Genome 112
Genomic Medicine 118
Genomics 120
Genomics Industry 123
Genotype and Phenotype 125
Growth Disorders 129

H

Hardy-Weinberg Equilibrium 133
Hemoglobinopathies 136
Hemophilia 141
Heterozygote Advantage 146
High-Throughput Screening 149
HIV 150
Homology 156
Hormonal Regulation 158
HPLC: High-Performance Liquid
   Chromatography 165
Human Disease Genes, Identification of . 167
Human Genome Project 171
Human Immunodeficiency Virus 178
Huntington's Disease 178
Hybrid Superiority 178

I

Immune System Genetics 178
Imprinting 183
In situ Hybridization 186
Inbreeding 189

Individual Genetic Variation 191
Information Systems Manager 192
Inheritance, Extranuclear 194
Inheritance Patterns 199
Intelligence 207
Internet 211

PHOTO CREDITS 215
GLOSSARY 219
TOPICAL OUTLINE 237
INDEX 243

VOLUME 3

FOR YOUR REFERENCE v
LIST OF CONTRIBUTORS xiii

L
Laboratory Technician 1
Legal Issues 3
Linkage and Recombination 4

M
Maize 8
Mapping 11
Marker Systems 15
Mass Spectrometry 18
McClintock, Barbara 21
McKusick, Victor 22
Meiosis 24
Mendel, Gregor 30
Mendelian Genetics 32
Metabolic Disease 37
Methylation 46
Microbiologist 50
Mitochondrial Diseases 51
Mitochondrial Genome 55
Mitosis 57
Model Organisms 60
Molecular Anthropology 62
Molecular Biologist 70
Morgan, Thomas Hunt 72
Mosaicism 76
Muller, Hermann 80
Multiple Alleles 82
Muscular Dystrophy 83
Mutagen 87
Mutagenesis 89

Mutation 93
Mutation Rate 98

N
Nature of the Gene, History 101
Nomenclature 106
Nondisjunction 108
Nucleases 112
Nucleotide 115
Nucleus 119

O
Oncogenes 127
Operon 131
Overlapping Genes 135

P
Patenting Genes 136
Pedigree 138
Pharmaceutical Scientist 142
Pharmacogenetics and
    Pharmacogenomics 144
Physician Scientist 147
Plant Genetic Engineer 149
Plasmid 150
Pleiotropy 153
Polymerase Chain Reaction 154
Polymorphisms 159
Polyploidy 163
Population Bottleneck 167
Population Genetics 171
Population Screening 175
Post-translational Control 178
Prenatal Diagnosis 182
Prion 187
Privacy 190
Probability 193
Protein Sequencing 196
Proteins 198
Proteomics 205
Pseudogenes 209
Psychiatric Disorders 213
Public Health, Genetic Techniques in .. 216
Purification of DNA 220

PHOTO CREDITS 223
GLOSSARY 227
TOPICAL OUTLINE 245
INDEX 251

VOLUME 4

FOR YOUR REFERENCE v

LIST OF CONTRIBUTORS xiii

Q

Quantitative Traits 1

R

Reading Frame 4
Recombinant DNA 5
Repetitive DNA Sequences 7
Replication 12
Reproductive Technology 19
Reproductive Technology: Ethical Issues . 26
Restriction Enzymes 31
Retrovirus 34
Reverse Transcriptase 39
Ribosome 42
Ribozyme 44
RNA 46
RNA Interference 54
RNA Processing 57
Rodent Models 60
Roundworm: *Caenorhabditis elegans* 62

S

Sanger, Fred 64
Science Writer 65
Selection 67
Sequencing DNA 69
Severe Combined Immune Deficiency ... 74
Sex Determination 78
Sexual Orientation 83
Signal Transduction 85
Speciation 91
Statistical Geneticist 93
Statistics 95

T

Tay-Sachs Disease 98
Technical Writer 102
Telomere 104
Transcription 106
Transcription Factors 112
Transduction 117
Transformation 121
Transgenic Animals 124
Transgenic Microorganisms 127
Transgenic Organisms: Ethical Issues .. 129
Transgenic Plants 132
Translation 135
Transplantation 139
Transposable Genetic Elements 143
Triplet Repeat Disease 148
Tumor Suppressor Genes 153
Twins 155

V

Viroids and Virusoids 162
Virus 164

W

Watson, James 171

X

X Chromosome 173

Y

Y Chromosome 176
Yeast 179

Z

Zebrafish 181

PHOTO CREDITS 185
GLOSSARY 189
TOPICAL OUTLINE 207
CUMULATIVE INDEX 213

genetics

Karyotype *See Chromosomal Banding; Nomenclature*

Knockout *See Gene Targeting*

Laboratory Technician

The technician in a molecular biology laboratory is a resourceful scientist who specializes in the various experimental techniques critical to the mission of the laboratory. The work offers many rewards beyond the financial ones. For example, one of the rewards of working as a laboratory technician lies in being part of a team dedicated to scientific discovery. Another rewarding aspect of the laboratory technician position is the need for continual learning, as new scientific techniques replace older ones.

Skills of the Laboratory Technician

Technicians must possess a variety of skills, depending on the work being done in the laboratory in which they work. For example, a technician in a laboratory that studies human genetic **polymorphisms** will be skilled in the techniques of DNA isolation and DNA sequencing. DNA may be isolated from cultured human cells, which the technician would grow, or from tissue biopsies or blood samples. The technician must exercise great care to prevent accidental contamination of the samples.

polymorphisms DNA sequence variants

Following (chemical) extraction of DNA from the material, the technician will **amplify** the region of the gene under investigation in an enzyme-catalyzed DNA sequencing reaction. The reaction products, which are pieces of DNA of various lengths, are loaded by the technician onto a thin gel and separated from one another according to size by applying an electric current to the gel. A computer-controlled laser excites fluorescent dye molecules that the technician has chemically attached to the DNA, and a photodetector records the color. The laboratory technician will operate the DNA sequencing machine, supervise the electronic data collection, and, in general, assure the accuracy of the DNA sequence obtained. Good communication skills are an important part of a technician's qualifications, as is the ability to follow (and give) instructions correctly.

amplify produce many copies of

In 2000, at Genetic ID, the nation's largest genetic testing lab in Fairfield, Iowa, a technician uses the triple check method to narrow results.

In a protein structure laboratory, on the other hand, the technician must have expertise in protein purification. Since relatively large amounts of protein may be required for structure determination, the gene encoding the protein may be placed into bacteria using recombinant DNA techniques. The bacteria can then be induced to produce (express) significant quantities of the protein. The technician is in charge of growing the bacteria and seeing whether the protein is properly expressed. Next, the technician breaks open the bacterial cells by mechanical means, and purifies (separates) the protein of interest away from contaminating proteins and nucleic acids using a series of chromatographic techniques. The technician must know how to detect and quantify proteins and **enzymes**.

enzymes proteins that control a reaction in a cell

Qualifications and Compensation

Often, a laboratory technician can demonstrate expertise in a wide variety of experimental techniques. This type of employee is highly sought after by the pharmaceutical and biotechnology industries. It is generally expected that a technician in a molecular biology work setting will have a bachelor's or master's degree in biology, biochemistry, or chemistry.

While many employers are willing to provide on-the-job training for laboratory technicians, a well-qualified technician will have at least some experience with techniques commonly used in biochemistry, genetics, and cell biology.

A laboratory technician who has just received a bachelor's or master's degree may receive an annual salary of $18,000 to $36,000 or more, depending on the employer (industry, academia, or government), the geographical location, and the supply of and demand for qualified technicians. Salaries increase with experience, technical expertise, and responsibility, particularly in industry, where many opportunities exist for the laboratory technician to climb the career ladder.

Samuel E. Bennett and Dale Mosbaugh

Legal Issues

The question of who owns tissues, DNA, and other biological materials raises numerous legal questions. One concern is that genetic information derived from someone's DNA sequences could be used to deny insurance coverage to people whose genes indicate that they have a disease or that they are at risk of contracting one.

Another concern is that the profits made by hospitals and transplant centers for transplantation procedures are unfair, as tissue donors and their families are typically not compensated, despite the fact that these donors often pay for the operations that provide the materials. There is a question of who should profit from such materials: those from whom the materials were originally derived or those who use the materials to treat other patients or conduct research.

In the criminal setting, genetic testing provides the opportunity to identify criminals. Through storage of **DNA** and DNA analysis data, old, unsolved cases can sometimes be resolved. DNA analysis is also useful for exonerating wrongly accused individuals, including those who have served significant jail time for crimes they did not commit. However, there is concern regarding the potential abuses of DNA data stored by law enforcement agencies. There is also concern that stored genetic material will be used to clone people. Additional concerns center on the safety and risks of genetically modified foods.

DNA deoxyribonucleic acid

Ownership of Tissues

The issue of ownership of tissues was addressed in California in the case *Moore* v. *Regents of the University of California*. Moore underwent treatment for leukemia at the University of California at Los Angeles Medical Center. His spleen was removed, and his cells were cultured without his consent. Eventually, he sued the medical center over the ownership of the cell line that was developed from his spleen cells.

The California Supreme Court refused to recognize Moore's ownership of the cell line, pointing to the investment the medical center made to develop it. The court did not place much weight on financial or other contributions Moore made to the development of the cell line.

The court indicated that recognizing a patient's right to own such cells would chill medical research, as scientists would be required to determine the originators of each cell culture they use. Because of the large number of cell cultures used, such a requirement would be burdensome and expensive, and it would potentially halt important research, the court said. The court also noted that researchers establishing cell lines are increasingly using contracts to clarify patent and ownership rights, though in Moore's case none was signed.

Criminal Law

DNA testing has proven to be a very valuable tool for convicting criminals, as well as for exonerating falsely accused individuals. The methods used to analyze DNA, as well as the implications of the results of such analyses, are still being standardized and are almost always questioned in court by at least one party, but they are becoming increasingly accepted and refined for use in criminal law.

Despite the usefulness of genetic testing, there are various concerns about privacy and the potential for discrimination. There are also some concerns about the consequences if insurance companies, employers, or other entities have access to such personal data.

Patenting Issues

Genetic material obtained from individuals is often used in developing patentable inventions. These patents are filed by the scientists who develop the materials and methods that utilize the genetic information. Cells obtained from a person with a rare disease, for example, might be used to develop tests to detect the disease as well as methods and materials for treatment. The patent rights are granted to the scientists who develop the tests, methods, and materials, rather than to the patient who was the source of the cells.

A patient might consider negotiating for an ownership interest in the cells. But very few patients are in a position to do so. They often are afraid that such negotiations would result in denial of treatment. Also, since it is illegal in the United States to sell organs and tissues, agreements involving ownership in such cases could be seen as falling afoul of the law. SEE ALSO DNA PROFILING; GENETIC COUNSELING; GENETIC DISCRIMINATION; GENETIC TESTING; GENETIC TESTING: ETHICAL ISSUES; PATENTING GENES; PRIVACY.

Kamrin T. MacKnight

Bibliography

Lewis, Ricki. *Human Genetics: Concepts and Applications*, 4th ed. Boston: McGraw Hill, 2001.

Internet Resource

Human Genome Project Information: Ethical, Legal, and Social Issues. U.S. Department of Energy. <http://www.ornl.gov/hgmis/elsi/elsi.html>.

Linkage and Recombination

Linkage refers to the association and co-inheritance of two DNA segments because they reside close together on the same chromosome. Recombination is the process by which they become separated during crossing over,

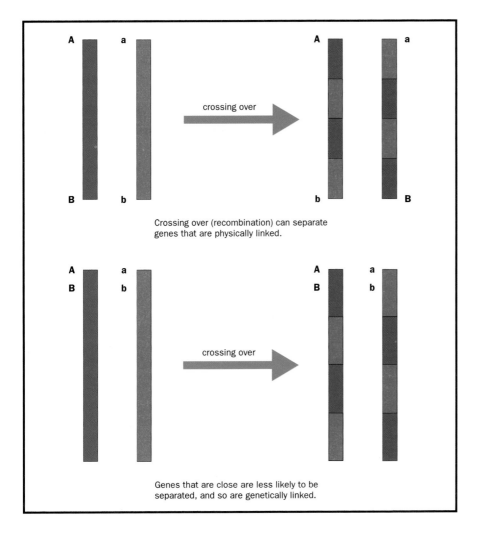

crossing over

Crossing over (recombination) can separate genes that are physically linked.

crossing over

Genes that are close are less likely to be separated, and so are genetically linked.

Physical linkage of genes simply means they are on the same chromosome. To be genetically linked, a pair of genes must be close enough that they are unlikely to be separated by crossing over.

which occurs during **meiosis**. The existence of linkage and the frequency of recombination allow chromosomes to be mapped to determine the relative positions and distances of the genes and other DNA sequences on them. Linkage analysis is also a key tool for discovering the location and ultimate identity of genes for inherited diseases.

meiosis cell division that forms eggs or sperm

Basic Concepts

Each individual inherits a complete set of twenty-three chromosomes from each parent, and chromosomes are therefore present in **homologous** pairs. The members of a pair carry the same set of genes at the same positions, or **loci**. The two genes at a particular locus may be identical, or slightly different. The different forms of a gene are called alleles.

homologous posessing the same set of genes

loci sites on a chromosome (singular, locus)

Genes or loci can be linked either physically or genetically. Genes that are physically linked are on the same chromosome and are thus syntenic. Only syntenic genes can be genetically linked. Genes that are linked genetically are physically close enough to one another that they do not segregate independently during meiosis.

Understanding independent segregation is crucial to understanding linkage. Independent segregation was first discovered by Gregor Mendel, who found that, in pea plants, the different forms of two traits found in the

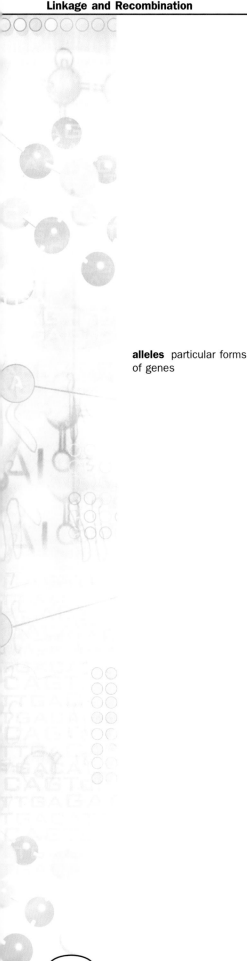

parents, such as color and height, could occur in all possible combinations in the offspring. Thus, a tall parent with green pods crossed with a short parent with yellow pods could give rise to offspring that were tall with yellow pods or short with green pods, as well as some of each parental type. Mendel concluded that the factors controlling height segregated independently from the factors controlling pod color. Later work showed that this was because these genes occurred on separate (nonhomologous) chromosomes, which themselves segregate independently during meiosis.

How is it possible for physically linked genes to nonetheless segregate independently? The answer lies in the events of crossing over. During crossing over, homologous chromosomes exchange segments at several sites along their length, in a process called recombination. Thus, two loci at distant ends of the chromosome are almost certain to have at least one exchange point occur between them. If only one exchange occurs, two alleles that began on the same chromosome will end up on different chromosomes. If there are two exchange points between them, they will end up together; if three, they end up apart, and so on. Over long distances, the likelihood of two **alleles** remaining together is only 50 percent, no better than chance, and, therefore, loci that are far apart on a large chromosome are not genetically linked. Conversely, loci that are close together will not segregate independently, and are therefore genetically linked. It is these that are most useful for mapping and discovering disease genes.

alleles particular forms of genes

The loci examined in linkage analysis need not be genes of functional significance; indeed, anonymous segments of DNA (stretches of DNA with no known function) called genetic markers are often more useful in genetic linkage analysis. In order for a genetic marker to be of benefit in a linkage analysis, the chromosomal location of the marker must be known and, most importantly, there must be some variation in the sequence or length of these markers among individuals. Nongene markers used in linkage analysis are classified into four broad categories: restriction fragment length polymorphisms (RFLPs), variable number of tandem repeat (VNTRs), short tandem repeat polymorphisms (STRPs), and single nucleotide repeats (SNPs).

Calculating Linkage and Map Distance

As noted above, when genes are not genetically linked, alleles at the loci segregate independently from one another. So, if locus 1 has alleles A and a, and if locus 2, not linked to locus 1, has alleles B and b, then four gametes can be formed (AB, Ab, aB, and ab). Each of these four will occur with equal frequency (a 1:1:1:1 ratio), and all possible offspring combinations are expected with equal frequency.

If locus 1 and locus 2 are genetically linked to one another, however, deviations from this 1:1:1:1 ratio will be observed. If A and B begin on the same chromosome, then AB and ab will be more common than either aB or Ab. By counting the number of each type and determining the extent of this deviation, one can estimate the extent of recombination between the two loci: A large deviation means little recombination. The "recombination fraction," expressed as a percentage, is an indirect measure of the distance between the loci and is the basis for the development of genetic maps.

Genetic maps order polymorphic markers by specifying the amount of recombination between markers, whereas physical maps quantify the distances among markers in terms of the number of base pairs of DNA. Although mapping in humans has a relatively recent history, the idea of a linear arrangement of genes on a chromosome was first proposed in 1911 by Thomas Hunt Morgan, who was studying the fruit fly, *Drosophila melanogaster*. The possibility of a genetic map was first formally investigated by the American geneticist Alfred H. Sturtevant in the 1930s, who determined the order of five markers on the X chromosome in *D. melanogaster* and then estimated the relative spacing among them.

For small recombination fractions (usually less than 10 percent to 12 percent), the estimate of the recombination fraction provides a very rough estimate of the physical distance. In general, 1 percent recombination is equivalent to about one million base pairs of DNA and is defined as one centimorgan. Physical measurements of DNA are often described in terms of thousands of **kilobases**. Crossing over does not occur equally at all locations, so estimates of distance from physical and genetic maps of the identical region may vary dramatically throughout the genome.

kilobases units of measure of the length of a nucleicacid chain; one kilobase is equal to 1,000 base pairs

Statistical Approaches

In experimental organisms, genetic mapping of loci involves counting the number of recombinant and nonrecombinant offspring of selected matings. Genetic mapping in humans is usually more complicated than in experimental organisms for many reasons, including researchers' inability to design specific matings of individuals, which limits the unequivocal assignment of recombinants and nonrecombinants. Therefore, maps of markers in humans are developed by means of one of several statistical algorithms used in computer programs.

Genetic maps can assume equal recombination between males and females, or they can allow for sex-specific differences in recombination, since it has been well established that there are substantial differences in recombination frequencies between men and women. Chromosomes recombine more often in females. On average, the female map is two times as long as the male map.

The complexity of the underlying statistical methods used to generate them renders genetic maps sensitive to marker genotyping errors, particularly in small intervals, and these maps are less useful in regions of less than about 2 centimorgans. While marker order is usually correct, genotyping errors can result in falsely inflated estimates of map distances.

Disease gene mapping is greatly facilitated by the availability of dense genetic maps. Linkage analysis for the mapping of disease genes boils down to the simple idea of counting recombinants and nonrecombinants, but in humans this process is complicated for a variety of reasons. The generation time is long in humans, so large, multigenerational **pedigrees** in which a disease or trait is segregating are rare. Scientists cannot dictate matings or exposures. They also cannot require that specific individuals participate in a study. Thus the process of linkage analysis in humans requires a statistical framework in which various **hypotheses** about the linkage of a trait locus and marker locus can be considered.

pedigrees sets of related individuals, or the graphic representation of their relationships

hypotheses testable statements

How far apart are the disease and marker, and how certain is the conclusion of linkage?

When the inheritance pattern for a disease is clearly known (e.g., autosomal dominant, sex-linked, etc.), the genetic data can be treated with a statistical approach that determines the likelihood that the gene is linked to a particular marker, at a particular position on a specific chromosome. This approach is often termed the "lod score approach," where "lod" is short for logarithm of the odds.

Lod score linkage analysis is used most frequently to consider diseases that follow a Mendelian pattern of transmission within families. Positive lod scores, especially those greater then 3.0, suggest evidence for linkage between a disease gene and a marker locus. Negative lod scores suggest that the disease gene and marker locus are unlinked to one another. SEE ALSO CROSSING OVER; GENE DISCOVERY; HUMAN DISEASE GENES, IDENTIFICATION OF; MAPPING; MEIOSIS; MORGAN, THOMAS HUNT; POLYMORPHISMS.

Marcy C. Speer

Bibliography

Strachan, Tom, and Andrew P. Read. *Human Molecular Genetics.* New York: Wiley-Liss Publishers, 1999.

Maize

Maize (*Zea mays* L.), otherwise known as corn, is a highly unusual, economically important, and genetically well-characterized member of the grass family. It is believed to have originated some 8,000 to 10,000 years ago in the fields of the first agriculturalists of Mexico and Central America. These early farmers carefully selected traits that would ultimately transform the tiny, sparsely seeded spike of a wild grass into the large cob bearing many rows of kernels that we recognize today as an ear of corn.

The success of these early plant breeders was manifested by the spread of corn cultivation throughout the New World, long before the arrival of Europeans. Today, maize is grown in more countries than any other crop, and is a major source of food for both humans and domesticated animals throughout the world. The world production of maize in 2000 exceeded 23 billion bushels, the largest producer being the United States (43 percent).

Early Studies of Maize

As a major crop plant, maize was already the subject of study by plant breeders at the time of the rediscovery of Mendel's laws of inheritance at the beginning of the twentieth century. The inheritance patterns of readily observed traits were uncovered through controlled crosses and the examination of progeny. In many respects, maize was an ideal model system for this early period in the study of genetics. Male and female flowers are borne separately and are easily manipulated for controlled crosses. Large amounts of pollen are produced in the tassels (male inflorescence) over a period of days, and one ear (female inflorescence) contains many seeds (kernels). Large progeny arrays could be produced in one season.

Mutations caused differences in these ears of corn stored in the Maize Genetics Cooperation Stock Center, a repository for mutated maize located at the University of Illinois, Urbana-Champaign.

The high genetic diversity of maize provided many interesting mutant **phenotypes** to study, many of which were recessive. These could be maintained in a **heterozygous** state by the outcrossed breeding system (most fertilizations are the result of pollen transfer among plants) and easily uncovered by selfing (fertilizations that result from a plant's own pollen). There was also ample scope for selection of extreme phenotypes in continuous (quantitative) traits. A drawback for maize, compared to short-lived fruit flies, is that it only produces one or two crops per year, depending on location. However, many early maize geneticists knew that kernel phenotypes, which were discernable at harvest time, often predicted phenotypes in the adult plants, and could be used to set up the following season's crosses.

phenotypes observable characteristics of an organism

heterozygous characterized by possession of two different forms (alleles) of a particular gene

One of the earliest breakthroughs in crop breeding was the detection of hybrid vigor in maize by George Harrison Shull in 1908. He found that the progeny of two inbred lines were more productive than their wind-pollinated progenitors. This discovery provided the stimulus for the commercial propagation of maize and made it one of the most productive food plants worldwide.

Later Maize Studies

Many important genetic discoveries were made in maize by a group of scientists brought together at Cornell University in the 1920s and 30s by R. A. Emerson, who is often referred to as the spiritual father of maize genetics. The Emerson group, which included the future Nobel laureates Barbara McClintock and George Beadle, laid the foundation of maize genetics. They assembled information on maize mutants and ultimately produced the first genetic map of maize, based on linkage studies, in 1935. McClintock's first major contribution occurred early in her career (1929), when she perfected the techniques used to visualize maize chromosomes under the microscope. This allowed individual chromosomes to be identified by size, form, and features such as the highly staining regions, called "knobs."

cytogenetics study of chromosome structure and behavior

trisomics mutants with one extra chromosome

This milestone allowed McClintock and other members of the Emerson group to make major advances in **cytogenetics**, which combines genetic crossing data and cytological landmarks to locate genes on chromosomes. Cytological landmarks include **trisomics**, reciprocal translocations, and deficiencies. Another of McClintock's breakthroughs, achieved with the collaboration of her colleague Harriet Creighton, was to establish the cytological proof of crossing over, which refers to the exchange of chromosomal segments during meiosis. Of course, McClintock's most famous discovery was that genetic elements within the genome can move (transpose) from one locus on the chromosome to another. These "jumping genes" (transposable genetic elements of transposons) were later discovered in bacteria, flies, and humans and eventually resulted in McClintock receiving a Nobel Prize in 1983.

In recent years, transposable elements have been exploited as tools for understanding the function of many maize genes. If a transposon inserts into a gene, it will disrupt the function of that gene. The disruption of gene function may result in a mutant phenotype affecting tissues or developmental stages of the plant that give some indication of the function of that gene. For instance, a transposon that inserts into a gene required for chlorophyll production would result in an albino seedling. Because the DNA sequences of many transposable elements in maize are known, they provide convenient molecular tags with which to clone and further characterize the gene into which they have inserted. Corn transposons have also been adapted to mutagenize and "tag" genes in the model plant *Arabidopsis thaliana*. SEE ALSO *ARABIDOPSIS THALIANA*; CROSSING OVER; HETEROZYGOTE ADVANTAGE; MCCLINTOCK, BARBARA; MODEL ORGANISMS; TRANSPOSABLE GENETIC ELEMENTS.

Denise E. Costich

Bibliography

Dold, Catherine. "The Corn War." *Discover* (December 1997): 109–113.

Fedoroff, Nina, and David Botstein, eds. *The Dynamic Genome: Barbara McClintock's Ideas in the Century of Genetics.* Cold Spring Harbor, NY: Cold Spring Harbor Laboratory Press, 1992.

Kass, Lee B. "Barbara McClintock: American Botanical Geneticist (1902–1992)." In *Plant Sciences for Students*, vol. 3. New York: Macmillan Publishing, 2000.

Keller, Evelyn Fox. *A Feeling for the Organism: The Life and Work of Barbara McClintock.* San Francisco: W. H. Freeman, 1983.

Rhoades, Marcus M. "The Early Years of Maize Genetics." *Annual Review of Genetics* 18 (1984): 1–29.

Mapping

Genetic mapping is the process of measuring the distance between two or more **loci** on a chromosome. In order to determine this distance, a number of things must be done. First, the loci (pronounced "low-sigh") have to be known, and **alleles** have to exist at each locus so that they can be observed. The specific pair of alleles that are present is usually referred to as a genotype. Second, there has to be a way to measure the distance between the loci.

loci sites on a chromosome (singular, locus)

alleles particular forms of genes

In genetic mapping, this distance is measured by the amount of meiotic recombination that occurs between the two loci. Meiotic recombination is the process in which the two chromosomes that are paired during meiosis each break apart and then reattach to each other, rather than back to themselves. These recombined chromosomes will end up in either eggs (for women) or sperm (for men).

Typically for any chromosome pair there will be only one or two such breaks per chromosome arm. The closer together two loci are, the less likely it is that such a break will occur. Thus, counting the number of breaks between two loci provides a good estimate of how far apart two loci are.

Genetic maps provide the order and distance between many markers all along the chromosome. In genetic maps, the loci that are used are called marker loci. Marker loci are almost always not in genes and serve only as signposts along the chromosome, "marking" a specific location. Thus genetic maps act much like road maps, and markers act much like mile markers or exit signs.

Why Create and Use Maps?

Genetic maps contain very important information and are used to help find the genes that can cause, or change the risk of developing, genetic diseases. For most diseases, the gene is not yet known and could be any one of the 30,000 to 70,000 genes that exist in the human genome. Since the disease gene is not known, its location is also not known. However, if the general location could be determined, then it would be much easier to figure out which of the genes near that location are the actual disease genes.

Genetic maps are very important for "disease-gene discovery," as they provide the reference locations for locating the disease gene. Finding the disease genes without a genetic map would be like trying to find a town by driving down a road without any mile markers or exit signs. There would be no clues as to where you are. The maps make it much easier to "navigate" the chromosomes.

Using Recombination and Map Functions

Genetic maps are created by measuring the amount of recombination that occurs between two or more loci. The easiest way to do this is to use families with a large number of children, since this provides a large number of recombination events to look at. Scientists have collected a panel of forty such families, called the CEPH families (pronounced "sef," from the French Centre d'Étude du Polymorphisme Humain—the Center for the Study of Human Polymorphisms). These families are measured (genotyped) for the variations at each locus, and the inheritance of each allele at each locus is compared.

An example of a CEPH family is shown in Figure 1. Using the father as an example (although in other families this could easily occur in the mother), allele a at locus 1 and allele b at locus 2 are always inherited together. Similarly, allele A at locus 1 and allele B at locus 2 are inherited together. There has been no recombination between locus 1 and locus 2, and therefore these loci are likely to be close together. In contrast, allele a at locus 1 and allele c at locus 3 are only inherited together half the time. There have been several recombination events between them, and therefore these loci are likely to be far apart.

The actual distance between two loci is measured using the recombination fraction, which is just the number of recombination events divided by the total number of events that are looked at. In the family diagrammed in the figure, the recombination fraction between locus 1 and locus 2 is 0 recombination events divided by 8 total events, or $0 \div 8 = 0$. The recombination fraction between locus 1 and locus 3 is 4 recombination events divided by 8 total events, or $4 \div 8 = 0.50$. Recombination fractions can vary between 0.00 and 0.50. To generate a complete genetic map of a chromosome, a large number of markers (between 50 and 200, depending on the size of the chromosome) are genotyped in many families, and more complex statistical analyses are used to compare the inheritance across all markers.

There is an additional complication in the analysis of recombination events. The further apart two loci are, the more likely it is that two recombination events could occur between them. The first event will shuffle the alleles, but the second event will reshuffle the alleles back to the way they were. Thus it will look like there were no recombination events when in fact there were two.

Another complication arises from the fact that the occurrence of one recombination event on a chromosome tends to inhibit the occurrence of a second recombination event, especially in regions close to the first one. This is called "interference" and will generally make the map smaller. To account for this, "map functions" have been created that are used to better estimate the true recombination distance between two markers.

Map functions are mathematical equations that are based on assumptions about how much recombination and how much interference exists on a chromosome. Map function distances are measured in units called centimorgans, named for Thomas Hunt Morgan, the first person to develop the techniques of genetic mapping. There are several map functions that have been proposed. Each is named for its originator. The most commonly used

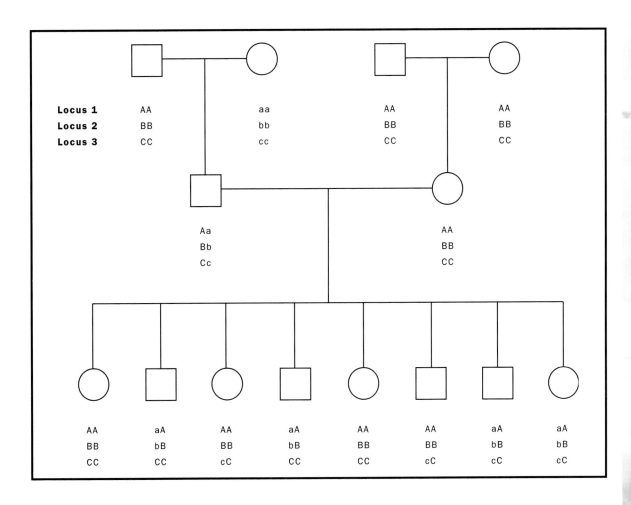

map function is the Haldane map function (named after John Burdon Sanderson Haldane), which assumes that there is no interference between loci. A second map function, the Kosambi map function (named after Damodar Kosambi), assumes a moderate level of interference and seems to more accurately reflect experimental data. Thus the recombination fraction is modified by the map function. Generally the recombination fraction and the centimorgans are very similar for distances from 0.00 to 0.10.

Figure 1. A typical CEPH family with genotypes from three genetic markers. Loci one and two are completely linked to each other, while loci one and three are not linked to each other.

Types of Markers, and Their Advantages and Disadvantages

There are four major kinds of genetic markers that have been used for genetic mapping. The oldest of these is the restriction fragment length polymorphism (RFLP) that was first proposed for genetic mapping in 1980. RFLPs arise from changes in a single **base pair** that can be detected by restriction endonuclease **enzymes**. These enzymes can cut the DNA at that locus if the right base pair is present. Many maps were made with these markers, but they are expensive and time-consuming to genotype, and they generally have only two alleles. Having only two alleles means that in many cases it is impossible to tell the two chromosomes in any person apart for that marker and makes that marker useless for genetic mapping in that family. In the figure, the mother of the eight children has the same alleles at locus 1, the same alleles at locus 2, and the same alleles at locus 3. Thus we cannot tell if there have been any recombination events coming from the

base pair two nucleotides (either DNA or RNA) linked by weak bonds

enzymes proteins that control a reaction in a cell

mother. RFLPs were the first type of marker known to occur almost everywhere, across all the chromosomes.

Variable number of tandem repeat (VNTR) markers were the next markers to be described. These result from the duplication of DNA sequences consisting of 50 to 5,000 base pairs each. The differences between the two **homologous** chromosomes are in the number of repeats present (and thus the length of the locus). These markers are expensive and time consuming to genotype but have the advantage of having many alleles (often more than twenty). Thus almost everyone in the world has a different allele on each paired chromosome at a VNTR locus. This allows more families to give recombination information. Having so many alleles, however, can cause problems, because it can be hard to tell many of the alleles apart during genotyping. VNTRs also tend to occur most often at the ends of chromosomes, not in the middle. This is unlike RFLPs, which occur at all locations on a chromosome.

Microsatellite markers—also known as simple tandem repeat polymorphisms (STRPs), simple sequence repeats (SSRs), or simple sequence length polymorphisms (SSLPs)—have become the most common type of marker for genetic maps. These markers are made of repeats of two, three, or four base pairs, with the variation being the number of repeats. For example, the most commonly used two-base-pair repeat is CA, and the most commonly used four-base-pair repeat is GATA. Thus a microsatellite marker actually varies in length between the paired chromosomes. On one chromosome, there might be eight repeats (CACACACACACACACA), while on the other chromosome there might be ten (CACACACACACACACACACA). Microsatellite markers are easy to genotype and have multiple (three to ten) but usually not large numbers (more than ten) of alleles. They also occur almost everywhere across the chromosome. Most of the genetic maps in use today are made with microsatellite markers.

The most recently described type of marker is the single nucleotide polymorphism (SNP, pronounced "snip"). As the name implies, these are variations at a single base on the chromosome. For example, on some chromosomes a locus might have a C, while on other chromosomes the same locus might have a T. These are the most common markers, with at least three million already described, and seem to occur across the entire genome. As with RFLPs, there are almost always only two alleles at a SNP locus. Individually they suffer the same problem as RFLPS of not being useful in many of the families. They are being used widely now because they are very easy to genotype, are very common (occurring at least ten times more frequently than the other types of markers) and thus can be used in combination with each other.

History of Genetic Mapping

The technique of genetic mapping was first described in 1911 by Thomas Hunt Morgan, who was studying the genetics of fruit flies. Morgan was able to study genetic mapping because he was able to actually see traits in the flies (like having white eyes instead of red) that were caused by mutations in single genes. He noticed that some traits violated Gregor Mendel's Law of Independent Assortment (which said that any two loci would segregate independently and thus have a recombination fraction of 0.50).

homologous carrying similar genes

Genetic mapping did not start being applied to humans until the 1950s, because it was hard to know what traits were caused by genetic mutations. When RFLPs were first described in 1980, a large effort was undertaken to generate maps of all the chromosomes. The first such maps were made in the early 1980s but covered only parts of chromosomes and had only a few markers. Maps of whole chromosomes were made by the late 1980s. By the mid-1990s, as the abilities of the research teams improved, and as the statistical methods of analysis were refined, a number of whole-genome (i.e., covering all the chromosomes) genetic maps were generated. These maps were updated and improved, and they were made available on the Internet.

The Comparison of Genetic and Physical Distance

Genetic maps are a measure of distance based on recombination, which is a biological process. A different way of measuring the distance between two loci is to measure the actual number of base pairs between the loci. This is known as the physical distance, and, when many such distances are put together, it makes a physical map.

Genetic maps and physical maps are similar in that the loci will be in the same order. There is also a general correspondence of distance, in that bigger genetic distances usually correspond to bigger physical distances. The overall rule of thumb is that one centimorgan of genetic distance is about one million base pairs of physical distance. However, this comparison can vary dramatically across certain parts of chromosomes. In some areas, one centimorgan might be only 50,000 base pairs (e.g., at the ends of chromosomes, where recombination seems to be increased). In other chromosomal areas (e.g., near the centromere), one centimorgan might be five million base pairs. SEE ALSO CROSSING OVER; GENE DISCOVERY; LINKAGE AND RECOMBINATION; MEIOSIS; MORGAN, THOMAS HUNT; POLYMORPHISMS; REPETITIVE DNA ELEMENTS.

Jonathan L. Haines

Bibliography

Bloom, Mark V., Greg A. Freyer, and David A. Micklos. *Laboratory DNA Science: An Introduction to Recombinant DNA Techniques and Methods of Genome Analysis.* Menlo Park, CA: Addison-Wesley, 1996.

Internet Resources

The Center for Medical Genetics. Marshfield Clinic. <http://research.marshfieldclinic.org/genetics>.

Thomas Hunt Morgan. Cold Spring Harbor Laboratory. <http://www.cshl.org/History/morgan.html>.

Marker Systems

Marker systems are tools for studying the transfer of genes into an experimental organism. In gene transfer studies, a foreign gene, called a transgene, is placed into an organism, in a process called transformation. A common problem for researchers is to determine quickly and easily if the target cells of the organism have actually taken up the transgene. A marker allows the researcher to determine whether the transgene has been transferred, where it is located, and when it is expressed (used to make protein).

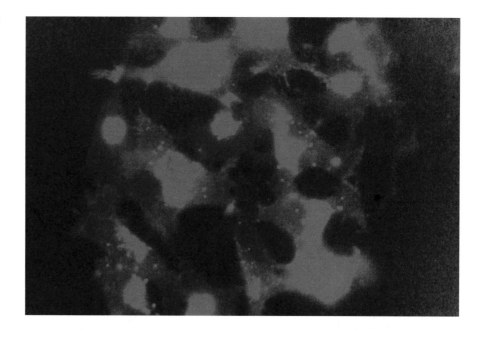

Green fluorescent protein (GFP) serves as a visible marker in these mouse cells.

The marker itself is also a gene. It is placed next to the transgene to make a single piece of DNA, which is then transferred. Markers are chosen because their gene products (proteins) have obvious effects on the **phenotype** of the organism. If the system is constructed properly, detection of the marker's product indicates that the transgene is present and functioning.

phenotype observable characteristics of an organism

Marker systems exist in two broad categories: selectable markers and screenable markers. Selectable markers are typically genes for antibiotic resistance, which give the transformed organism (usually a single cell) the ability to live in the presence of an antibiotic. Screenable markers, also called reporter genes, typically cause a color change or other visible change in the tissue of the transformed organism. This allows the investigator to quickly screen a large group of cells for the ones that have been transformed. Selectable and screenable markers are essential to genetic engineering in both **prokaryotes** and **eukaryotes**, and are often built into engineered DNA plasmids used for genetic transformation.

prokaryotes single-celled organisms without a nucleus

eukaryotes organisms with cells possessing a nucleus

Selectable Markers

Selectable markers are said to cause either negative or positive selection. Negative selection kills cells that do not have the marker gene, while positive selection kills those that have it but not in the correct place in the chromosome.

Negative selection is most commonly used in the transformation of bacterial cells. A gene for resistance to an antibiotic such as kanamycin is placed on a **plasmid** with the transgene (such as an insulin gene). Resistance genes often code for an enzyme that phosphorylates (adds a phosphate to) the antibiotic, thereby inactivating it. Cells that take up the plasmid can thus tolerate an otherwise lethal exposure to the antibiotic. The researcher exposes the entire group of cells, and harvests those that remain alive.

plasmid a small ring of DNA found in many bacteria

Positive selection is often performed in mammalian cells grown in cell culture. Because of the complexity of the mammalian cells, it is important

that a transgene not only enter the cell, but also be integrated into the correct place in the chromosome. If it does not, it is unlikely to be regulated properly. The "correct place" is the site on the chromosome where the normal gene is found. For example, if the researcher is inserting a human nerve cell gene into a mouse, it should be inserted at the site where the corresponding mouse nerve cell gene sits. Selection of cells with the properly located transgene is accomplished by killing off transformed cells in which the gene is in the wrong place.

This system, an example of positive selection system, has three parts. The first is an antibiotic, the second is an **enzyme** that acts on the antibiotic, and the third is an enzyme that cuts and splices DNA.

enzyme a protein that controls a reaction in a cell

The antibiotic ganciclovir is used to kill cells. Ganciclovir is a "nucleotide analog," meaning it is structurally similar (but not identical) to the building blocks of DNA. It must be phosphorylated before it can be incorporated into DNA in the target cell. Once it is incorporated, it acts like a monkey wrench in the machinery, preventing normal DNA function and thus killing the cell. The enzyme that acts on the ganciclovir is called thymidine kinase (TK). It adds a phosphate on the antibiotic, inactivating the antibiotic. Mammalian TK does not phosphorylate ganciclovir very efficiently, so mammalian cells are not normally killed by it. TK from the *Herpes simplex* virus (HSV) does phosphorylate it efficiently, and any mammalian cell transformed with an active HSV TK enzyme will be killed.

In this system, a plasmid is constructed with the **transgene**, the HSV TK gene, and a "recombination site," a stretch of DNA that is recognized by the cellular recombinase enzymes that cut and splice DNA. If the transgene is integrated into the chromosome at the site of the normal gene, then the HSV TK gene is eliminated by the cellular "recombinase" enzymes, and the cells are not sensitive to ganciclovir. In improperly transformed cells, the recombinase can't remove the HSV TK gene, and so those cells will be killed when exposed to ganciclovir.

transgenes genes introduced into an organism

Screenable Markers

Screenable marker systems employ a gene whose protein product is easily detectable in the cell, either because it produces a visible pigment or because it fluoresces under appropriate conditions. Visible markers rarely affect the studied trait of interest, but they provide a powerful tool for identifying transformed cells before the gene of interest can be identified in the culture. They can also identify the tissues that have (and have not) been transformed in a multicellular organism such as a plant.

Green fluorescent protein (GFP) is used as a screenable marker or a reporter gene in a variety of cells. GFP is a small protein that is isolated from jellyfish. It possesses a trio of amino acids that absorb blue light and fluoresce yellow-green light that is detectable using a fluorescence microscope or other means. Using GFP as a reporter has the enormous advantage that transgenic cells can be located noninvasively, simply by illuminating with blue light and observing the fluorescence. It is a simple protein, and it works in many different model systems (plants, mammalian cell culture, and the like) because it requires no post-translational processing of the protein to make it active. This is helpful, because processing enzymes are typically specific to each type of organism, thus limiting the usefulness of transgenes

that require such modifications. In addition, whereas some reporter products are toxic to the cell, GFP is not, and the intensity of the fluoresced light can be used to quantify gene expression.

The *Escherichia coli* bacterium provides another reporter gene system commonly used in plants. The bacterium makes an enzyme, called B-glucuronidase gus A (uid A), that cleaves a group of sugars called B-glucouronides. This enzyme will also cleave a chemical that is added to the culture such that the cleaved chemical is converted into an insoluble, visible blue precipitate at the site of enzyme activity. Many plants lack their own B-glucuronidase enzymes, so it is easy to determine if the plant has been transformed. Enzyme activity can be easily, sensitively and cheaply assayed **in vitro**, and can also be examined in tissues to identify transformed cells and tissues. The level of gene expression can be measured by the intensity of the blue color produced. SEE ALSO CLONING GENES; MODEL ORGANISMS; PLASMID; POST-TRANSLATIONAL CONTROL; RECOMBINANT DNA.

Linnea Fletcher

in vitro "in glass"; in lab apparatus, rather than within a living organism

Bibliography

Bloom, Mark V., Greg A. Freyer, and David A. Micklos. *Laboratory DNA Science: An Introduction to Recombinant DNA Techniques and Methods of Genome Analysis.* Menlo Park, CA: Addison-Wesley, 1996.

Ponder, Bruce A. "Cancer Genetics." *Nature* 411 (2001): 336–341.

Risch, Neil J. "Searching for Genetic Determinants in the New Millenium." *Nature* 405 (2001): 847–856.

Mass Spectrometry

Mass spectrometry is a technique for separating and identifying molecules based on mass. It has become an important tool for proteomics, the analysis of the whole range of proteins expressed in a cell. Mass spectrometry is used to identify proteins and to determine their amino acid sequence. It can also be used to determine if a protein has been modified by the addition of phosphate groups or sugars, for example. The technique also allows other molecules, including DNA, RNA, and sugars, to be identified or sequenced.

The use of mass spectrometry has greatly aided proteomics. Whereas DNA sequencing is simple and straightforward, protein sequencing is not. The ability to quickly and accurately identify proteins being expressed in a cell allows a range of **hypotheses** to be tested that cannot be approached by simply looking at DNA. For instance, it is possible with mass spectrometry to determine what proteins are expressed in cancer cells that are not expressed in healthy cells, possibly leading to further understanding of the disease and to development of drugs that target these proteins.

hypotheses testable statements

Data derived from mass spectrometry is usually analyzed by computer programs that search databases to help identify the analyzed protein. Such tools are the province of **bioinformatics**. The databases are usually located at a centralized institution and are searched via the Internet.

bioinformatics use of information technology to analyze biological data

Ionize, Accelerate, Detect

gel electrophoresis technique for separation of molecules based on size and charge

Proteins to be analyzed, such as those from a cell, are first separated and purified. One technique for this is two-dimensional **gel electrophoresis**.

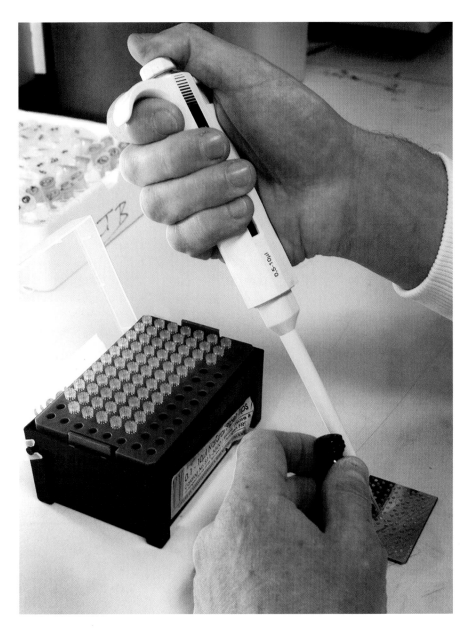

At the Manchester Metropolitan University in 2001, a technician prepares samples to be analyzed by a mass spectrometry machine. Combined with a database of controlled spectra, this machine can aid in the detection of anthrax spores.

Individual proteins form spots on the gel, which can then be cut out individually. Chromatography can also be used. In this technique, a mixture of proteins is separated by being passed through a column containing inert beads, which slow the proteins to different extents based on their chemical properties. Unlike the two-dimensional gel method, chromatography allows continuous (versus batch) processing of cellular samples, which reduces the requirement for handling of samples and speeds up analysis.

Mass spectrometry begins by ionizing the molecules in the target sample—removing one or more electrons to give them a positive charge. Molecules must be charged so they can be accelerated. The principle is the same as that used in a television or fluorescent light bulb: Charged particles are accelerated by being pulled toward something of the opposite charge. In the mass spectrometer, the speed the molecules attain during acceleration is proportional to their mass (actually, their mass-charge ratio). By determining the speed of the molecules, researchers can calculate their mass.

Proteins are ionized in one of two common ways. The first is matrix-assisted laser desorption ionization, or MALDI. The "matrix" that is used is a crystalline structure of small organic molecules in which the protein is suspended. When excited by a laser, the protein is vaporized ("desorbed") and ionized to a +1 charge. The second method is electrospray ionization (ESI). In this process, the protein is dissolved in a solution, which is sprayed to form a fine mist (it is ionized at the same time). Evaporation of the surrounding solvent eventually leaves the protein by itself. A benefit of the solution method of ESI is that a mixture of proteins can first be separated by chromatography or capillary gel electrophoresis, and then passed on to the ionizer without additional handling, avoiding the labor-intensive two-dimensional gel method.

Following ionization, the protein is accelerated. The most common way to determine mass is with a "time-of-flight" (TOF) tube. Just as its name implies, this tube is used to determine the time of flight of the protein, allowing a simple determination of velocity (velocity = distance / time). The accelerator imparts a known amount of kinetic energy to the molecule. Since kinetic energy = $1/2$(mass) (velocity)^2, the determination of mass is straightforward.

Applications

Identifying Unknown Proteins. Since several different proteins may have the same mass, simply obtaining the mass of the whole protein is not enough to identify it. However, if it is broken into a characteristic set of fragments (called **peptides**), and the mass of each of these is determined, it is usually possible to identify the protein based on its "peptide fingerprint."

peptides amino acid chains

Sequencing Peptides. Peptides can be sequenced by generating multiple sets of fragments and analyzing the differences in masses among them. Removing a single amino acid from a peptide, for instance, will decrease its mass by a specific amount and at the same time create a new, detectable particle with the same mass. Individual amino acids can be identified by their characteristic molecular masses. Mass spectrometry has made protein sequencing much easier than it had been. The traditional method required about twelve hours to sequence a ten-amino acid peptide. Mass spectrometry can do the same job in about one second. The entire protein need not be sequenced to be identified. Often four to five amino acids are enough.

Identifying Chemical Modifications. Chemical modifications to proteins after they are synthesized (called post-translational modifications) are important for regulation. For instance, the addition of a phosphate group (PO_4) is used to turn on or turn off many enzymes. The presence of such groups can be detected by the additional weight they bring. Sugar groups can be detected in the same fashion. SEE ALSO BIOINFORMATICS; HPLC: HIGH-PERFORMANCE LIQUID CHROMATOGRAPHY; INTERNET; POST-TRANSLATIONAL CONTROL; PROTEINS; PROTEOMICS.

Richard Robinson

Bibliography

Perkel, Jeffrey M. "Mass Spectrometry Applications for Proteomics." *The Scientist* 15, no. 16 (2001): 31–32.

Internet Resource

Mass Spectrometry. Richard Caprioli and Marc Sutter, eds. Vanderbilt University Mass Spectrometry Research Center. <http://ms.mc.vanderbilt.edu/tutorials/ms/ms.htm>.

McClintock, Barbara

Geneticist
1902–1992

Barbara McClintock was one of the most important geneticists of the twentieth century and among the most controversial women in the history of science. She made several fundamental contributions to our understanding of chromosome structure, put forward a bold and incorrect theory of gene regulation, and, late in her career, developed a profound understanding of the interactions among genes, organisms, and environments. She was born on June 16, 1902, in Hartford, Connecticut, the third of four children and the youngest daughter. She grew up in Brooklyn, New York, and in 1919 she enrolled in the agricultural college of Cornell University, where she received all her post-secondary education. She took a bachelor's degree in 1923, a master's in 1925, and a Ph.D., under the direction of the **cytologist** Lester Sharp, in 1927.

McClintock gravitated toward the cytology and genetics of maize, or Indian corn, and by 1929 she was a rising star in her field. Not quite single-handedly, she made possible the "golden age of maize genetics," from 1929 to 1935. During those years, McClintock published a string of superb papers identifying novel cytological phenomena and linking them to genetic events. Working with Harriet Creighton, she confirmed that chromosomes physically exchange pieces during the genetic phenomenon known as "crossing over." She was supported by a series of prestigious fellowships, from the National Research Council, the Guggenheim Foundation, and others, that took her from Cornell to the California Institute of Technology, and to Berlin and back. In 1935 she took a faculty position at the University of Missouri in Columbia. She was not happy there, however, and resigned in 1939, despite the apparent imminence of a promotion with tenure.

In 1941 she took a summer position at Cold Spring Harbor on New York's Long Island, at the Carnegie Institution of Washington's Department of Genetics. It was an ideal position for her, with no teaching or administrative duties. Within a year the post became permanent, and she remained there until her death. On arrival, she continued work that she had begun while at Missouri, investigating a phenomenon called the breakage-fusion-bridge (BFB) cycle. This is a repeating pattern of chromosome breakage she had discovered among strains of maize plants grown from X-rayed pollen. In 1944, during an experiment designed to use the BFB cycle to create new mutations, she discovered numerous "mutable" genes: genes that turned on and off spontaneously during development. In the cells of some of these new mutants lay her most important discovery, chromosome segments that move from place to place on the chromosome. That same year, the National Academy of Sciences honored a woman for only the third time in its eighty-year history when it elected McClintock a member.

During the rest of the 1940s McClintock developed a novel theory of how genes could control the development and differentiation of organisms.

Barbara McClintock, having just received the prestigious $15,000 Lasker Award in 1981 for her many contributions to the field of genetics.

cytologist a scientist who studies cells

The key to the theory was a new type of genetic element, not a gene but a gene-controller, that first appeared in her 1944 BFB experiment. These "controlling elements," she argued, inhibited or modulated the effects of the genes near them. She proposed that through coordinated movement from gene to gene (transposition) controlling elements executed the genetic program of development, much as the hammers on a player piano execute the program encoded on a piano roll.

Transposition in maize was confirmed immediately and repeatedly by other researchers. Few scientists, however, could accept her notion that the movements were coordinated. After about 1954, McClintock did little more with transposition, but she continued to work on genetic control for the rest of her long career. Her systems grew increasingly complex, and her thinking led her to comparisons between embryology and evolution.

During the 1970s transposition was discovered in bacteria, and its biochemistry was explained in terms of DNA sequences and enzymatic action. Soon transposition was found to be nearly universal in the living world and was linked to medical fields such as cancer, virology, and immunology. McClintock experienced a rare scientific renaissance. Her theories of genetic control, never widely accepted and by this time rejected outright, were forgotten, and she was reborn as the discoverer of transposition. She won, unshared, the 1983 Nobel Prize in physiology or medicine "for her discovery of transposable genetic elements."

genome the total genetic material in a cell or organism

Since then, the experiments of other researchers have provided at least qualified support for even some of her wilder ideas, such as her conception of the **genome** as a "sensitive organ of the cell"; and the idea that any organism has the genetic instructions to make any other. Some of these findings had appeared by the time she died, on September 2, 1992, but it has been the various genome projects, human and otherwise, that have lent the strongest support to McClintock's dynamic, interactive vision of nature. Had she lived to be one hundred, she might well have been considered for a second Nobel, this time for her insights into the workings of the genome. SEE ALSO CHROMOSOMAL ABERRATIONS; CHROMOSOMAL THEORY OF INHERITANCE, HISTORY; GENE; MAIZE; MUTATION; TRANSPOSABLE GENETIC ELEMENTS.

Nathaniel Comfort

Bibliography

Comfort, Nathaniel C. *The Tangled Field: Barbara McClintock's Search for the Patterns of Genetic Control.* Cambridge, MA: Harvard University Press, 2001.

———. "Two Genes, No Enzyme: A Second Look at Barbara McClintock and the 1951 Cold Spring Harbor Symposium." *Genetics* 140, no. 4 (1995): 1161–1166.

Keller, Evelyn Fox. *A Feeling for the Organism: The Life and Work of Barbara McClintock*, 10th Anniversary Edition. New York: W. H. Freeman, 1993.

McClintock, Barbara. *The Discovery and Characterization of Transposable Elements: The Collected Papers of Barbara McClintock.* New York: Garland, 1987.

McKusick, Victor

Geneticist
1921–

Victor Almon McKusick was born in the little town of Parkman in central Maine, on October 21, 1921. He went to Tufts College in Boston from

1940 to 1943, and then received his medical degree from Johns Hopkins University in 1946. He has remained at Johns Hopkins ever since, rising from a medical student to physician-in-chief of the entire Johns Hopkins hospital and on the way creating the field of medical genetics and changing forever the way medicine is practiced. Due to his interest, energy, determination, and discoveries, McKusick is often called the Father of medical genetics.

McKusick was first interested in heart diseases, but he discovered that many of the patients he saw had family members who also had heart problems, piquing his interest in the possibility that genes might be causing these diseases. His interest in genetic disorders continued, and in 1957 he established the first Department of Medical Genetics in the country. He started studying many different diseases, including those related to heart defects, blood problems, and dwarfism. He helped found the American College of Medical Genetics and the American Board of Medical Genetics.

Victor McKusick

Soon after, McKusick realized that many physicians and scientists knew little about genetics and helped to start one of the first courses in genetics which quickly became known as the "short course in experimental mammalian genetics." This was a two-week course taught by McKusick and his colleagues from Johns Hopkins and the Jackson Laboratories, a research institute studying the genetics of mice. The course started in 1960 and was taught in Bar Harbor, near the Jackson Laboratories, in McKusick's home state of Maine. This course quickly became the most well-known and highly respected course in genetics, and brought in students, researchers, and doctors from all over the country. It remains the premier course of its kind today.

When McKusick started the field of medical genetics, there was no list of genetic diseases, nor any one place to go to find information about any of these disorders. He started keeping track of the diseases, and this led to the first edition of his book *Mendelian Inheritance in Man*, which was published in 1966. Even then, when computers were room-sized boxes kept behind closed doors, McKusick knew their value, and he maintained his list using one of these computers. In that first edition, McKusick listed 1,466 different genetic diseases. In the most current edition, over 12,000 diseases are listed. *Mendelian Inheritance in Man* is available on the Internet.

McKusick was one of the first to realize the potential of the Human Genome Project. In 1973 he helped to organize the first Human Gene Mapping Workshop, the predecessor of the Human Genome Project. He helped found the Human Genome Organization (HUGO) in 1988. He has won over twenty prestigious awards, including the Albert Lasker Medical Research Award in 1997, and he has received over twenty honorary degrees from colleges and universities around the world.

McKusick's accomplishments go well beyond all the papers he has written, all the books he has authored, and all the awards he has won. He has inspired two generations of medical geneticists with his dedication, rigorous scientific approach, warmth, and good humor. SEE ALSO GENETICIST; HUMAN DISEASE GENES, IDENTIFICATION OF; HUMAN GENOME PROJECT.

Jonathan L. Haines

Bibliography

McKusick, V. A., et al. "40 Years of the Annual Bar Harbor Course (1960–1999): A Pictorial History." *Clinical Genetics* 55, no. 6 (1999): 398–415.

McKusick, V. A. "The Anatomy of the Human Genome: A Neo-Vesalian Basis for Medicine in the Twenty-first Century." *Journal of the American Medical Association.* 286 (2001): 2289–2295.

Internet Resource

Online Mendelian Inheritance in Man. Johns Hopkins University, and National Center for Biotechnology Information. <http://www.ncbi.nlm.nih.gov/entrez/Omim>.

Meiosis

Meiosis is a type of cell division that, in humans, occurs only in male testes and female ovary tissue, and, together with fertilization, it is the process that is characteristic of sexual reproduction. Meiosis serves two important purposes: it keeps the number of chromosomes from doubling each generation, and it provides genetic diversity in offspring. In this it differs from mitosis, which is the process of cell division that occurs in all **somatic** cells.

somatic nonreproductive; not an egg or sperm

Overview

All of our somatic cells except the egg and sperm cells contain twenty-three pairs of chromosomes, for a total of forty-six individual chromosomes. This number, twenty-three , is known as the diploid number. If our egg and sperm cells were just like our somatic cells and contained twenty-three pairs of chromosomes, their fusion during fertilization would create a cell with forty-six chromosome pairs, or ninety-two chromosomes total. To prevent that from happening and to ensure a stable number of chromosomes throughout the generations, a special type of cell division is needed to halve the number of chromosomes in egg and sperm cells. This special process is meiosis.

Meiosis creates haploid cells, in which there are twenty-three individual chromosomes, without any pairing. When **gametes** fuse at conception to produce a zygote, which will turn into a fetus and eventually into an adult human being, the chromosomes containing the mother's and father's genetic material combine to form a single diploid cell. The specialized diploid cells that will eventually undergo meiosis to produce the gametes are called primary **oocytes** in the female ovary and primary spermatocytes in the male testis. They are set aside from somatic cells early in the course of fetal development.

gametes reproductive cells, such as sperm or egg

oocyte egg cell

Even though meiosis is a continuous process in reality, it is convenient to describe it as occurring in two separate rounds of nuclear division. In the first round (meiosis I), the two versions of each chromosome, called homologues or homologous chromosomes, pair up along their entire lengths and thus enable genetic material to be exchanged between them. This exchange process is called crossing over and contributes greatly to the amount of genetic variation that we see between parents and their children. Subsequently, the two homologues are pulled toward opposite ends of their surrounding cell, thus creating a haploid cluster of chromosomes at each pole, at which point division occurs, separating the two clusters. Meiosis I is therefore the actual reduction division. At the end of meiosis I, each chromosome is still composed of two sister strands (chromatids) held together by a

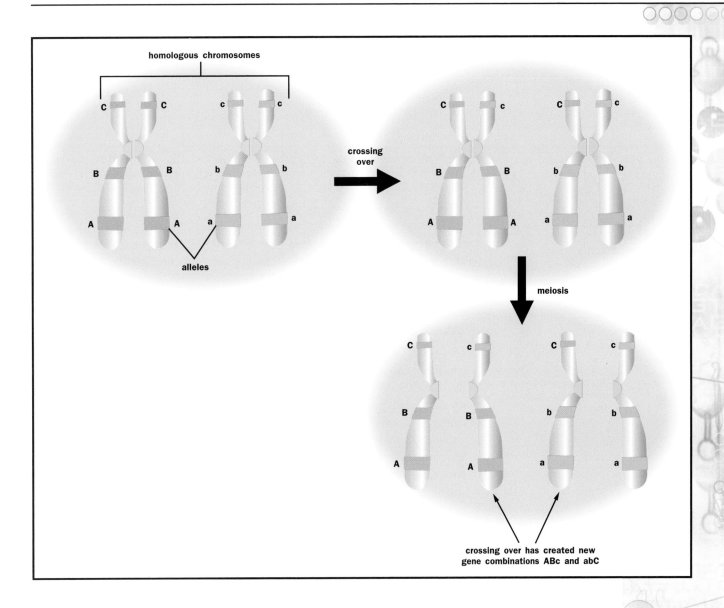

crossing over has created new
gene combinations ABc and abC

particular DNA sequence of about 220 nucleotides, called the centromere. The centromere has a disk-shaped protein molecule (kinetochore) attached to it that is important for the separation of the sister chromatids in the second round of meiosis (meiosis II). Meiosis II is essentially the same division process as mitosis. Through the separation of the two sister chromatids, a total of four daughter cells, each with a haploid set of chromosomes, are created.

New combinations of alleles result from the crossing over between homologous chromosomes. Adapted from Robinson, 2001.

Meiosis I

Meiosis must be preceded by the S phase of the cell cycle. This is when DNA replication (the copying of the genetic material) occurs. Thus, each chromosome enters meiosis consisting of two sister chromatids joined at the centromere. The first stage of meiosis is a stage called prophase I. First, the DNA of individual chromosomes coils more and more tightly, a process called DNA condensation. The sister chromatids then attach to specific sites on the nuclear envelope that are designed to bring the members of each homologous pair of chromosomes close together. The sister chromatids line up in a fashion that is precise enough to pair up each gene on the DNA

molecule with its corresponding "sister gene" on the homologous chromosome. This four-stranded structure of maternal and paternal homologues is also called a bivalent.

Next in prophase I is the process of crossing over, in which fragments of DNA are exchanged between the homologous sister chromatids that form the paired DNA strands. Crossing over involves the physical breakage of the DNA double helix in one paternal and one maternal chromatid and joining of the respective ends. Under the light microscope, the points of this exchange can often be seen as an X-shaped structure called a chiasma.

The exchange of genetic material means that new combinations of genes are created on two of the four chromatids: Stretches of DNA with maternal gene copies are mixed with stretches of DNA with paternal copies. This creation of new gene combinations is called "recombination" and is very important for evolution, since it increases the amount of genetic material that evolution can act upon. A statistical technique known as linkage analysis uses the frequency of recombination to infer the location of genes, such as those that increase a person's risk for certain diseases.

At the beginning of metaphase I, the nuclear envelope has dissolved, and specialized protein fibers called **microtubules** have formed a spindle apparatus, as also occurs in the metaphase of mitosis. These microtubules then attach to the kinetochore protein disks on the two centromeres of the homologous pair of chromosomes. However, there is an important difference between mitosis and meiosis in the way this attachment occurs. In mitosis, microtubules attach to both faces of the kinetochore and thus separate sister chromatids when they pull apart. In meiosis, because the chiasma structures still hold the homologous sister chromatids together, only one face of each kinetochore is accessible to the microtubules. Since the microtubules can only attach to one face of the kinetochore, the sister chromatids will be drawn to opposite poles as a pair, without separation of the individual chromatids.

At the end of metaphase I, the pairs of homologues line up on the metaphase plate in the center of the cell, the spindle apparatus is fully developed, and the microtubules of the spindle fibers are attached to one side of each of the two kinetochores. In anaphase I, the microtubules begin to shorten, thus breaking apart the chiasmata and pulling the centromeres with their respective sister chromatids toward the two cell poles. The centromeres do not divide, as they do in mitosis. At the final stage of meiosis I, called telophase I, each cell pole has a cluster of chromosomes that corresponds to a complete haploid set, one member of each homologous chromosome pair.

The Sources of Genetic Diversity

It is completely random whether the maternal or paternal chromosome of each pair ends up at a particular pole. The orientation of each pair of homologous chromosomes on the metaphase plate is random, and a mixture of maternal and paternal chromosomes will be drawn toward the same cell pole by chance. This phenomenon is often called "independent assortment," and it creates new combinations of genes that are located on different chromosomes. Thus, we have two levels of gene reshuffling occurring in meiosis I. The first occurs during recombination in prophase I, which creates new combinations of genes on the same chromosome. In contrast to mitosis, the sister chromatids of a chromosome are not genetically identical because of the

microtubules protein strands within the cell, parts of the cytoskeleton

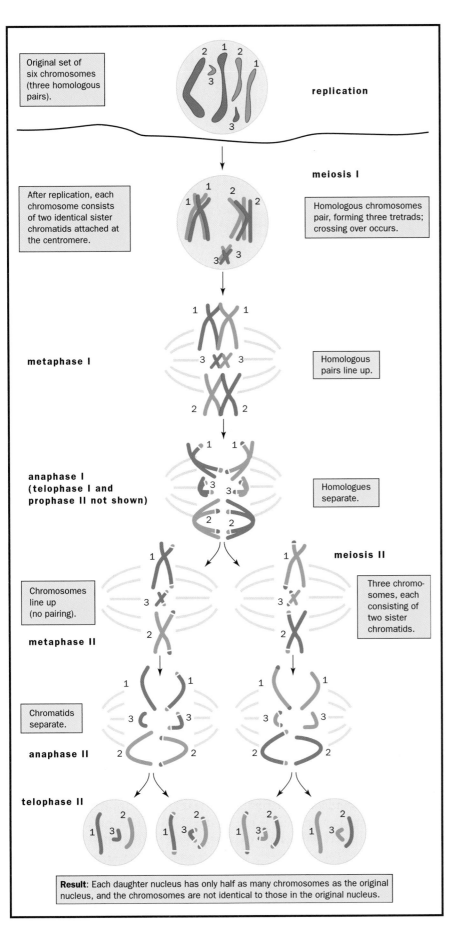

Original set of six chromosomes (three homologous pairs).

replication

After replication, each chromosome consists of two identical sister chromatids attached at the centromere.

meiosis I

Homologous chromosomes pair, forming three tretrads; crossing over occurs.

metaphase I

Homologous pairs line up.

anaphase I
(telophase I and
prophase II not shown)

Homologues separate.

meiosis II

Three chromo-somes, each consisting of two sister chromatids.

Chromosomes line up (no pairing).

metaphase II

Chromatids separate.

anaphase II

telophase II

Result: Each daughter nucleus has only half as many chromosomes as the original nucleus, and the chromosomes are not identical to those in the original nucleus.

Meiosis in an organism with six chromosomes. Replication precedes meiosis. Adapted from Curtis, 1994.

Kinetochore fibers separate homologues in Meiosis I, and sister chromatids in Meiosis II. Adapted from Alberts, 1994.

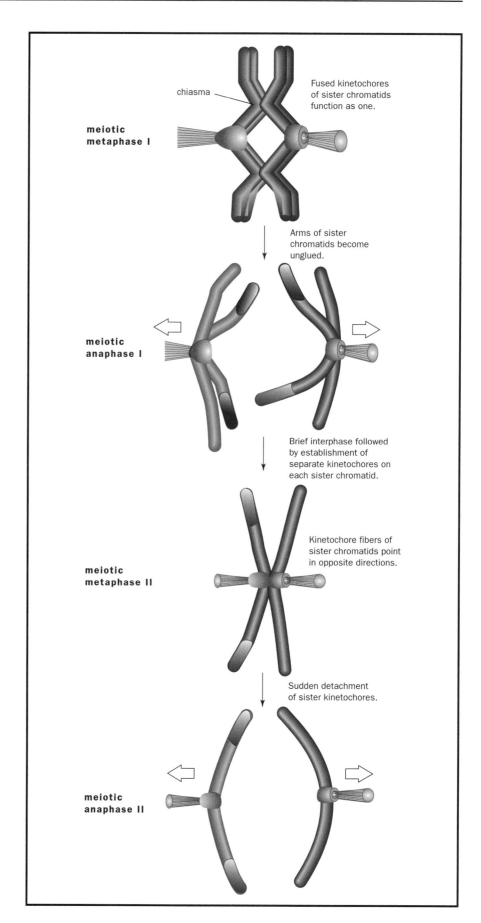

chiasma

Fused kinetochores of sister chromatids function as one.

meiotic metaphase I

Arms of sister chromatids become unglued.

meiotic anaphase I

Brief interphase followed by establishment of separate kinetochores on each sister chromatid.

meiotic metaphase II

Kinetochore fibers of sister chromatids point in opposite directions.

Sudden detachment of sister kinetochores.

meiotic anaphase II

crossing-over process. Anaphase I then adds the independent assortment of chromosomes to create new combinations of genes on different chromosomes. A total of 2^{23} (8.4 million) possible combinations of parental chromosomes can be produced by one person, and recombination further increases this to an almost unlimited number of genetically different gametes.

Meiosis II

Once both cell poles have a haploid set of chromosomes clustered around them, these chromosomes divide mitotically (without reshuffling or reducing the number of chromosomes during division) during the second part of meiosis. This time, the spindle fibers bind to both faces of the kinetochore, the centromeres divide, and the sister chromatids move to opposite cell poles. At the end of meiosis II, therefore, the cell has produced four haploid groups of chromosomes. Nuclear envelopes form around each of these four sets of chromosomes, and the cytoplasm is physically divided among the four daughter cells in a process known as **cytokinesis**.

cytokinesis division of the cell's cytoplasm

In males, the four resulting haploid sperm cells all go on to function as gametes (spermatozoa). They are produced continuously from puberty onwards. In females, all primary oocytes enter meiosis I during fetal development but then arrest at the prophase I stage until puberty. During infancy and early childhood, the primary oocytes acquire various functional characteristics of the mature egg cell. After puberty, one oocyte a month completes meiosis, but only one mature egg is produced, rather than the four mature sperm cells in males. The other daughter cells, called polar bodies, contain little **cytoplasm** and do not function as gametes.

cytoplasm the material in a cell, excluding the nucleus

Comparison with Mitosis

In summary, the main differences between meiosis and mitosis are that meiosis occurs only in specialized cells rather than in every tissue; it produces haploid gametes rather than diploid somatic cells; and each daughter cell is genetically different from the others due to recombination and independent assortment of homologues, rather than genetically identical. The pairing of homologous chromosomes and crossing over occur only in meiosis.

Chromosomal Aberrations

Meiosis is a very intricate process that requires, among other things, the precise alignment of homologous chromosome pairs and correct attachment of microtubules. During meiosis, errors in chromosome distribution may occur and lead to chromosomal aberrations in the offspring. One example is Down syndrome, where affected children carry three copies of chromosome 21 (trisomy 21). This may be explained by the failure of paired chromosomes or sister chromatids to separate in either sperm or egg, leading to the presence of two copies of chromosome 21. After fertilization with a normal gamete, the zygote will carry three copies, which leads to several **phenotypic** abnormalities, including mental retardation. SEE ALSO CELL, EUKARYOTIC; CHROMOSOMAL ABERRATIONS; CROSSING OVER; DOWN SYNDROME; FERTILIZATION; LINKAGE AND RECOMBINATION; MITOSIS; REPLICATION.

phenotypic related to the observable characteristics of an organism

Silke Schmidt

Bibliography

Alberts, Bruce, et al. *Molecular Biology of the Cell*, 3rd ed. New York: Garland Publishing, 1994.

Curtis, Helena, and Susan Barnes. *Invitation to Biology*, 5th ed. New York: Worth Publishers, 1994.

Raven, Peter H., and George B. Johnson. *Biology*, 2nd ed. St. Louis, MO: Times Mirror/Mosby College Publishing, 1989.

Robinson, Richard. *Biology*. Farmington Hills, MI: Macmillan Reference USA, 2001.

Mendel, Gregor

Natural Scientist
1822–1884

Gregor Mendel.

Gregor Mendel laid the foundation for the modern understanding of inheritance with his experiments on transmission of traits in garden peas. The ideas he developed are still in use today, and his essential insights into the physical nature of inheritance led directly to the understanding of the gene as a physical entity within the cell.

Education and Training

Mendel was born into a farming family in Heinzendorf, Austria (now Hyncice, Czech Republic). He attended university in Olmutz, but financial difficulties soon persuaded him to enter the Augustinian monastery in Brno, where he received both theological and agricultural training. Mendel remained affiliated with the monastery for the rest of his life. He served briefly as a parish chaplain in the region, and for many years served as a popular and successful teacher at the technical school in Brno. His training in agricultural experimentation, obtained at the University of Vienna, beginning in 1851, prepared him for the experiments that he began in 1856 on peas.

Experiments on Peas

Mendel's experiments were designed to investigate the most widely accepted model of inheritance, blending, which held that the traits of an offspring would be a blend of the parental traits. For example, the theory of blending predicts that a tall and short parent would give rise to a medium-height offspring. Mendel's results showed that for many simple traits, at least, this model was wrong. Instead, the offspring displayed traits in exactly the same form as they appeared in one or the other of the parents.

Mendel chose to study a small group of traits that occur in either of two forms, such as round versus wrinkled pea shape. He began by developing "pure-breeding" lines of each form. In a pure-breeding line, crossing two members gives only offspring that are identical to the parents for that trait. Mendel then crossed pure-breeding parents who had different forms of a trait. For example, he crossed a pea plant that produced only round peas with one that produced only wrinkled peas. All the offspring from this cross developed only round peas; no wrinkled peas were found. When these offspring were crossed among themselves, however, both round and wrinkled were observed, in a numerical ratio of three round-pea plants for every one wrinkled-pea plant.

Mendel explained these results by proposing that each visible trait is governed by the presence of two "factors," which may be the same or different in any individual. One of these factors is "dominant," while the other is "recessive." In the above example, the round-producing factor is dominant, and the wrinkled-producing factor is recessive. If two recessive factors are present, the organism will display the recessive trait. If the organism has two dominant factors, or one dominant and one recessive, the dominant trait will be displayed.

Laws of Inheritance

To explain the numerical relationships he obtained, Mendel developed the Law of Segregation. He proposed that during the process of egg and sperm formation, the two factors separate, or segregate, so that each egg or sperm contain only one factor. For a parent containing one of each type of factor, this means that half the sperm (or eggs) will contain the dominant factor, and half the recessive factor. During fertilization, these randomly pair up, so that some offspring will have two dominants, some two recessives, and some one of each. Simple algebra shows that the ratio of offspring in such a cross will be 3:1, just as Mendel found.

To show how this works, let $0.5D$ be the proportion of dominant factors and $0.5r$ be the proportion of recessive factors. Multiplying $(0.5D + 0.5r)$ times itself gives the offspring ratios, $0.25D^2 + 0.5Dr + 0.25r^2$. In this expression, $0.25D^2$ indicates that one-quarter of the offspring will have both dominant factors, $0.5Dr$ means half will have one of each type, and $0.25r^2$ means one-quarter will have both recessive factors. Since both the D^2 and Dr organisms will show the dominant trait, the ratio of dominant to recessive traits in the offspring will be $0.75:0.25$, or 3:1.

Mendel went on to study crosses between peas with multiple sets of traits, such as round seeds plus tall plants crossed with wrinkled seeds plus short plants. He found that the factors for each trait acted independently, so that the offspring of these crosses showed all possible combinations of traits. From the results of these experiments, he formulated his second principle, known as the Law of Independent Assortment, which states that the members of factor pairs assort (segregate) independently of each other during sperm and egg formation, and combine again randomly.

Mendel's Scientific Legacy

While neither Mendel nor anyone else in his day knew anything about chromosomes or genes, the laws of inheritance he discovered predicted exactly how genes behave on chromosomes during the reproductive process. Indeed, the factors he discovered are genes, which come in pairs and segregate on separate chromosomes during sperm and egg production, just as he suggested. Gene pairs located on different sets of chromosomes will assort independently during the process. While most genes do not exhibit simple dominance-recessiveness relations, and most traits are governed by more than one gene, it is to Mendel's credit that he began by trying to understand simple systems in order to develop generalizable laws.

Mendel published the results of his experiments, "Versuche über Pflanzenhybriden" ("Attempts at Plant Hybridization") in 1866. He did little scientific work after he became abbot of the monastery two years later.

His work was ignored by the larger scientific community, in part because it was not published in a widely read journal, and in part because it tackled a problem, the physical basis of heredity, that few other scientists were thinking deeply about at that time.

That changed shortly afterward, when microscopic studies of cells revealed that chromosomes divided when cells divided, provoking speculation that they might be involved in inheritance. Mendel's studies were rediscovered in 1900, sixteen years after his death, by three biologists studying similar phenomena. The importance of his theory of inheritance was immediately recognized and widely accepted, and became the starting point for further investigations of the nature of inheritance that were carried out by Thomas Hunt Morgan, Alfred Sturtevant, and other twentieth-century geneticists. Mendelism, as the theory was called, was merged with Darwinism in the 1930s to form the "New Synthesis," which explained evolutionary theory in modern genetic terms. SEE ALSO CHROMOSOMAL THEORY OF INHERITANCE, HISTORY; INHERITANCE PATTERNS; MENDELIAN GENETICS; NATURE OF THE GENE, HISTORY; PROBABILITY.

Richard Robinson

Bibliography

Henig, Robin Marantz. *The Monk in the Garden: The Lost and Found Genius of Gregor Mendel, the Father of Genetics.* Boston: Houghton Mifflin, 2000.

Mendelian Genetics

hybrids combinations of two different types

Gregor Mendel (1822–1884), an Austrian monk and botanist, was curious and loved nature. He grew plants with diverse flower colors, and he cross-pollinated plant varieties to create **hybrids**. Mendel's fascination with "the striking regularity with which the same hybrid forms always reappeared," broadened his quest into discovering laws for inheriting any trait, not just flower color, from one generation to the next.

Mendel designed a series of experiments to learn the statistical rules governing the features that appeared in hybrids and in their offspring. Mendel identified plant varieties that exhibited the same features over many generations when the plants were allowed to self-pollinate or cross-pollinate with plants from the same variety. He chose hybrids that were fertile, so that their inherited characteristics, or traits, could be passed on to their offspring. He also made sure to exclude foreign pollen, so that outside plants did not get mixed up in his breeding experiments. Mendel chose peas as an ideal plant that had these characteristics.

Mendel obtained thirty-four varieties of peas from seedsmen, and, after two years of preparative work, he selected for study seven traits exhibited by the peas. The seven traits were: color of the seed coats (white or non-white); form of the ripe seeds (round or wrinkled); color of the seeds (yellowish orange or green); form of the ripe pods (inflated or constricted); color of the unripe pods (dark green or vivid yellow); position of the flowers (axial or terminal); and length of the stems (long or short).

Mendel carefully avoided choosing any traits, such as size and form of leaves, length of flower stalk, or size of pods, that would have generated a

paternal gametes \ maternal gametes	AB	Ab	aB	ab
AB	AABB	AABb	AaBB	AaBb
Ab	AABb	AAbb	AaBb	Aabb
aB	AaBB	AaBb	aaBB	aaBb
ab	AaBb	Aabb	aaBb	aabb

Figure 1. This Punnett square shows the possible outcomes, in genotypes, of a dihybrid cross (AaBb x AaBb). Reginald Punnett designed this layout to illustrate how alleles will assort according to Mendel's laws (segregation and independent assortment). The two parents are heterozygous for two traits, A and B, on non-homologous chromosomes and carry the genotype AaBb. The four gamete types from the male and female are shown in boldface. Lowercase letters represent recessive alleles; uppercase letters represent dominant alleles.

chaotic mix of forms. He chose traits that would allow plants and their offspring to be distinctly classified.

Instead of looking at all seven traits at once, Mendel focused on one at a time. For each trait, he crossed two plant varieties to make hybrid plants. This was a monohybrid cross, because only one of the plant's many traits was studied. Mendel crossed the two chosen forms for each of the seven traits, using several hundred plants in each cross.

He found that in each case, all the first-generation offspring exhibited the same form as one of the parents, despite the hybrid having received input from two different parental varieties. Mendel called the form of the trait that appeared in these first-generation offspring **dominant**, as it was able to hide the other form during that generation. When the first-generation hybrid plants were allowed to self-pollinate, the hidden feature resurfaced in the next generation. Mendel called the hidden feature **recessive**. He further discovered that, on average, for every four offspring in the second generation, three displayed the dominant form of the trait, and one displayed the recessive form. He used these observations to suggest that each trait was governed by two "factors," one dominant and one recessive.

dominant controlling the phenotype when one allele is present

recessive requiring the presence of two alleles to control the phenotype

Mendel concluded that each plant carried two factors for every trait—either two dominant factors, two recessive factors, or one dominant and one recessive factor.

He proposed that his true-breeding parents carried two factors of the same kind. This is now defined as being homozygous. One parent plant was homozygous dominant, and the other homozygous recessive. When the parents were crossed, each offspring plant inherited one factor from each parent, but exhibited only the dominant form of the trait, even though they had received both a dominant and a recessive form. The offspring plants were hybrids, now called **heterozygotes**.

heterozygotes individuals whose genetic information contains two different forms (alleles) of a particular gene

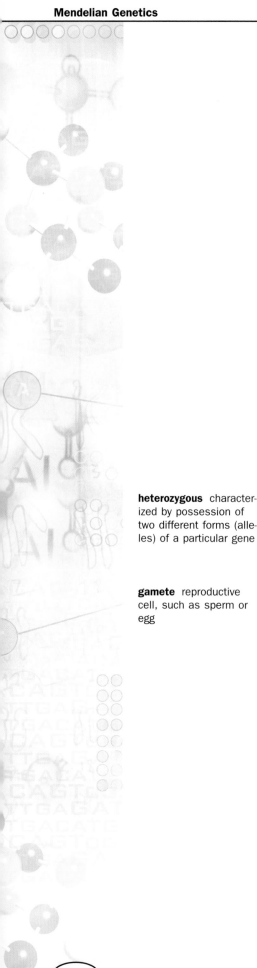

When these heterozygous plants self-pollinated, their offspring had an equal chance of receiving either two recessive factors, two dominant factors, or a dominant and a recessive factor. One quarter of these offspring were homozygous recessive, one quarter were homozygous dominant, one-half, were heterozygotes. Except for the one quarter that were homozygous recessives, the rest had at least one dominant factor and showed the dominant form of the trait. This explained Mendel's observation that three of every four plants showed the dominant form, and one in four the recessive.

Mendel also allowed the offspring of the heterozygous plants to self-pollinate. When he let plants with recessive features self-pollinate, only recessive features developed in their descendants, supporting the theory that they all contained only recessive factors.

When he let plants with dominant features self-pollinate, one-third gave rise to descendants that exhibited only dominant features. The other two-thirds gave rise to progeny with both dominant and recessive features, and therefore had to contain both dominant and recessive factors. Mendel tested six generations of plants and got similar results for each generation. Each generation of self-pollinating heterozygotes bore offspring, of which half were heterozygotes and half were homozygotes.

Mendel also did reciprocal crosses for each of the seven traits, switching the egg-bearing and the pollen-bearing variety to transmit the dominant and recessive features. The same ratio—three plants with dominant features for every one with recessive features—emerged from all the reciprocal crosses. Mendel concluded that a descendant had an equal chance of getting a dominant or a recessive factor (now called alleles) from either **heterozygous** parent, regardless of sex.

heterozygous characterized by possession of two different forms (alleles) of a particular gene

The Principle of Segregation

Mendel used his observations to formulate his First Law, the Principle of Segregation. According to this principle, each **gamete** receives from a parent cell only one of the two alleles the parent cell carries for each trait, and the gamete has an equal chance of getting either allele. (Exceptions to the principle were described later by Thomas Hunt Morgan, an American geneticist.) When the two gametes unite during fertilization, the resulting cell contains two alleles, either identical or different, for each trait. These two alleles are referred to as the individual's genotype for the trait.

gamete reproductive cell, such as sperm or egg

An alternative idea that other scientists during Mendel's time had was that two parental characteristics fused and blended into a single hybrid characteristic. Mendel's results showed this was not the case. His results showed, instead, that individuals inherit from their parents intact units that can leap through time. Mendel's discovery that inheritance had a particulate nature set the stage for modern advances in genetics.

The Principle of Independent Assortment

Mendel began a numerical evaluation of, respectively, two and three traits simultaneously, because he also wanted to know how different traits sorted themselves during gamete formation. He began a numerical evaluation of how two and then three traits were inherited simultaneously. The dihybrid cross involved two traits: the form of the plant's ripe seeds and the color of

its interior seeds. He crossed one true-breeding variety that had wrinkled seeds and a green interior, with another that had round seeds and a yellow interior. The dihybrid cross generated offspring that all had round, yellow seeds, but the seeds' outward appearance, or phenotype, hid the offspring's heterozygous nature. The offspring contained recessive alleles for making wrinkled, green seeds, as well as the dominant alleles that generated the seeds' round and yellow appearance.

The round, yellow seeds, which were the seeds of the first filial generation, or F1, were planted, raised, and made to self-pollinate. Their progeny, the second filial generation, or F2, had four phenotypes for seed form and color, in a ratio of 9:3:3:1 (nine round and yellow, to three wrinkled and yellow, to three round and green, to one wrinkled and green).

To unmask the F2 genotypes, the next generation's wrinkled, round, yellow, or green seeds were collected. The seeds showed that there were nine different genotypes among the F2 plants. If Y represents yellow, y represents green, R represents round, and r represents wrinkled, the nine genotypes were: YyRr, YyRR, Yyrr, YYRr, YYRR, YYrr, yyRr, yyRR, and yyrr. Four of the genotypes were homozygous for both traits, four were homozygous for one trait and heterozygous for the other, and one was heterozygous for both traits.

Mendel's trihybrid cross included the trait for the color of the seed coats, which could be white or non-white, in addition to the same two traits used in the dihybrid cross. In this cross, the F2 generation had eight different combinations of seed shape, seed coat color, and interior seed color and twenty-seven different genotypes.

The existence of all these allelic combinations revealed that chance had a lot to do with what ended up in the same gamete. The chance of a descendent getting a specific seed shape and color depended on straight math, not on interaction between shape and color or another unknown influence. A ratio of three dominant to one recessive phenotype appeared for each trait, as if the other traits' alleles did not exist. The arrival of one allele inside a gamete was unaffected by the entry of another trait's allele. Mendel described this formally as "each pair of different characters in hybrid union is independent of the other differences."

The chance of a descendant getting a specific trait depends on probability, not on the interaction between traits. This is formally stated as Mendel's Second Law, or the Principle of Independent Assortment: Different traits assort (i.e. are included in gametes) independently of one another.

A Punnett square, designed by English geneticist Reginald Punnett (1875–1967) and shown in Figure 1, shows the outcomes of crosses that follow Mendel's laws. The capital letters A and B represent dominant alleles, and the lowercase letters a and b represents recessive alleles. A genotype that is heterozygous for both traits in a dihybrid cross is represented as AaBb.

Exceptions to Mendel's Laws

The seven traits that Mendel evaluated all assort independently, but not all sets of traits do. Independent assortment is true for the seven traits that Mendel evaluated, and holds generally true for traits (genes) found

on non-homologous chromosomes. Any chromosome carries a collection of traits located on a long string of DNA, and the traits are therefore physically linked in a series or sequence. A non-homologous chromosome carries a unique collection of traits on a long string of DNA, that is different from the gene collection of an other non-homologue. Normal non-homologous chromosomes are not attached to each other during meiosis, and move independently of one another, each carrying their own gene collection. Each chromosome, composed of a long string of DNA, carries a collection of genes, with each gene showing up in a particular form or type. Seven chromosomes reside within a pea gamete, and each of the traits Mendel chose to study lie on a different (non-homologous) chromosome.

Independent assortment is not true for the collection of traits that are located on a homologous chromosome. In eukaryotes, homologues come in pairs, one donated from each parent. Two homologous chromosomes carry the same collection of genes, but each gene can be represented by a different allele on the two homologues (a heterozygous individual). A gamete will receive one of those homologues, but not both. Genes or alleles that travel together on a chromosome do not show independent assortment, because they do not move independently of each other into a gamete.

Punnett and William Bateson (1861–1926), an English biologist, published the first report of gene linkage in peas. A comparison between the ratios at which certain genes were inherited and the expected Mendelian ratios showed that the traits did not assort independently.

Sometimes two traits on non-homologous chromosomes affect each other's phenotypic expression. Purple flowers, for example, occur only with the presence of at least one dominant allele from two different genes. Offspring of two parents who are heterozygotes for both genes produce flowers that are purple or white at a ratio of nine to seven. The genotypes of the purple offspring are either aaB– or A–bb. (A dash indicates that the individual could have either a dominant or a recessive allele.)

Incomplete dominance occurs when a heterozygote has a unique phenotype. Pink flowers, for example, result when one parent is homozygous white and the other homozygous red. Neither allele hides the other, and their appearance together creates a unique intermediate phenotype.

Alleles are said to be codominant when heterozygotes express both alleles but neither affects the other's character. Individuals who have the allele for blood type A and the allele for blood type B, for example, have the characteristics of both blood types and are referred to as being of blood type AB.

Modern studies of genetic diseases use Mendel's ratios to determine whether or not genes are linked to certain chromosomes. Family histories are converted into pedigrees to help understand inheritance patterns. A disease might skip generations as expected of recessive alleles, or be linked to other traits. Deafness, white hair, and blue eyes are linked in cats, for example. A disease's symptoms might also become more severe with successive generations, as is the case with some dominant alleles.

Mendelian genetics and molecular biology together can elucidate the function of genes that are critical for development and life, in both experimental animals and human beings. Understanding of genetic processes can help to cure diseases. SEE ALSO CHROMOSOMAL THEORY OF INHERITANCE, HISTORY; INHERITANCE PATTERNS; MEIOSIS; MENDEL, GREGOR; MORGAN, THOMAS HUNT; NATURE OF THE GENE, HISTORY; PROBABILITY.

Susanne D. Dyby

Bibliography

Internet Resource

Mendel, Gregor. Trans. C. T. Druery, and William Bateson. "Experiments in Plant Hybridization." (1866). MendelWeb. <http://www.netspace.org/MendelWeb>.

Metabolic Disease

Metabolism is the sum of the chemical processes and interconversions that take place in the cells and the fluids of the body. This includes the absorption of nutrients and minerals, the breakdown and buildup of large molecules, the interconversion of small molecules, and the production of energy from these chemical reactions. Virtually every chemical step of metabolism is catalyzed by an **enzyme**. Disorders of these enzymes that result from abnormalities in their genes are known as inborn errors of metabolism.

Inborn errors of metabolism were first recognized by Sir Archibald Garrod, a British physician who noted in 1902 that the principles of Mendelian inheritance applied to certain examples of human metabolic variation. He perceived the genetic basis for a particular metabolic condition that leads to visible effects—alkaptonuria, which results in a black pigment in the urine. Since then, more advanced chemical methods have allowed the discovery of hundreds of enzyme defects that cause metabolic diseases.

enzyme a protein that controls a reaction in a cell

Enzymes Control Metabolic Reactions

Enzymes are proteins that control the rate of chemical reactions in the cell. In general, each enzyme controls the rate of only one or a few reactions. Enzymes function by binding to the molecules to be reacted (called substrates or precursors) and altering their chemical bonds, producing products. The binding occurs on the surface of the enzyme, usually in a pocket or groove, called the active site. The enzyme releases the products after reaction. The active site has a specific three-dimensional structure that is required for binding substrates. In addition, it may have other sites that bind regulatory molecules or cofactors. Some cofactors are vitamins, which perform some accessory function critical for enzyme action.

Enzymes are often linked in multistep pathways, such that the product of one reaction becomes the substrate for another. In this way, a simple molecule can be changed step by step into a complex one, or vice versa. In addition, the multiple steps provide additional levels of regulation, and intermediates can be shunted into other pathways to make other products. For instance, some intermediates in the breakdown of sugar can be shunted to make amino acids. When all the enzymes in a pathway are functioning properly, intermediates rarely build up to high concentrations.

William Lucas, shown in 1998, when he was twelve, must wear a protective suit while riding his bicycle because he suffers from a metabolic disease called Erythropoietic Protoporphyria (EPP). The disorder is characterized by extreme sunlight sensitivity that can cause exposed skin to swell and blister.

Enzyme Defects Cause Metabolic Disorders

The causes of enzyme defects are genetic mutations that affect the structure or regulation of the enzyme protein or create problems with the transport, processing, or binding of cofactors. In general, the consequences of an enzyme deficiency are due to perturbations of cellular chemistry, because of either a reduction in the amount of an essential product, the buildup of a toxic intermediate, or the production of a toxic side-product, as shown in Figure 1.

Except as noted below, most metabolic disorders are inherited as autosomal recessive conditions. In this inheritance pattern, two defective gene copies are needed (one from each parent) to develop the disease. The parents, each of whom almost always has only one gene copy, will not have the disease but are **carriers**. The chance that two carrier parents will have a child who inherits two defective gene copies is 25 percent for each birth.

Metabolic disorders tend to be recessive, because they are due to inactivating, or "loss-of-function," mutations. One working copy of the gene is

carriers people with one copy of a gene for a recessive trait, who therefore do not manifest the trait

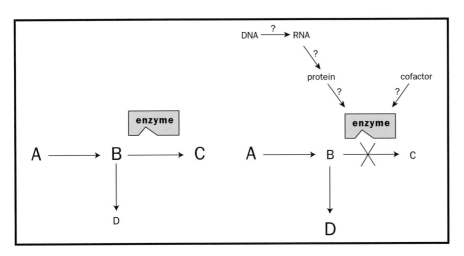

Figure 1. *Left:* Normal sequence of metabolism, in which precursor A is converted to intermediate B, and then into final product C. The B-to-C conversion is catalyzed by the enzyme. The side-product D is made in very small amounts. *Right:* Metabolic disease, resulting in altered enzyme*, resulting in decrease of product C, and/or buildup of precursor A and/or side-product D. The enzyme defect may be due to either an abnormality of the gene (DNA), a problem with production of messenger RNA (mRNA), a defect in the production or stability of the protein, or a deficiency or abnormal interaction with an enzyme cofactor.

usually enough to maintain sufficient levels of the enzyme, and so with one copy present, no disease develops.

Approaches to Treatment

Treatment approaches for metabolic disorders include (a) modifying the diet to limit the amount of a precursor that is not metabolized properly; (b) using cofactors or vitamins to enhance the residual activity of a defective enzyme system; (c) using detoxifying agents to provide alternative pathways for the removal of toxic intermediates; (d) enzyme replacement, to provide functional enzymes exogenously (from the outside); (e) organ transplantation, which in principle allows for endogenous (internal) production of functional enzymes; and (f) gene therapy, or replacement of the defective gene.

Gene therapy is expected to become the most important approach. It offers the potential for definitive treatment, and it is being actively investigated as a treatment for virtually every one of the metabolic disorders. Most of the genes for the enzymes involved in metabolic diseases have been identified and cloned, and in many cases the genes can be replaced in experimental systems. Genetic approaches have been used to produce mass quantities of enzymes to use for enzyme replacement, but as of 2002, gene therapy has not yet been used successfully to provide the stable expression of active enzymes in the human body.

This chapter will summarize classes of inborn errors of metabolism based upon the type of chemical process involved, and individual disorders will be discussed that illustrate the various disease mechanisms and treatment approaches.

Major Classes of Metabolic Disorders

Cells are constructed from four major types of molecules: carbohydrates, proteins, fats, and nucleic acids. The metabolic pathways involving each are

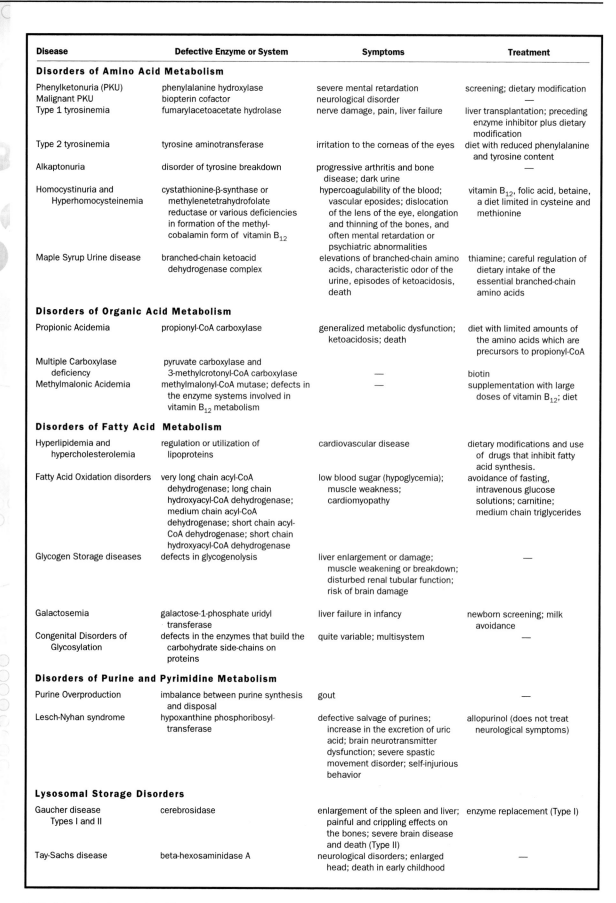

Disease	Defective Enzyme or System	Symptoms	Treatment
Disorders of Amino Acid Metabolism			
Phenylketonuria (PKU)	phenylalanine hydroxylase	severe mental retardation	screening; dietary modification
Malignant PKU	biopterin cofactor	neurological disorder	—
Type 1 tyrosinemia	fumarylacetoacetate hydrolase	nerve damage, pain, liver failure	liver transplantation; preceding enzyme inhibitor plus dietary modification
Type 2 tyrosinemia	tyrosine aminotransferase	irritation to the corneas of the eyes	diet with reduced phenylalanine and tyrosine content
Alkaptonuria	disorder of tyrosine breakdown	progressive arthritis and bone disease; dark urine	
Homocystinuria and Hyperhomocysteinemia	cystathionine-β-synthase or methylenetetrahydrofolate reductase or various deficiencies in formation of the methyl-cobalamin form of vitamin B_{12}	hypercoagulability of the blood; vascular eposides; dislocation of the lens of the eye, elongation and thinning of the bones, and often mental retardation or psychiatric abnormalities	vitamin B_{12}, folic acid, betaine, a diet limited in cysteine and methionine
Maple Syrup Urine disease	branched-chain ketoacid dehydrogenase complex	elevations of branched-chain amino acids, characteristic odor of the urine, episodes of ketoacidosis, death	thiamine; careful regulation of dietary intake of the essential branched-chain amino acids
Disorders of Organic Acid Metabolism			
Propionic Acidemia	propionyl-CoA carboxylase	generalized metabolic dysfunction; ketoacidosis; death	diet with limited amounts of the amino acids which are precursors to propionyl-CoA
Multiple Carboxylase deficiency	pyruvate carboxylase and 3-methylcrotonyl-CoA carboxylase	—	biotin
Methylmalonic Acidemia	methylmalonyl-CoA mutase; defects in the enzyme systems involved in vitamin B_{12} metabolism	—	supplementation with large doses of vitamin B_{12}; diet
Disorders of Fatty Acid Metabolism			
Hyperlipidemia and hypercholesterolemia	regulation or utilization of lipoproteins	cardiovascular disease	dietary modifications and use of drugs that inhibit fatty acid synthesis.
Fatty Acid Oxidation disorders	very long chain acyl-CoA dehydrogenase; long chain hydroxyacyl-CoA dehydrogenase; medium chain acyl-CoA dehydrogenase; short chain acyl-CoA dehydrogenase; short chain hydroxyacyl-CoA dehydrogenase	low blood sugar (hypoglycemia); muscle weakness; cardiomyopathy	avoidance of fasting, intravenous glucose solutions; carnitine; medium chain triglycerides
Glycogen Storage diseases	defects in glycogenolysis	liver enlargement or damage; muscle weakening or breakdown; disturbed renal tubular function; risk of brain damage	—
Galactosemia	galactose-1-phosphate uridyl transferase	liver failure in infancy	newborn screening; milk avoidance
Congenital Disorders of Glycosylation	defects in the enzymes that build the carbohydrate side-chains on proteins	quite variable; multisystem	—
Disorders of Purine and Pyrimidine Metabolism			
Purine Overproduction	imbalance between purine synthesis and disposal	gout	—
Lesch-Nyhan syndrome	hypoxanthine phosphoribosyl-transferase	defective salvage of purines; increase in the excretion of uric acid; brain neurotransmitter dysfunction; severe spastic movement disorder; self-injurious behavior	allopurinol (does not treat neurological symptoms)
Lysosomal Storage Disorders			
Gaucher disease Types I and II	cerebrosidase	enlargement of the spleen and liver; painful and crippling effects on the bones; severe brain disease and death (Type II)	enzyme replacement (Type I)
Tay-Sachs disease	beta-hexosaminidase A	neurological disorders; enlarged head; death in early childhood	—

Table 1 (continued on next page).

Disease	Defective Enzyme or System	Symptoms	Treatment
Lysosomal Storage Disorders [CONTINUED]			
Fabry disease	α-galactosidase	severe pain; renal failure; heart failure	enzyme replacement
Hurler syndrome, Hunter syndrome	α-iduronidase (Hurler syndrome); iduronate sultatase (Hunter syndrome)	enlargement of the liver and spleen; skeletal deformities; coarse facial features; stiff joints; mental retardation; death within 5–15 years	enzyme replacement
Sanfilippo syndrome	enzymes for heparan sulfate degradation	enlargement of the liver and spleen	enzyme replacement
Maroteaux-Lamy syndrome	arylsulfatase B	progressive, crippling and life-threatening physical changes similar to Hurler syndrome, but generally with normal intellect	—
Morquio syndrome	galactose 6-sulfatase; β-galactosidase	truncal dwarfism; severe skeletal deformities; potentially life-threatening susceptibility to cervical spine dislocation; valvular heart disease	—
Disorders of Urea Formation			
	carbamyl phosphate synthetase deficiency; ornithine transcarb-amylase deficiency, citrullinemia, argininosuccinic aciduria	hyperammonemia; mental retardation; seizures; coma; death	limitation of dietary protein; phenylacetate; liver transplantation
Disorders of Peroxisomal Metabolism			
Refsum disease	branched-chain fatty acid buildup	neurologic symptoms	—
Alanine-glyoxylate transaminase defect	alanine-glyoxylate transaminase	oxalic acid increase; organ dysfunction; renal failure	liver transplantation

Table 1, continued.

the basis for classification of many of the metabolic disorders. The mitochondria in cells are organelles that play a major role in most metabolic pathways, and mitochondrial disorders are one of the most significant and common types of metabolic disorders. Defects in the storage and disposal of molecules also give rise to metabolic disorders.

Carbohydrates are used primarily as fuel and can be built and broken down rapidly. The major storage form is glycogen. They are also added to proteins to make glycoproteins. Fatty acids are long-chain molecules that are used to construct membranes. Fatty acids are derived from dietary fats. Excess fat is used as fuel by mitochondria. Proteins are made of amino acids.

Humans must eat eight kinds of amino acids and then convert these into twelve other types to make the twenty amino acids found in our proteins. Excess amino acids in the diet are used for fuel by mitochondria. Along the way, they generate organic acids. Nucleic acids—DNA and RNA—are the molecules that store and process genetic information. They must be built from smaller units, called nucleotides. The storage and interconversion of different types of nucleotides assures a steady supply.

Below, representative disorders of each system are discussed. Other disorders are listed in Table 1. Many of the disease names end in "emia." This suffix indicates a blood disorder, and the names are derived from the fact that most metabolic disorders are diagnosed by detecting abnormal levels of intermediates or other substances in the blood.

Disorders of Mitochondrial Oxidative Metabolism

Most cellular energy is derived from the mitochondrial electron transport chain, which reduces oxygen to water in a series of steps to drive the

formation of the high-energy compound ATP. The Krebs cycle creates high-energy intermediates that it feeds to the electron transport chain, the energy of which ultimately is derived from a two-carbon compound called acetate, which is broken down successively to carbon dioxide. Acetate is derived from several pathways of amino acid, carbohydrate, and fat metabolism.

Thus, many pathways of metabolism feed into the Krebs cycle to drive oxidative metabolism in a web of processes requiring hundreds of enzymes. When there are defects in the Krebs cycle or the electron transport chain, one result may be ketoacidosis, which is due to the accumulation of lactic acid and ketone bodies.

neurological related to brain function or disease

The lack of cellular energy may be manifest in many cellular processes and can affect several tissues and organ systems, particularly those that are most dependent upon oxidative metabolism for energy. The brain and muscles are generally affected first, which can cause developmental delay, **neurological** crises—including episodes of coma, stroke-like events, and seizures—and muscle weakness or cardiomyopathy. Kidney function—most often the tubular function required for retention of electrolytes—may also be affected. Endocrine (hormone) systems may also be affected, resulting in conditions such as diabetes mellitus (caused by effects on the pancreas or by sensitivity to insulin in muscle and fat cells) or adrenal insufficiency (from effects on the adrenal glands).

Disorders of mitochondrial oxidative metabolism are very variable in terms of age of onset, severity, specific symptoms, and clinical course. Even the inheritance patterns of mitochondrial diseases are heterogeneous. Most are inherited in the usual autosomal recessive manner (although the chromosomal locations of only a few of the relevant genes are known). A few are inherited from defects in the mitochondrial DNA, which is passed on in the maternal line.

The mitochondrion contains a circular chromosome of about 16,500 bases. It codes for thirteen components of the electron-transport chain, as well as transfer RNA molecules and ribosomal RNAs required for their expression. Since there are multiple copies of mitochondrial DNA and there may be mixtures of normal and abnormal mitochondrial DNA (a phenomenon known as heteroplasmy), the precise proportion of mutated mitochondrial DNA may vary in an unpredictable manner from individual to individual within a family, and from tissue to tissue within an individual. There may also be variations within an individual tissue over time, adding to the unpredictability of mitochondrial disease and the difficulty in the diagnosis.

Disorders of Amino Acid Metabolism

Phenylketonuria. Phenylketonuria (PKU) is the most common disorder of amino acid metabolism, and it is a paradigm for effective newborn screening. Phenylalanine is an essential amino acid (meaning that it cannot be synthesized but must be taken in through the diet). The first step to its breakdown is the phenylalanine hydroxylase reaction, which converts phenylalanine to another amino acid, tyrosine. A genetic defect in the phenylalanine hydroxylase enzyme is the basis for classical PKU. Untreated PKU results in severe mental retardation, but PKU can be detected by

screening newborn blood spots, and the classical form can be very effectively treated by using medical formulas that are limited in their phenylalanine content.

The hydroxylase enzyme requires a cofactor called biopterin, which is also a cofactor for other enzymes. Defects affecting the production of biopterin result in another form, so-called malignant PKU. In this form, the other biopterin-dependent hydroxylases are also affected, resulting in deficient neurotransmitter synthesis and significant neurological symptoms.

Alkaptonuria. Alkaptonuria is a disorder of tyrosine breakdown. The intermediate that accumulates, called homogentisic acid, can **polymerize** to form pigment that binds to cartilage and causes progressive arthritis and bone disease and that also is excreted to darken the urine—the effect that allowed Garrod to recognize the genetic inheritance of this inborn error of metabolism.

polymerize to link together similar parts to form a polymer

Disorders of Organic Acid Metabolism

Propionic Acidemia. Propionyl-CoA is formed mainly from the breakdown of four essential amino acids (isoleucine, valine, threonine, and methionine). Defects of the enzyme propionyl-CoA carboxylase result in propionic acidemia, a life-threatening disease characterized by episodes of generalized metabolic dysfunction and ketoacidosis. The basis of treatment is a carefully applied diet containing limited amounts of the amino acids that are precursors to propionyl-CoA.

Methylmalonic Acidemia. Methylmalonyl-CoA is the product of propionyl-CoA carboxylase. There are a variety of metabolic defects in the further metabolism of this compound, resulting in methylmalonic acidemia. The best-known of these conditions arises from a defect in methylmalonyl-CoA mutase, the vitamin B_{12}-dependent enzyme that converts methylmalonyl-CoA to succinyl-CoA, which enters the Krebs cycle. There are other conditions resulting in methylmalonic acidemia that are due to defects in the enzyme systems involved in vitamin B_{12} metabolism. In some cases, supplementation with large doses of vitamin B_{12} is effective, but in most cases of methylmalonic acidemia, a special diet is required, similar to that used to treat propionic acidemia.

Disorders of Fatty Acid Metabolism

Hyperlipidemia and Hypercholesterolemia. Dietary fats are distributed through the body attached to proteins, in lipoprotein complexes. There are a number of disorders involving the regulation or utilization of lipoproteins, which result in hyperlipidemia and/or hypercholesterolemia, including the common conditions in adults that are associated with cardiovascular disease. Standard treatment approaches include modifying the diet and administering drugs that inhibit fatty acid synthesis.

Disorders of Carbohydrate Metabolism

The most active pathways in carbohydrate metabolism are glycogenolysis (the breakdown of glycogen, a polymerized form of carbohydrate, which is stored primarily in the liver and muscles), which produces glucose and distributes it through the bloodstream, and **glycolysis**, which releases energy

glycolysis the breakdown of the six-carbon carbohydrates glucose and fructose

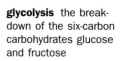

and produces pyruvate. Pyruvate is a three-carbon molecule that can be converted to acetate and enter the Krebs cycle or form several building-block molecules. The reverse processes are referred to as glycogen synthesis and gluconeogenesis, respectively.

Glycogen Storage Diseases. A number of defects may occur in glycogenolysis, giving rise to the disorders known as glycogen storage diseases. Glycogen storage diseases may affect the liver (enlarging it or damaging it due to increased amounts of glycogen) or muscle (weakening muscle or causing breakdown during times of exercise, due to inadequate glucose production). There may be additional problems, including disturbed kidney tubular function (which causes loss of nutrients and minerals), and there is a risk of brain damage in cases that result in critically low blood sugar.

Galactosemia. Another common disorder of carbohydrate metabolism is galactosemia, which is due to the inability to form glucose from galactose, the sugar that is found in milk. The classic form of galactosemia is due to a deficiency of the enzyme galactose-1-phosphate uridyl transferase, and, if untreated, it presents in the infant with fatal liver failure. Galactosemia is important because newborn screening (conducted by most developed countries on blood spots collected in the first days of life) has been very successful, and simple alteration of the diet (replacing milk with formulas that contain glucose or glucose polymers) has permitted a generation of individuals to survive with quite normal lives and, in general, normal intellect.

Disorders of Purine and Pyrimidine Metabolism

Purines and pyrimidines are chemicals that form the nucleic acids (DNA and RNA). An important purine compound is adenosine triphosphate (ATP), which is used to transfer chemical energy for processes such as biosynthesis and transport. There are several rare defects in the synthesis of purines and pyrimidines. The most common symptom of purine overproduction is gout, which arises for several reasons, often not associated with an identifiable enzyme defect but rather due to an imbalance between purine synthesis and disposal. Gout manifests when the ultimate product of purine degradation, uric acid, accumulates and crystallizes in the joints.

A very dramatic disorder of purine metabolism is Lesch-Nyhan syndrome, which is due to a defect in the enzyme hypoxanthine phosphoribosyltransferase (HPRT), resulting in defective salvage of purines and, accordingly, in an increase in the excretion of uric acid. For reasons that are still incompletely understood, a severe defect of HPRT also causes brain-neurotransmitter dysfunction, resulting in a severe spastic form of movement disorder and also a stereotypical compulsion for self-injurious behavior. The concentration of uric acid can be reduced by using the drug allopurinol, but there is no satisfactory treatment for the neurological symptoms associated with Lesch-Nyhan disease.

Lysosomal Storage Disorders

macromolecules large molecules such as proteins, carbohydrates, and nucleic acids

Lysosomes are intracellular compartments in which **macromolecules** are broken down in an acidic environment. Various classes of lysosomal storage disorders arise when there are defects in specific enzymes, and the manifestations of these disorders depend upon the class of macromolecule whose breakdown is affected.

Gaucher's Disease. The most common lysosomal storage disorder is Gaucher's disease, caused by a deficiency of the enzyme cerebrosidase, which is needed to break down cerebroside, a component of the cell membrane in blood cells and **neurons**. Partial defects of cerebrosidase cause Type 1 Gaucher's disease, in which material accumulates in the lysosomes of **macrophage** cells in the spleen, liver, and bone marrow, where most of the cell-turnover takes place. Significant accumulation usually occurs by childhood or early adulthood, resulting in dramatic enlargement of the spleen and liver. Later there may be painful and crippling effects on the bones. Type 1 Gaucher's disease can be effectively treated with enzyme replacement, but the enzyme must be infused intravenously approximately every two weeks for life. More severe defects of cerebrosidase cause Type 2 Gaucher's disease, which is rare, appears in infancy, and presents with the same problems as in Type 1 disease as well as severe brain disease that progresses to death. Very rarely, defects of intermediate severity can give rise to Type 3 Gaucher disease, which is a chronic neuronopathic form.

neurons brain cells

macrophage immune system cell that consumes foreign material and cellular debris

Tay-Sachs Disease. Tay-Sachs disease is due to a defect in the beta-hexosaminidase A enzyme, which removes a sugar from certain lipids called gangliosides, which build up in the lysosome. The disease causes neurological symptoms, an enlarged head, and death in early childhood.

Mucopolysaccharidosis. Mucopolysaccharidoses are lysosomal storage disorders affecting the breakdown of mucopolysaccharides, which are carbohydrate-protein macromolecules found on several cell types. Hurler syndrome (α-iduronidase deficiency) and Hunter syndrome (iduronate sultatase deficiency) are two disorders that affect the breakdown of the mucopolysaccharides dermatan sulfate and heparan sulfate, which are components of connective tissues throughout the body. The usual clinical manifestations of these syndromes are enlargement of the liver and spleen, skeletal deformities, coarse facial features, stiff joints, and mental retardation. Most cases are severe and progress to death within five to fifteen years, but there are exceptions. By 2002, there were several experimental approaches with enzyme replacement for mucopolysaccharidoses.

Disorders of Urea Formation

The urea cycle is a series of enzyme reactions that removes waste nitrogen from the body, allowing it to be excreted in the urine as urea. Disorders of the enzymes of the urea cycle disrupt this pathway, increasing blood ammonia (hyperammonemia). Hyperammonemia results in mental retardation, and acute episodes can progress to seizures, coma, and death. These conditions are inherited in an autosomal recessive pattern, except for ornithine transcarbamylase deficiency, which is X-linked, affecting males more severely than females. Treatment for these disorders includes limiting dietary protein (the major source of nitrogen intake) and using agents (such as phenylacetate) that provide an alternate mechanism to remove waste nitrogen (through excretion of phenylacetyl-glutamine in urine). Liver transplantation may also be effective in controlling blood ammonia in these conditions.

Disorders of Peroxisomal Metabolism

Several specialized metabolic functions are performed in the subcellular organelles known as peroxisomes. Severe defects in the biogenesis of

peroxisomes result in Zellweger syndrome, which is characterized by structural and developmental abnormalities and which is generally fatal in infancy. Defects in individual peroxisomal enzymes are also encountered, including Refsum disease, which results in the buildup of a branched-chain fatty acid (phytanic acid) and progressive problems in the nervous system. A defect in the enzyme alanine-glyoxylate transaminase causes an increase in the production of oxalic acid, an insoluble chemical that is progressively deposited in the tissues of the body and, over years, causes organ dysfunction, including renal failure. Renal transplantation does not prevent recurrence, but liver transplantation is effective in preventing the progression of the disease in the kidneys and other organs. SEE ALSO CELL, EUKARYOTIC; INHERITANCE PATTERNS; MITOCHONDRIAL DISEASES; POPULATION SCREENING; PROTEINS; TAY-SACHS DISEASE.

Bruce A. Barshop

Bibliography

Berg, Jeremy, John Tymoczko, and Lubert Stryer. *Biochemistry*, 5th ed. New York: W. H. Freeman, 2001.

Internet Resource

Online Mendelian Inheritance in Man. Johns Hopkins University, and National Center for Biotechnology Information. <http://www.ncbi.nlm.nih.gov/Omim>.

Methylation

DNA methylation is a mechanism used to regulate genes and protect DNA from some types of cleavage. It is one of the regulatory processes that are referred to as epigenetic, in which an alteration in gene expression occurs without a change in the nucleotide sequence of DNA. Defects in this process cause several types of disease that afflict humans.

Biochemical Features

In methylation, a methyl group ($-CH_3$) is added to position five of the cytosine ring in a DNA molecule (see diagram), attaching itself there by means of a chemical bond. For methylation to occur in DNA, certain conditions must be met. The cytosine must be linked to guanine, with the guanine occurring at the 3' ("three prime") end of the DNA molecule, in a formation that, in scientific notation, is expressed as 5'-CG-3' and is referred to as a CpG dinucleotide (with the "p" representing a phosphate group). It occurs in many eukaryotic organisms, including mammals, and was recently found to occur in *Drosophila* (fruit fly), but does not occur in yeast.

The methylation process is performed by enzymes called DNA methyltransferases (DNMTs). Currently, five DNA methyltransferase members have been identified in humans (DNMT1, 2, 3A, 3B, 3L). The precise function of many of these proteins is not yet known. The most well-characterized DNA methyltransferase is DNMT1. This **enzyme** is required for proper embryonic development in mammals, and is involved in copying the methylation pattern from an existing DNA strand to the newly synthesized DNA strand following DNA **replication**. For this reason, DNMT1 is called the maintenance DNA methyltransferase. In contrast, DNMT3A and DNMT3B are believed to be de novo methyltransferases, or proteins that can add a

enzyme a protein that controls a reaction in a cell

replication duplication of DNA

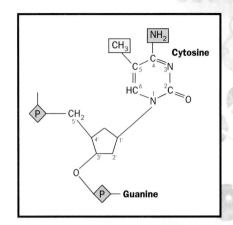

methyl group to a cytosine at a new location in the DNA strand, instead of just copying one that already exists. It is not yet known what determines which cytosines in the DNA will have a methyl group added by a de novo methyltransferase.

CpG Islands

The frequency of occurrence of the CpG dinucleotide in the genome is not random, as would be expected. Instead, the CpG dinucleotide is greatly under represented in eukaryotic genomes, occurring at approximately 5 to 10 percent of its predicted frequency, according to some estimates. Of these occurrences, it is further estimated that 70 to 80 percent are methylated. This under representation of CpG dinucleotides in the genome may result from a spontaneous conversion of methyl cytosine to thymine in DNA by a process known as deamination, in which an amino group (in this case, NH_2) is removed from 5-methylcytosine. For this reason, methylated cytosines represent potential sites of spontaneous DNA mutation in the genome.

There are, however, small regions of DNA that are very rich in linked cytosines and guanines, but which are unmethylated. These regions, which can consist of from 500 to 5,000 base pairs of unmethylated DNA, are referred to as CpG islands. These "islands" commonly occur in **promoter** regions of genes (regions where RNA polymerase binds to start transcription), which are located at the 5′ ("five prime") end of the genes. In fact, about 50 percent of all genes contain a CpG island in their promoter regions. The lack of methylation in CpG islands leads to a less compact chromatin structure, and generally allows for active gene expression. The methylation of unmethylated CpG islands leads to the silencing of genes required for proper cell growth control and is a common mechanism in the development of many types of cancer.

Host Defense

The process of methylation was first described in bacteria in 1948. Most bacterial strains contain enzymes called restriction **endonucleases**. These restriction enzymes recognize certain short sequences of DNA, and cleave the DNA strand at these sites. By modifying its DNA with a pattern of methylation specific to its strain, a bacterium can use this system of modification and restriction to distinguish its own DNA from invading foreign DNA. Methylation serves to protect the bacterial DNA from digestion by its own restriction enzymes.

In mammals, methylation has also been proposed to be a genome defense system against foreign DNA such as viruses. Viruses that infect cells and integrate into the host cell DNA frequently become methylated. While methylation in eukaryotes does not mark DNA for digestion, methylation can inactivate a promoter and thereby silence gene expression from a viral promoter. Evidence in support of this comes from the fact that most methylated cytosines in the mammalian genome lie within viral and **transposon** DNA. In addition to silencing gene expression from foreign DNA promoters, methylation has also been shown to prevent DNA sequences such as transposons from moving to a new site in the DNA. In this way,

Methylation is the addition of a –CH_3 group at position 5 on the cytosine ring (box). Removal of the –NH_2 group by deamination converts this nucleotide into thymine. Methylation occurs on cytosines linked to guanines at their 3′ end.

promoter DNA sequence to which RNA polymerase binds to begin transcription

endonucleases enzymes that cut DNA or RNA within the chain

transposon genetic element that moves within the genome

High densities of methylated cytosines occur in gene promoter regions.

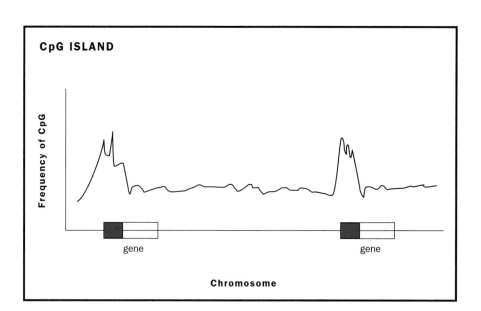

CpG ISLAND

Frequency of CpG

gene   gene

Chromosome

methylation can limit the spread of infectious virus from cell to cell, and prevent the damaging spread of transposon sequences.

Gene Repression

The addition of methyl groups to DNA can repress, or silence, gene expression by leading to a more compact DNA structure that excludes the binding of most proteins. Because of this, regions of DNA that are heavily methylated are not usually accessible to the binding of proteins needed for gene expression, such as **transcription factors**. Transcription repression is also aided by proteins that specifically bind to methylated DNA and contribute to the more compact DNA structure. These methyl-binding proteins contain a methyl-binding domain (MBD) that specifically recognizes methylated DNA. MeCP2, which causes a genetic disorder known as Rett syndrome, is one of these methyl-binding proteins that can bind to a single methylated cytosine in DNA and prevent the binding of other proteins like transcription factors. If it appears in a gene promoter region, it can prevent transcription from occurring.

transcription factors proteins that increase the rate of gene transcription

Gene Imprinting

Methylation can also function in the process of genomic imprinting, which is found in sexually reproducing species. During sexual reproduction, each parent contributes one allele for each gene to the offspring. Genomic imprinting is a difference in gene expression that depends on whether the gene allele originated from the mother or the father. This differential pattern of gene expression occurs as the result of differential methylation in the gene promoter. One example of an imprinted gene is the insulin-like growth factor II (*IGF2*) gene. There are two copies (or alleles) of this gene, but only one is expressed. For *IGF2*, the maternal allele is methylated and silent, whereas the paternal allele is unmethylated and expressed. The opposite situation may occur in other genes. In this way, only one copy of an imprinted gene is expressed, and this provides a mechanism for a cell to determine the parental origin of certain genes.

DNA Methylation and Human Disease

Changes in DNA methylation patterns have been implicated in the development and progression of many types of cancers. Additionally, defects in DNA methylation have been associated with several genetic diseases, including ICF (Immunodeficiency, Centromere Instability, and Facial Anomalies), Rett, and Fragile X syndromes, all of which result in variable degrees of mental retardation. This common effect on neurological function may result from the fact that DNA methylation occurs at high levels in the brain, and implies that the brain requires DNA methylation for proper development.

ICF syndrome is a rare recessive disease characterized by variable immunodeficiency, developmental delays, distinctive facial features, and mental retardation. In 1999 it was found that patients with ICF syndrome have mutations in the DNA methyltransferase gene *DNMT3B*, located on human chromosome 20q11. These mutations impair the function of *DNMT3B*, resulting in an overall reduction in DNA methylation, or hypomethylation. This, in turn, leads to destabilization of the **centromeres** of chromosomes 1, 9, and 16. The alteration in chromosome structure leaves these chromosomes susceptible to DNA breakage and possibly alters the expression of genes located in these regions.

centromeres regions of the chromosome linking chromatids

Rett syndrome and Fragile X syndrome are other genetic disorders that result from a disruption in the function of methylated DNA. Rett patients, who are almost all young females, at first develop normally. Later on, however, they develop mental retardation, autism, and movement disorders. These patients have a mutation in the gene for the methyl-binding protein MeCP2. This protein usually represses gene expression by binding tightly to methylated DNA and causing repression.

Fragile X syndrome is the most common form of inherited mental retardation. Fragile X results from an increase in the number of CGG repeats in the promoter region of the *FMR1* gene on the X chromosome. When the number of repeated sequences reaches the 200 to 600 copy range, the repeat itself becomes very heavily methylated, leading to silencing of the *FMR1* gene. The critical importance of DNA methylation in mammalian development is obvious, given the diseases that result when this process is improperly regulated. SEE ALSO GENE EXPRESSION, OVERVIEW OF CONTROL; IMPRINTING; NUCLEOTIDE.

Theresa M. Geiman and Keith D. Robertson

Bibliography

Baylin, Stephen B., and James G. Herman. "DNA Hypermethylation in Tumorigenesis: Epigenetics Joins Genetics." *Trends in Genetics* 16 (2000): 168–174.

Hendrich, Brian. "Human Genetics: Methylation Moves into Medicine." *Current Biology* 10 (2000): R60–R63.

Jones, Peter A., and Peter W. Laird. "Cancer Epigenetics Comes of Age." *Nature Genetics* 21 (1999): 163–167.

Robertson, Keith D., and Peter A. Jones. "DNA Methylation: Past, Present and Future Directions." *Carcinogenesis* 21 (2000): 461–467.

Robertson, Keith D., and Alan P. Wolffe. "DNA Methylation in Health and Disease." *Nature Reviews Genetics* 1 (2000): 11–19.

Wolffe, Alan P. "The Cancer-Chromatin Connection." *Science and Medicine* (1999): 28–37.

Microbiologist

A microbiologist studies living organisms that are invisible to the naked eye. Microbiologists study bacteria, fungi, and other one-celled organisms (microbes), as well as viruses. A microbiologist may study a single molecule isolated from a bacterium, or a complex ecosystem with many microbial species.

In the course of their work, microbiologists do a variety of tasks. These include inoculating microbes using sterile techniques, viewing microbes under the microscope, purifying DNA or other molecules from microbes, counting microbes, preparing sterile **media**, and developing experiments to better understand these organisms, often using the techniques of genetic analysis. Many microbiologists supervise other employees, and they frequently work with a group of people to plan experiments or validate laboratory procedures. Some microbiologists teach or are engaged in sales, so their duties may include presenting information. As they gain experience, microbiologists working in a laboratory frequently assume more responsibility to supervise others and to plan a research program.

In most cases, microbiologists have training beyond high school. This training could consist of a two-year associate degree with coursework in biology, chemistry, physics, mathematics, and microbiology. Many microbiologists have a four-year bachelor's degree in microbiology, biology, or chemistry. This degree includes coursework in all the sciences, as well as several courses in microbiology. A master's degree usually requires an additional two years of coursework and research, and prepares individuals for additional supervisory or teaching positions. A doctoral degree (Ph.D.) typically requires an additional four to seven years of schooling, and opens up a wide range of teaching, research, and executive opportunities. In many cases, microbiologists who earn a Ph.D. do a further two to four years of postdoctoral research to gain additional research experience before they assume an independent position.

Some microbiologists work in hospital or clinical laboratories, where they identify disease-causing bacteria. Others work in the food industry, checking to make sure that food products do not contain **pathogenic** organisms. Environmental microbiologists study the role of bacteria and other organisms in ecological systems. Some microbiologists work in research laboratories, helping to make fundamental discoveries about microbes. Microbiologists also teach about these organisms in community colleges and universities.

The rewards of a career in microbiology are many, and may vary depending upon the career path taken. Medical microbiologists gain satisfaction from the knowledge that they assist in helping to prevent and treat disease. Food microbiologists are critical for maintaining the safety of our food supply. Research microbiologists are rewarded by the knowledge that they may, in the course of their work, learn something that no one else has ever understood. Teaching microbiologists gain satisfaction in helping others learn about a discipline they find fascinating.

Starting salaries for microbiologists also vary a great deal, depending on whether a person is employed in industry or in the public sector. In general, positions in industry command higher salaries than positions in

media (bacteria) nutrient source

pathogenic disease-causing

educational institutions, and individuals with many years of experience can command salaries at the high end of the range offered in their particular career path and at their educational level. Salaries for microbiologists with a bachelor's degree start at about $18,000 per year, and range up to $50,000 per year. For microbiologists with a master's degree, the starting salary is typically $25,000 per year, and can rise to $80,000. For microbiologists with a doctoral degree, the starting salary is about $35,000 per year, and can range up to $200,000 or more. SEE ALSO BIOTECHNOLOGY ENTREPRENEUR; COLLEGE PROFESSOR; LABORATORY TECHNICIAN; MOLECULAR BIOLOGIST.

Patrick G. Guilfoile

Bibliography

American Society for Microbiology. *Your Career in Microbiology: Unlocking the Secrets of Life.* Washington, DC: Author, 1999.

Internet Resource

Careers in the Microbiological Sciences. American Society for Microbiology. <http://www.asmusa.org/edusrc/edu21.htm>.

Mitochondrial Diseases

Mitochondria are intracellular organelles that play a critical role in cellular metabolism. Mitochondria contain the electron transport chain, which transfers electrons to oxygen by means of a process called oxidative **phosphorylation**. This process releases energy for the production of adenosine triphosphate (ATP) by forming a pH and electrical **gradient** (called the chemiosmotic gradient) across the inner mitochondrial membrane. In addition to oxidative phosphorylation, the mitochondria fulfill a number of other functions, including the following:

phosphorylation addition of the phosphate group PO_4^{3-}

gradient a difference in concentration between two regions

- Make ATP for cellular energy
- Metabolize fats, carbohydrates, and amino acids
- Interconvert carbohydrates, fats, and amino acids
- Synthesize some proteins
- Reproduce themselves (replicate)
- Participate in apoptosis
- Make free radicals

Of these functions **apoptosis** is particularly important in development and disease. However, human disease may result from impairment of any of these functions.

apoptosis programmed cell death

Mitochondria are inherited from the mother, but not from the father. In the process of egg formation, there is thought to be a "bottleneck" in mitochondrial number, such that the unfertilized egg may have as few as 1,000 mitochondria. This number increases 100-fold after the ovum is fertilized. The mitochondria contain their own DNA, mitochondrial or mtDNA, and during development there may be selective **amplification** of some of these mtDNA molecules, leading to increases or decreases in the presence of mutated mtDNAs.

amplification multiplication

ELECTRON TRANSPORT CHAIN DEFECTS OF LEIGH SYNDROME

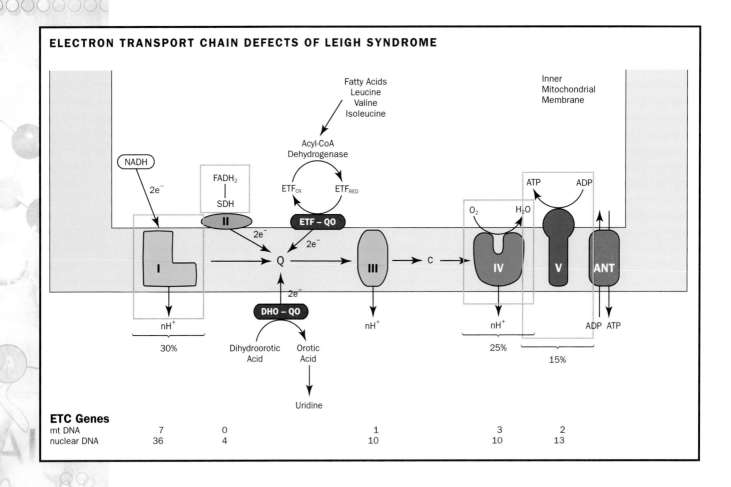

ETC Genes					
mt DNA	7	0	1	3	2
nuclear DNA	36	4	10	10	13

Figure 1. This figure provides an overview of the electron transport chain (ETC). Three defects causing Leigh's disease are highlighted above (purple boxes), with the percentage of cases due to these defects indicated. Numbers of nuclear and mitochondrial DNA (mtDNA) genes encoding the human ETC complexes are listed.

symbiotic describes a close relationship between two species in which at least one benefits

anaerobic without oxygen or not requiring oxygen

glycolysis the breakdown of the six-carbon carbohydrates glucose and fructose

The Importance of the Electron Transport Chain

The origins of mitochondria are unknown, but the likely explanation, called the endosymbiont hypothesis, holds that they arose as free-living bacteria that colonized proto-eukaryotic cells, thereby establishing a **symbiotic** relationship. Primitive eukaryotic cells with intracellular mitochondria capable of metabolizing oxygen would have had an advantage in an oxygen-rich environment. The electron transport chain produces far more energy for each molecule of glucose consumed than is produced by **anaerobic** respiration. The oxidative phosphorylation process conducted by the mitochondria produces thirty-eight molecules of ATP, compared to two molecules of ATP produced by anaerobic **glycolysis**. Oxidative phosphorylation allows the conversion of toxic oxygen to water, a protective biological advantage.

A disadvantage of oxidative phosphorylation, however, is the formation of reactive oxygen species, such as singlet oxygen and hydroxyl radicals, which damage such cellular components as lipids, proteins, and DNA. A normally functioning electron transport chain produces reactive oxygen species from about 2 percent of the electrons that it transports. In disease states and in aging, larger quantities of reactive oxygen species are generated, and this may be a significant factor in cellular deterioration as well as a major contributor to the aging process.

Mitochondrial Genes and Disease

Mitochondrial DNA encodes approximately 3 percent of mitochondrial proteins. The relative contribution of the mitochondrial and nuclear genomes

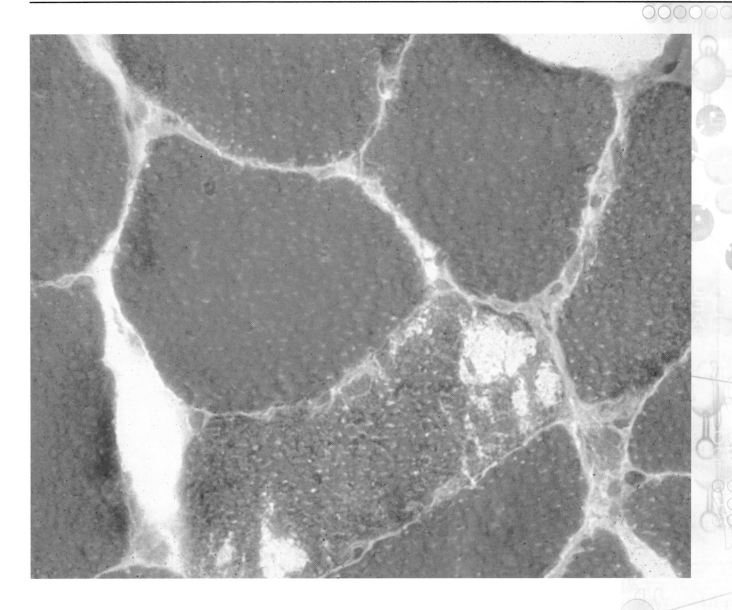

in coding for electron transport chain subunits is detailed in Figure 1. Human mtDNA contains 16,569 nucleotide bases and encodes thirteen polypeptides of the electron transport chain, twenty-two transfer RNAs (tRNAs) and two ribosomal RNAs (rRNAs). In addition, mtDNA has a control region (termed the D-loop), which contains considerable genetic variation. The D-loop forms the basis of forensic medicine DNA identification and has been very useful in the molecular anthropological study of human origins.

In 1988 the first human disease associated with mtDNA deletions was reported. These patients suffered from muscle and brain diseases with ragged red fiber muscle disease (myopathy), with or without progressive neurological deterioration. Ragged red fibers are muscle fibers, that have a disorganized structure and an excess of abnormal mitochondria and that stain red when treated with a histochemical stain called modified Gomori trichrome (Figure 2). In 1988 Kearns-Sayre syndrome, which primarily affects the muscles, heart, and brain, was found to be due to mtDNA deletions or duplications. About the same time, the maternally inherited disorder Leber's hereditary optic atrophy was traced to point mutations in mitochondrial DNA encoding subunits of complex I of the electron transport chain.

Figure 2. Muscle biopsy of mitochondrial myopathy patient (modified Gomori trichrome stain) with ragged red fiber showing disorganization and mitochondrial proliferation. Mitochondria stain red. Other muscle fibers also have increased numbers of mitochondria.

Table 1.

MULTISYSTEM MITOCHONDRIAL DISEASES	
Organ or System Diseased	**Symptoms**
brain	stroke, seizures, dementia, ataxia, developmental delay
muscle	weakness, pain, fatigue
nerve	neuropathy
heart	cardiomyopathy, heart failure, heart block, arrhythmia
pancreas	diabetes, pancreatitis
eye	retinopathy, optic neuropathy
hearing	sensorineural deafness
kidney	renal failure
GI system	diarrhea, pseudo-obstruction, dysmotility

Mitochondrial diseases tend to affect multiple organ systems. The cells and organs most severely affected are those most heavily dependent on ATP, such as those listed in Table 1. Patients will frequently have multiple symptoms or signs, a circumstance that often causes confusion in diagnosis and treatment.

One of the more common presentations of mitochondrial disease in infants and young children is Leigh's disease, first described by the pathologist Dennis Leigh in 1951. This progressive disease primarily affects the brain, with episodic deterioration that is often triggered by mild viral illnesses. Other organ systems are often involved, and there is often high blood or brain lactic acid as a result of a failure in oxidative metabolism (lactic acid is formed from glucose in the absence of oxygen). Figure 1 details the sites of metabolic defect and the percentages of cases affected in cases of Leigh's syndrome. Complex I and IV defects are autosomal recessive diseases, with the culprit genes residing on the nuclear chromosomes. Complex V mutations are mtDNA inherited, and another 25 percent of cases are X-linked, due to pyruvate dehydrogenase deficiency (another mitochondrial enzyme, not shown in Figure 1).

phenotypes observable characteristics of an organism

One of the most common mtDNA diseases seen is due to a single point mutation at position 3,243, with an adenine to guanine mutation in a tRNA leucine gene. Patients with this mutation may have **phenotypes** ranging from asymptomatic (that is, having no visible effects) to diabetes mellitus (with or without deafness). It is estimated that 1 to 2 percent of all diabetics have the A3243G mutation as the cause, affecting 200,000 people in the United States alone. The most severe phenotype to occur from this mutation has been given the acronym MELAS, for mitochondrial encephalomyopathy, with lactic acidosis and stroke-like episodes. The variability of disease phenotype or heterogeneity of disease due to mtDNA mutations arises in part because of variations in the amount of mutated mtDNA within different tissues. This mixture of wild type and mutant DNA within a cell is called heteroplasmy. In many mtDNA diseases, heteroplasmy changes over time, so that there is an increase in mutant DNA in nondividing cells and tissues such as muscle, heart, and brain, with a decrease over time in rapidly dividing tissues such as bone marrow. SEE ALSO APOPTOSIS; DIABETES; INHERITANCE, EXTRANUCLEAR; METABOLIC DISEASE; MITOCHONDRIAL GENOME; MOLECULAR ANTHROPOLOGY.

Richard Haas

Bibliography

Johns, D. R. "Mitochondrial DNA and Disease." *New England Journal of Medicine* 333, no. 10 (1995): 638–644.

Raha, S., and B. H. Robinson. "Mitochondria, Oxygen Free Radicals, and Apoptosis." *American Journal of Medical Genetics* 106, no. 1 (2001): 62–70.

Wallace, D. C. "Mitochondrial DNA in Aging and Disease." *Scientific American* 277, no. 2 (1997): 40–47.

Mitochondrial Genome

One of the defining features of eukaryotic cells is the presence of membrane-enclosed organelles. Two of these organelles, the mitochondria and chloroplast, are unique in that they contain their own genetic material necessary for proper functioning. These organelle genomes are evolutionary relics of free-living bacteria that entered into a **symbiotic** relationship with a host cell. Through the process of cellular respiration, mitochondria produce about 90 percent of the chemical energy that a cell needs to survive. The discovery that mutations in the mitochondrial genome can cause a variety of human diseases has increased our interest in this "other" human genome.

symbiotic describes a close relationship between two species in which at least one benefits

Organelle Structure and Energy Production

The mitochondria (singular: mitochondrion) are enclosed by two membranes, each a phospholipid bilayer with a unique collection of embedded proteins. The outer membrane is smooth, but the inner membrane contains extensive folds called cristae. The cristae provide a means of packing a relatively large amount of the inner membrane into a very small container, thus enhancing the productivity of cellular respiration. The number of mitochondria per cell is correlated with the cell's level of metabolic activity, with a typical cell containing hundreds to thousands of these organelles. Time-lapse photography of living cells reveals mitochondria as very dynamic structures, moving around, changing shape, and dividing.

Often described as the "power plant" of the cell, mitochondria generate ATP by extracting energy from sugar, fats, and other fuels with the help of oxygen. Mitochondria generate most of the energy in animal cells through a process called oxidative phosphorylation. In this process, electrons are passed along a series of protein complexes that are located in the inner mitochondrial membrane. The passage of electrons between these protein complexes releases energy that is stored in the membrane, and is then used to make ATP from ADP.

Mitochondrial DNA: Function and Replication

Scientists have known since the early 1960s that the nucleus is not the only location for DNA in a **eukaryotic** cell. The mitochondria (and the chloroplasts in plant cells) harbor their own small genome. The genes found on the circular 16,569 **base-pair** piece of mitochondrial DNA (mtDNA) in human cells code for thirteen proteins, two ribosomal RNAs (rRNA), and twenty-two transfer RNAs (tRNAs), all of which are essential for the production of ATP by the mitochondria. Each individual organelle contains several copies of the mitochondrial genome. Although they comprise only

eukaryotic describing an organism that has cells containing nuclei

base-pair two nucleotides (either DNA or RNA) linked by weak bonds

Structure of a mitochondrion.

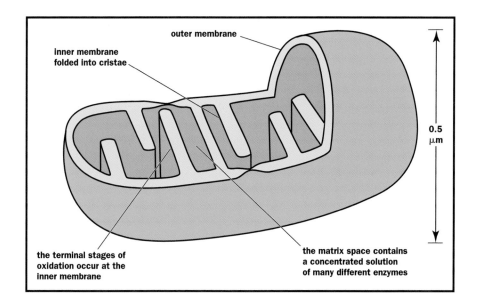

inner membrane folded into cristae

outer membrane

0.5 μm

the terminal stages of oxidation occur at the inner membrane

the matrix space contains a concentrated solution of many different enzymes

introns untranslated portions of genes that interrupt coding regions

a small portion of the proteins found in the mitochondrion, all thirteen proteins encoded by the mtDNA are essential, because they are necessary for oxidative phosphorylation and the production of cellular ATP. All of the remaining mitochondrial components are encoded by nuclear genes and are imported into the organelle.

The mitochondrial genome in mammals is extremely compact, with essentially no **introns** and very little DNA sequence between genes. Each of the protein and rRNA genes is immediately flanked by tRNA genes. Initial transcription of mtDNA produces large RNA molecules that are then processed into smaller units to generate mature tRNAs, rRNAs, and mRNAs.

The two mtDNA strands in the circular molecule can be separated based on their density (due to their differing nucleotide compositions), and are thus designated as the heavy strand (H-strand) and the light strand (L-strand). Both strands are transcribed completely, making two long RNA molecules. Since the two strands are complementary (not identical), they do not each code for the same genes. Instead, the H-strand transcript codes for most of the proteins and tRNAs, while the L-strand codes for most of the rRNAs. Ninety percent of the L-strand does not code for useful products, and is degraded after it is transcribed. The processed L-strand transcript also functions as the starting point for replication of the mitochondrial chromosome.

Endosymbiosis and Genome Reduction

Given the bacterial-like features of mitochondria and chloroplasts (small size, circular genome, and ability to divide on their own), it is believed that each organelle traces its evolutionary history to a free-living bacterial ancestor that was engulfed by a larger cell and then entered into a symbiotic relationship with the host cell. This "serial endosymbiotic theory" proposes that the evolution of the modern eukaryotic cell was a step-wise association, with the acquisition of the mitochondria preceding that of the chloroplast. The most compelling evidence for the endosymbiosis theory has come from the

analysis of complete genome sequences. Comparison of DNA sequence data has identified two specific groups of bacteria, α-Proteobacteria and Cyanobacteria, as the closest living relatives of mitochondria and chloroplasts, respectively. The mtDNA sequence information from numerous organisms has revealed remarkable similarity, reinforcing the idea of a single primary ancestor for the organelle originating very early in the evolution of the eukaryotic cell.

During the course of evolution, a large portion of mitochondrial genes were either lost or transferred to the nuclear **genome**. Elimination of genes from mtDNA is an ongoing evolutionary process made possible because their functions either become dispensable or can be replaced by nuclear functions. A comparison of the complete mtDNA sequence and the working draft of the human nuclear genome project reveals numerous areas of similarity. These regions represent mtDNA sequences that have been transferred from the **cytoplasm** to the nucleus over the course of mammalian evolution. This transfer accounts for the current nuclear location of most of the genes that encode mitochondrial proteins, including most of the proteins required for oxidative phosphorylation. Even eukaryotes that lack mitochondria (such as some protists) contain nuclear genes that encode typical mitochondrial proteins, implying that these eukaryoytes once had mitochondria but subsequently lost them. SEE ALSO CELL, EUKARYOTIC; EUBACTERIA; INHERITANCE, EXTRANUCLEAR; MITOCHONDRIAL DISEASES; MOLECULAR ANTHROPOLOGY.

Stephan Zweifel

genome the total genetic material in a cell or organism

cytoplasm the material in a cell, excluding the nucleus

Bibliography

Wallace, Douglas C. "Mitochondrial DNA in Aging and Disease." *Scientific American* 277, no. 2 (1997): 40–47.

"Special Section: Mitochondria." *Science* 283 (1999): 1475–1497. (A series of articles on the mitochondrial genome, mtDNA diseases, and the evolution of the organelle.)

Internet Resource

MITOMAP (a human mitochondrial genome database). Center for Molecular Medicine, Emory University. <http://www.gen.emory.edu/mitomap.html>.

Mitosis

Mitosis is the process by which all cells divide. Many cells have a limited life span, and mitosis allows them to be renewed on a regular basis. Mitosis is also responsible for generating the many millions of cells that are needed for an embryo to develop into a fetus, an infant, and finally an adult.

Most human cells continually undergo a cycle of different phases. The phases have distinct names but flow smoothly into one another. The mitotic (M) phase is the phase in which the cell's genetic material is split in two. Once the phase is completed, the cell is physically divided into two daughter cells, in a process called cytokinesis.

Before entering the M phase, cells are in interphase, the phase between two cell divisions. Interphase is itself divided into three phases: G_1, S, and G_2, where G stands for gap or growth, and S for synthesis. During the G_1

Various stages of mitosis are evident in this magnification of an onion root tip. The chromosomes are stained red.

phase, daughter cells formed in the previous M phase undergo active cell growth. During the S phase, the genetic material (DNA) contained in the chromosomes is duplicated so that both of the future daughter cells receive the same set of chromosomes. This ensures that they will be genetically identical to each other and to the cell from which they originated.

In human **somatic** cells, each of the forty-six chromosomes replicates to produce two daughter copies that are called sister chromatids. These two copies remain attached to each other at a single point, the centromere, which is a DNA sequence of about 220 nucleotides. The centromere has a disk-shaped protein molecule, called a kinetochore, attached to it. In interphase, the chromosomes are not visible as discrete entities under the light micro-

somatic nonreproductive; not an egg or sperm

scope. Interphase chromosomes are uncoiled threads composed of DNA and protein molecules. This noncondensed form of chromosomes is also called chromatin.

During the G_2 phase of the cell cycle, the chromatin fibers start to condense, eventually turning into tightly coiled, compact bodies, visible as chromosomes. The cell also begins to manufacture protein fibers called microtubules, which will be used later to move the chromosomes to opposite poles of the cell, so two new daughter cells can form. Chromosome condensation and microtubule formation both begin in the G_2 phase but occur mostly during the first stage of the M phase, which is called prophase. The microtubules are organized into a three-dimensional spindle apparatus, where each fiber of the spindle apparatus connects one cell pole to the other like a bridge.

During the next stage of mitosis, called prometaphase, the envelope surrounding the cell nucleus breaks down so that the chromosomes are free to migrate to the central plane of the spindle apparatus. A second group of microtubules grows out, to connect the two opposite sides of the kinetochore to the two poles of the spindle. This arrangement is crucial for ensuring that the two sister chromatids end up in two separate daughter cells rather than being pulled into the same cell.

In the next stage of mitosis, metaphase, the chromosomes become maximally condensed and line up in an imaginary plane, called the metaphase plate, in the center of the cell and perpendicular to the spindle apparatus. All the centromeres are neatly arranged in a circle, about halfway between the two cell poles.

In human cells, at this point the twenty-three pairs of chromosomes, each made up of two condensed sister chromatids held together by a centromere, are visible under the microscope. Unlike in meiosis, the paternal and maternal copies in each pair of chromosomes align independently in the metaphase plate and are not associated with each other. At the end of metaphase, the centromeres that hold the two sister chromatids together all divide simultaneously.

During the next stage, anaphase, microtubules that are attached to the sister chromatids' kinetochores draw the chromatids quite rapidly to opposite poles of the spindle. The separation of sister chromatids completes the partitioning of the replicated genetic material.

The only task remaining during the final phase of mitosis, telophase, is to disassemble the spindle apparatus and re-form the nuclear envelope around each set of sister chromatids. The chromatids can be called chromosomes again, because they each have their own centromere. The chromosomes begin to uncoil, and the genes they carry begin to be expressed again. This is the end of mitosis.

The cell cycle is completed by cytokinesis, the physical division of the cytoplasm into two daughter cells. By the time cytokinesis occurs, other cytoplasmic organelles, such as mitochondria, already have been replicated during the S or G_2 phases, and they have also been directed to the areas around the cell poles that will become the daughter cells. Cytokinesis is followed by the G_1 phase, with active cell growth occurring in each of the two daughter cells.

Differences between Mitosis and Meiosis

Mitosis occurs in all eukaryotic cell tissues and produces genetically identical daughter cells with a complete set of chromosomes. In humans, mitosis produces somatic cells that are diploid, which means they contain two non-identical copies of each of the twenty-three chromosomes. One copy is derived from the person's mother and the other from the person's father.

Meiosis, on the other hand, occurs only in testis and ovary tissues, producing sperm and ova (eggs). The gametes that are produced by meiosis in humans are haploid, containing only one copy of each of the twenty-three chromosomes. Because of recombination and independent assortment of parental chromosomes, the daughter cells produced by meiosis are not genetically identical.

In mitosis, one round of DNA replication occurs per cell division. In meiosis, one round of DNA replication occurs for every two cell divisions. Prophase in mitosis typically takes about thirty minutes in human cells. Prophase in meiosis I can take years to complete. SEE ALSO CELL, EUKARYOTIC; CELL CYCLE; MEIOSIS; NUCLEUS; REPLICATION.

Silke Schmidt

Bibliography

Nasmyth, Kim. "Segregating Sister Genomes: The Molecular Biology of Chromosome Separation." *Science* 297, no. 5581 (2002): 559.

Raven, Peter H., and George B. Johnson. *Biology*, 5th ed. New York: McGraw-Hill, 1999.

Model Organisms

A model organism is a species that biologists choose to study, not necessarily because it has any inherent medical, agricultural, or economic value, but because it has certain traits that make it easy and convenient to work with. Studying model organisms enables researchers to perform experiments that might be impossible to carry out, due to logistical, financial, or ethical constraints, on organisms of more practical interest, such as humans.

This approach has been tremendously successful in the fields of genetics and molecular and cellular biology because, at their most fundamental levels, biological processes are remarkably similar across species. For example, the genetic code and much of the cellular machinery responsible for **replication, transcription, translation,** and gene regulation are essentially identical in all eukaryotic organisms. In many cases, genes have even been demonstrated to be functionally interchangeable between humans and baker's yeast.

Among the most commonly studied model organisms are: *Escherichia coli* (a bacterium), *Saccharomyces cerevisiae* (baker's yeast), *Dictyostelium discoideum* (slime mold), *Drosophila melanogaster* (a fruit fly), *Caenorhabditis elegans* (a soil roundworm), *Brachydanio danio* (zebrafish), *Xenopus laevis* (African clawed frog), *Arabidopsis thaliana* (a mustard weed), *Zea mays* (maize, or corn), and *Mus musculus* (mouse). Others include sea slugs, sea

replication duplication of DNA

transcription messenger RNA formation from a DNA sequence

translation synthesis of protein using mRNA code

This albino African clawed frog is one of the most commonly studied model organisms.

urchins, cyanobacteria, *Chlamydomonas* (an alga), puffer fish, *Tetrahymena* (a protozoan), and rats.

Useful Traits and Attributes

Most model organisms share a set of common features that make them amenable to study in the laboratory: They are generally small, easy, and inexpensive to rear in the lab, and reproduce quickly and prodigiously. In addition, the best genetic model organisms have small genome sizes, and many can reproduce sexually, allowing researchers to cross-breed individuals of different genotypes.

Beyond these common traits, most model organisms have one or several unique attributes that make them ideal for a particular line of research. For instance, zebrafish readily produce many large, transparent embryos, and are therefore a favorite research subject for developmental biologists. The roundworm, *Caenorhabditis elegans*, has a simple but nonetheless sophisticated nervous system, and displays simple behaviors, such as movement, feeding, and mating. These properties make it well suited for neurobiology and behavioral genetics.

The Model Mouse

Besides certain primates (such as monkeys and chimpanzees), which are costly and difficult to rear in the laboratory, the model organism most closely related to humans is the mouse, *Mus musculus*. The mouse genome is about the same size as the human genome, and the organization of genes (the order of genes on chromosomes) is strikingly similar between the two species. Findings from the nearly complete mouse genome sequencing project indicate that mice and humans share about 95 percent DNA sequence similarity. This means that any gene in humans is likely to have an identical or very similar counterpart (homologue) in the mouse genome.

In addition, it is much easier and less expensive to study genes in mice. A technology that has made the mouse an invaluable genetic model system is the ability to engineer "knockout" strains. These are mutant strains in which a single known gene has been selectively deleted from the genome of every cell. For human genes implicated in diseases, knocking out the homologous gene in mice can provide an excellent model system for studying the disease. The knockout mouse may show disease conditions similar to those of the human disease. Learning how the elimination of the gene in the mouse contributes to the mouse disease may then give important clues about the involvement of the **homologous** gene in the human disease. "Disease model" mouse strains are available for such disorders as cancer, Alzheimer's disease, arthritis, diabetes, heart disease, cystic fibrosis, and obesity. SEE ALSO *ARABIDOPSIS THALIANA*; *ESCHERICHIA COLI*: (*E. COLI* BACTERIUM); FRUIT FLY: *DROSOPHILA*; MAIZE; RODENT MODELS; ROUNDWORM: *CAENORHABDITIS ELEGANS*; TRANSGENIC ANIMALS; YEAST; ZEBRAFISH.

Paul J. Muhlrad

homologous carrying similar genes

Bibliography

Alberts, Bruce, et al. *Molecular Biology of the Cell*, 3rd ed. New York: Garland Publishing, 2002.

Pines, Maya, ed. *The Genes We Share with Yeast, Flies, Worms, and Mice*. Chevy Chase, MD: Howard Hughes Medical Institute, 2001. (Available from the Howard Hughes Medical Institute Web site: <www.hhmi.org>.)

Watson, James D., et al. *Recombinant DNA*, 2nd ed. New York: Scientific American Books, 1992.

Internet Resources

euGenes: Genomic Information for Eukaryotic Organisms. <http://iubio.bio.indiana.edu>.

WWW Virtual Library: Model Organisms. <http://ceolas.org/VL/mo/>.

Molecular Anthropology

Anthropology is the study of the origin and development of the human species. Molecular anthropology uses the tools and techniques of molecular genetics to answer anthropological questions, especially those concerning the origins and spread of humans across the globe. These questions mainly fall under the heading of physical or biological anthropology, as opposed to cultural anthropology, which studies social relationships, rituals, and other aspects of culture.

Tracing Human Origins through Genetic Data

Molecular anthropology attempts to answer such questions as whether humans are more genetically similar to chimpanzees than to gorillas; in what region or regions modern humans first developed; what the patterns are of migration and mixture of early human populations; and whether Neandertals were a different species, and whether they died out or mixed in with modern humans. Molecular anthropology is perhaps best known for the studies that surround the discovery of "mitochondrial Eve" (discussed below), although the meaning of that discovery is often misunderstood.

Two major approaches are used in addressing these questions, both of which involve analyzing DNA. The first and most common approach is

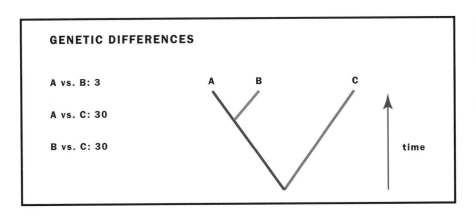

GENETIC DIFFERENCES

A vs. B: 3

A vs. C: 30

B vs. C: 30

A   B   C

time

The fundamental postulate of molecular anthropology is that closer genetic similarity indicates a more recent common ancestry. Here, the similarity of A and B vs. C indicate A and B are more closely related, as shown in the tree diagram.

to compare the DNA of groups of *living* organisms, for example, comparing humans to humans or humans to primates. The second approach relies on isolating and analyzing DNA from an *ancient* source, and comparing it to other ancient DNA or to modern DNA. In both cases, the number of differences between the DNA sequences of the two groups are determined, and these are used to draw conclusions about the relatedness of the two groups, or the time since they diverged from a common ancestor, or both.

The results of molecular anthropological studies are rarely used alone. Instead, the data are combined with information from fossils, archaeological excavations, linguistics, and other sources. Sometimes the data from these different sources conflict, however, and much of the controversy in anthropology centers around how much weight to give each when this occurs.

Advantages of DNA Comparisons

The essential postulate on which molecular anthropology is based is that closer genetic similarity indicates a more recent common ancestry. All organisms are believed to have evolved from a single ancestor. As different life forms evolved, their DNA began to diverge through the processes of mutation, natural selection, and genetic drift. Even within the same species, populations that do not interbreed will accumulate genetic differences, which increase over time. The number of these differences is proportional to the amount of time since the two groups diverged.

There are several advantages to comparing DNA data instead of external physical characteristics (collectively called the phenotype). Environmental factors can shape the phenotype to make two individuals with the same genetic makeup look different. For instance, nutrition has a profound effect on height, and if we used average height to classify humans, we might mistakenly conclude that medieval humans represented a different subspecies because they were significantly shorter than modern humans. DNA comparisons, on the other hand, would show no significant difference between these groups.

Another advantage is that DNA sequence differences can be easily quantified—two base changes in a gene are more different than one. Despite being random events, **mutations** occur at a fairly steady rate, constituting a "molecular clock," and so the number of differences can be use to estimate the time since the two organisms shared a common ancestor.

mutations changes in DNA sequences

Molecular anthropologic data has been used to argue that modern humans arose in Africa and migrated out to replace archaic populations elsewhere. The details and the timing are still controversial.

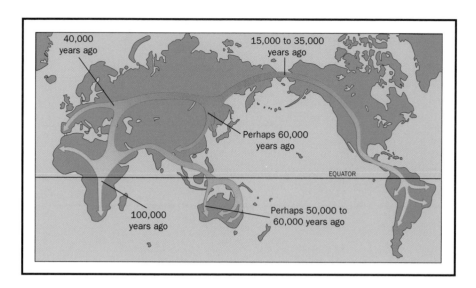

Finally, since all organisms contain DNA, the sequences of any two organisms can be compared. The same techniques used in molecular anthropology can also be applied to evolutionary questions in other species, to determine the evolutionary relations between different animal species, for instance, or even between bacteria and humans.

Caveats About Sequence Comparisons

On the other hand, the simplicity and power of sequence comparisons can lead too easily to an oversimplified interpretation of results, and to conclusions that may sound more significant than they are. A prime example is the often-repeated statement that humans and chimpanzees share 98 percent of their DNA.

It may be true that 98 out of 100 bases are the same in the two genomes, but what is the significance of this fact? It does not mean that 98 percent of our genes are identical. In fact, almost all of them differ slightly, some dramatically. It also does not tell us whether the significant differences between humans and chimps arise from a few very different genes, or many slightly different ones. Moreover, there are significant differences in genome structure not accounted for by the sequence comparison. For instance, humans have forty-six chromosomes, whereas chimpanzees have forty-eight; they have about 10 percent more DNA than humans do; and humans have more copies of a certain kind of **transposable genetic element** than they do.

transposable genetic element DNA sequence that can be copied and moved in the genome

Most importantly, the sequence similarity certainly does not tell us that humans "are" 98 percent chimpanzee—we are two entirely different species, as is obvious from differences in anatomy and behavior. If the profound differences between humans and chimps are not reflected in the sequence data, it may be that this simple tabulation of difference does not adequately summarize the ways in which DNA can cause two organisms to differ.

The 98 percent figure, therefore, may be used to say that chimps and humans are closely related, and are more closely related to each other than either is to an organism with a greater number of sequence differences, such as the orangutan. However, it may not be used to draw conclusions about the similarity of humans and chimps as organisms.

Types of DNA Comparisons

The human genome is much too large to sequence all of it to make comparisons, using current technology. Instead, much smaller portions of it are used. One strategy is to compare gene sequences, such as the sequence for hemoglobin. A potential problem with this is that most mutations in such useful genes are harmful, and so the few harmless mutations they accumulate may be similar between two individuals, despite a long evolutionary separation. Nonetheless, gene comparisons are useful for distantly related species, such as humans and yeast.

An alternative is to look at noncoding regions of DNA. These include microsatellite DNA sequences, a type of repetitive DNA element found throughout the genome. Because these sequences do not code for protein, most mutations in them do not affect the viability of the organism in which they occur. Thus they accumulate mutations more quickly. Another option is single nucleotide polymorphisms. These are sequences which differ among individuals or groups by a single nucleotide. There are millions of such sequences in the genome. Because there are so many different forms, these noncoding sequences are especially useful for determining kinship among closely related individuals, such as members of a tribe or extended family.

One potential problem with sequence comparisons is back mutation, in which a base mutates to another, and then reverts to the original (for example, $C \rightarrow T \rightarrow C$). When this occurs, two sequences may appear to be more closely related (less separated in time) than they really are, since the intervening mutation (the change from C to T, in this case) may not be apparent. Because of back mutation, the observed number of differences between sequences represents the minimum actual difference. Correction factors can be applied to estimate the true difference.

Another potential problem with any sequence on a chromosome, whether or not that sequence codes for a protein, is that most chromosomes do not remain intact during **meiosis**. This is because crossing over occurs, in which homologous chromosomes recombine (exchange segments). After a few generations, it becomes very difficult to track individual sequences and compare them with any confidence to similar sequences in another person. To avoid this problem, molecular anthropologists focus on two sources of DNA that do not recombine: the Y chromosome and mitochondrial DNA.

meiosis cell division that forms eggs or sperm

The Y Chromosome

The Y chromosome, which determines male sex, does not undergo recombination along most of its length. Instead, it passes intact from father to son. A man's Y chromosome, therefore, is a more-or-less exact copy of the one possessed by his father, grandfather, great-grandfather, and so on back through time. Like any other DNA segment, it may mutate, and any changes it accumulates are faithfully passed along as well. Two brothers are likely to have exactly the same Y chromosome sequence. Two men whose last common male ancestor was ten generations ago, however, are likely to have slightly different sequences. Comparison of the sequences of two Y chromosomes, therefore, can show how closely related two males are.

Analysis of Y chromosome microsatellite sequences was used to show that Thomas Jefferson was an ancestor of some of the male descendants of Sally Hemings, a slave owned by Jefferson.

Skulls from a Neandertal (left) and an anatomically modern human Cro-Magon (right). Molecular anthropologists analyze modern and ancient DNA to determine how the Neandertal is related to modern humans.

Y chromosome analysis has been used to track migration of human populations, and to study the relatedness of modern populations. For instance, Jews and Palestinian Arabs derive from a common ancestral population that lived in the Middle East about 4,000 years ago. Recent studies have linked the ancestors of American Indians to several small populations in Siberia, confirming the predominantly Asian origin of American Indians and refining the understanding of their migration history. Many other similar studies have been performed, providing an increasingly clear (and complex) picture of human migration and mixture.

Mitochondrial DNA and the Origin of Modern Humans

Mitochondria are energy-harvesting organelles in the cell. They are inherited only from the mother, and so track maternal inheritance in the same way that the Y chromosome tracks paternal inheritance. Like microsatellite DNA, mitochondrial DNA accumulates mutations faster than chromosomal coding DNA.

One of the earliest and most famous mitochondrial studies was used to address a central question in anthropology: Where and how did modern humans originate?

The *Homo* genus itself is universally believed to have originated in Africa. Groups of *Homo erectus* are known to have migrated out of Africa, populating Europe and Asia between one and two million years ago. *H. erectus* gradually changed in character, so that by about half a million years ago, it had taken on some more modern characteristics. Anthropologists call these groups "archaic" modern humans. They include the Neandertals, who lived in Europe and the Middle East from 150,000 to 28,000 years ago. Did modern humans evolve from these older populations in several different regions simultaneously? Or did they arise from a small group in Africa, and spread out from there? If so, did they mix with less advanced local populations (such as Neandertals), or replace them entirely?

The scientists who performed the mitochondrial DNA study (Rebecca Cann, Mark Stoneking, and Allan Wilson) reasoned that populations that had been in one place for only a short period of time would show very little variation in their mitochondrial DNA, since they all shared a relatively recent common ancestor. This would be the case in a modern human population if it had only recently migrated into the area in which it is found. (Such relative genetic homogeneity in newly formed populations is known as the founder effect.) In contrast, populations that have remained in place for long periods have much more ancient common ancestors, and therefore have more mitochondrial DNA variations.

To perform their analysis, the scientists collected samples from different ethnic groups from all over the world. They found that the populations with the greatest amount of sequence variation were in sub-Saharan Africa, indicating these were the groups with the most ancient ancestry. All other groups had much less variation, indicating more recent arrivals of those groups in those regions. Cann, Stoneking, and Wilson went on to estimate the date at which all these groups had their most recent common ancestor. Using a figure of 2 to 4 percent sequence divergence per million years, they estimated that the most recent common ancestor lived approximately 200,000 years ago.

The simplest explanation, they argued, was that ancestors of the non-African *Homo sapiens* migrated out of Africa about 200,000 years ago to populate other regions, over time replacing the nonmodern humans (*H. erectus*, Neandertals, and possibly others) already living in these regions. They argued that the relatively short time since the divergence of all modern humans was too brief to support the alternative hypothesis, that each local group of archaic humans had independently evolved modern traits, a model called multiregional evolution.

The conclusions drawn in this study are still controversial. Numerous other studies have been done since, and the data have been subjected to multiple different analyses. Some studies suggest differing dates for the most recent common ancestor (ranging from 100,000 to 400,000 years ago), and others suggest that an exclusive African origin is not the only possible interpretation of the data.

It is important to keep in mind that the vast number of comparisons that must be made in such studies require computer programs, not only to make the comparisons, but to draw from them the simplest "family tree" that fits. Much of the controversy surrounds the assumptions that must be built into these programs in order to generate results. The mutation rates by which events are timed (the "molecular clock") are also not known with precision, leading to further uncertainties about the exact timing of migrations.

Mitochondrial Eve

In their study, Cann, Stoneking, and Wilson pointed out that the patterns of mitochondrial variations they saw suggested that all the mitochondria of all living groups could be traced back to a single woman who lived in Africa approximately 200,000 years ago. Many people at the time of the original study and since have misinterpreted the results to claim there was a single female ancestor for all modern humans, dubbed "Eve."

It is true the study showed that the mitochondrial DNA in all living humans probably derives from this single woman. However, our nuclear DNA certainly does not derive exclusively (perhaps even at all) from this woman, and the thirty thousand or more genes in our nuclear DNA are far, far more important in determining our characteristics than the thirty-seven mitochondrial genes. Because of recombination, our nuclear DNA cannot be traced back to any single person. Rather, it is an amalgam of countless ancestors through time.

Mitochondrial Eve was also not the first modern human woman, nor the only woman in existence at the time she lived. She was not even the only woman in her local population; it is estimated that Eve was one of about 10,000 people in her population. There was really nothing particularly special about her, except that, by chance, the descendants of her mitochondria happen to have ended up in the cells of every living human. Even this, which sounds remarkable, is pretty much what we should expect from small populations.

To understand why, consider four couples, each of which has two children. Remember that mitochondria are passed from the mother to each child. One couple has two boys. Each boy inherits the mother's mitochondria, but neither passes them on to his children. The mother's mitochondrial type thus becomes extinct in one generation. Two of the couples have a boy and a girl, while the fourth has two girls. These four daughters go on to have children of their own, each with the same distribution according to sex. Whenever a family has only boys, a mitochondrial type becomes extinct. Any time a family has only girls, the mitochondrial type handed down from the mother becomes more common in the next generation. In a small population, over time, it is highly likely that one type will become most prevalent, ultimately becoming the one type found in all the members of the population. Looking back, we would give the name "Eve" to the original mother of that line of mitochondrial genetic inheritance.

A similar phenomenon occurs with the Y chromosome, for exactly the same reasons: Any family with only girls extinguishes that Y chromosome type. The "Y chromosome Adam" lived 60,000 to 150,000 years ago. There is no reason to expect that "Y chromosome Adam" would know "mitochondrial Eve"; indeed, even without the dates to make it impossible, it would be a remarkable coincidence if they had.

Neandertal DNA

DNA can be extracted from some archaeological samples, allowing direct sequencing and comparison with modern DNA. This has so far been possible with specimens up to about 40,000 years old (the dating of such samples is often inexact). DNA is isolated, purified, amplified with the polymerase chain reaction, and sequenced. By this technique, DNA from extinct animals such as the woolly mammoth has been obtained, but not dinosaur DNA, which is millions of years old. The DNA that can be isolated is typically highly fragmented and incomplete, and unsuitable for cloning the whole organism. One application is to analyze the DNA from plant and animal material at camp sites to determine the diet of early humans.

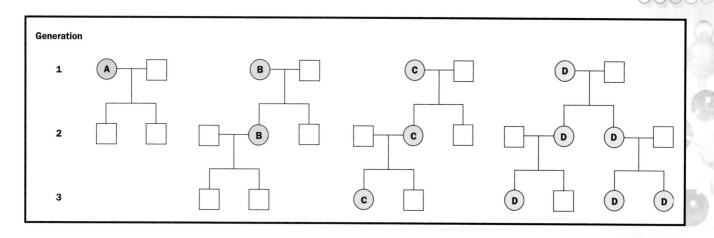

DNA can also be extracted from ancient human remains. As of summer 2002, mitochondrial DNA from two Neandertal skeletons had been extracted, sequenced, and compared. The first was from Germany, and was approximately 35,000 to 70,000 years old. A 378-base pair sequence was determined, and compared to almost one thousand different modern humans. On average, it differed at twenty-seven locations, while modern humans differed among each other at an average of only eight locations. There was some overlap, however, with the least number of differences between Neandertals and modern humans being twenty-two, and the greatest difference noted between modern humans being twenty-three.

The second skeleton was from Russia, and was 29,000 years old. A 345-base pair sequence was determined. It differed at twenty-three locations from a standard modern human sequence, but at only twelve locations compared to the German Neandertal DNA.

Keeping in mind that only two Neandertal sequences have been studied so far, some tentative conclusions have been offered from these data. The amount of difference between the two Neandertal sequences is similar to the amount found between randomly selected modern humans, suggesting that these two specimens, despite being separated by thousands of years, were indeed part of the same lineage.

The amount of difference between the Neandertal skeletal DNA and modern humans suggests that Neandertals were genetically distinctly different from modern humans in their mitochondrial DNA. Were they different enough to constitute a separate species? That is much less clear, and is a source of disagreement among anthropologists. The difference is much less than that between modern humans and chimpanzees, for instance, which suggests that they were not separate species, but it is greater than the differences among subspecies of chimpanzees, which suggests that perhaps they were. Scientists have not been able to compare Neandertal sequences to sequences from anatomically modern humans living at the same time as the Neandertals. It may be that those sequences would be more similar. At present, the relationship of Neandertals to modern humans has still not been conclusively determined.

Conclusion

Using the tools of molecular genetics, DNA sequences can be compared among groups to test hypotheses about the evolutionary relatedness of

Mitochondrial inheritance is through the mother (circles) only. After only three generations, mitochondrial type D is present in 75 percent of the women in this hypothetical population. Over time, it is quite likely that only one type will be left, simply by chance.

organisms, and about the time that has elapsed since divergence. Molecular anthropology has made major contributions to understanding the migration and mixture patterns of human groups. It has also provided significant new insights into the rise and spread of modern humans and their relation to earlier human groups. As more data becomes available and better models are devised for their interpretation, the results are likely to become less provisional and more certain. SEE ALSO INHERITANCE, EXTRANUCLEAR; FOUNDER EFFECT; MITOCHONDRIAL GENOME; MUTATION RATE; POLYMORPHISMS; POPULATION BOTTLENECK; REPETITIVE DNA ELEMENTS; SEQUENCING DNA; Y CHROMOSOME.

Richard Robinson

Bibliography

Avise, John C. *Molecular Markers, Natural History, and Evolution.* New York: Chapman and Hall, 1994.

Hammer, M. F., et al. "Jewish and Middle Eastern Non-Jewish Populations Share a Common Pool of Y Chromosome Biallelic Haplotypes." *Proceedings of the National Academy of Sciences* 97, no. 12 (2000): 6769–6774.

Marks, Jonathan. *What It Means to Be 98% Chimpanzee: Apes, People, and Their Genes.* Berkeley, CA: University of California Press, 2002.

Relethford, John H. *Genetics and the Search for Human Origins.* New York: Wiley-Liss, 2001.

Vigilant, L., et al. "African Populations and the Evolution of Human Mitochondrial DNA." *Science* 253 (1991): 1503–1507

Internet Resource

Y Chromosome Links. <http://john.hynes.net/y.html>.

Molecular Biologist

Biologists study life on many different levels. For example, a cellular biologist is concerned with the most basic unit of life, the single cell, whereas an evolutionary biologist may investigate the origin and genealogical history of a particular species of plant or animal life. The molecular biologist is concerned with understanding the biological phenomena of life at the molecular level. Molecular biology is a multifaceted discipline of recent origin, having emerged in the 1980s from the related fields of biochemistry, genetics, and cell biology.

Basic and Applied Research

A molecular biologist might investigate the genetic basis of a disease, analyzing the gene or genes suspected of causing the disease at the molecular level by using the biochemical technique of DNA sequencing. Genes code for proteins; that is, a particular gene contains the molecular information for producing one particular protein. A gene is expressed through the process of **transcription**. The DNA of the gene is transcribed by a protein known as RNA polymerase. Some genes are expressed frequently, others rarely or only during special times in development. Thus, a molecular biologist might also seek to understand the regulation of gene expression by studying how and when a gene's RNA message (mRNA) appears.

The mRNA resulting from gene expression is the blueprint for the protein. Ribosomes, the cell's protein synthesis factories, translate the mRNA

transcription messenger RNA formation from a DNA sequence

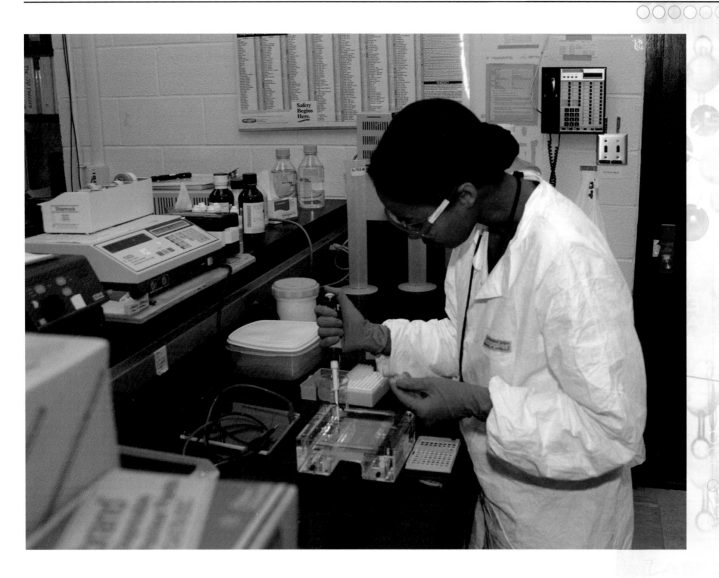

(read the message) and assemble the protein. After **translation**, a protein may be modified by covalent attachment of carbohydrates and lipids to particular amino acids. A molecular biologist might seek to determine the three-dimensional structure of a modified protein using techniques like X-ray diffraction and nuclear magnetic resonance.

In addition to carrying out basic research, molecular biologists may also work in applied research. Using recombinant DNA technology, for example, molecular biologists have created economical vaccines against deadly diseases. The molecular biologist often works at the frontier or cutting edge of a discipline. The rewards of such work include the thrill of intellectual discovery and the opportunity to conduct independent research. Also, the efforts of molecular biologists can bring great benefits to society.

Career Preparation

To prepare for a career as a molecular biologist, the student should begin by taking a broad selection of science and math courses in high school. Typically, such a course of study would include biology, chemistry, and physics, as well as geometry, algebra, and calculus. Good communication skills are

At the Los Alamos National Laboratory, molecular biologist Susana Delano performs an analysis of anthrax DNA utilizing gel electrophoresis. This New Mexico lab has performed research and DNA analysis to aid investigations of anthrax exposures in the United States.

translation synthesis of protein using mRNA code

very important for a scientist, so English and public speaking should not be neglected. At college, the student may wish to pursue a field of study involving biology or chemistry as a major, with an emphasis on laboratory training.

Usually about four to five years of study are needed to satisfy the requirements for an undergraduate (bachelor's) degree. Time for specialization comes at the graduate level. At this point, students should begin to focus their curiosity and choose a major area of interest, such as biochemistry, biophysics, genetics, or cell biology. They should also begin to look for a university that has an excellent reputation for research in the area of choice. Then, the student must seek out a particular laboratory and research advisor for real hands-on training and experimentation.

Two degree tracks are typically offered in graduate school: a master's and a doctoral (Ph.D.) program. On average, a student may earn a master's degree in molecular biology in roughly three years; earning the Ph.D. degree may require four to six years of work. Many pharmaceutical and biotechnology companies actively recruit scientists at the master's level who have training in molecular biology techniques. Students who have obtained a Ph.D. degree in molecular biology commonly undertake postdoctoral training: two to four years of additional study and research. Traditionally, only students interested in academic or government careers pursued postdoctoral studies; however, today many private companies offer one- and two-year postdoctoral positions as a means of attracting top scientific talent.

The annual salaries of molecular biologists can range from $20,000 to $150,000 or more, and are influenced by many factors, such as education (master's versus doctoral degree), experience (just beginning or a seasoned veteran), field of expertise ("hot" fields pay better), employer location (big city or small town), and the local supply of and demand for trained life scientists. Typically, industry positions come with somewhat higher salaries than academic or government positions; however, job security in industry may be tied to the financial success of the company. Academic and government positions may offer more intellectual independence, but sometimes lower salaries. The demand for well-trained, creative molecular biologists in government, industry, and academia continues to grow as our knowledge of life's basic processes deepens. SEE ALSO GENETICIST; LABORATORY TECHNICIAN.

Samuel E. Bennett and Dale Mosbaugh

Molecular Clock *See Mutation Rate*

Morgan, Thomas Hunt

Geneticist
1866–1945

Thomas Hunt Morgan proved the validity of the chromosomal theory of heredity and led a research group whose insights into the physical nature of inheritance propelled genetics into the center of biology in the twentieth century.

Training and Early Interests

Morgan was born and raised in Kentucky, and received his bachelor's degree from the State College of Kentucky in 1886. He pursued graduate study at

Johns Hopkins University in Baltimore, and eventually became a professor of biology at Bryn Mawr College in 1891. His early interests were in developmental biology and evolution. After moving to Columbia University in 1901 and coming under the influence of the great cell biologist Edwin Wilson, Morgan turned his attention to understanding the physical basis of inheritance, which he saw as a means to test theories about the role of mutation in evolution.

At the time Morgan began his work, chromosomes had been seen in cells, but their significance was unknown and not widely considered. A student of Wilson's, Walter Sutton, had recently proposed that chromosomes carried the genetic material, but had little evidence to support this important hypothesis. At the time, the gene itself was an abstract concept with no known physical correlate, and many scientists thought it was not a physical entity at all, but only a convenient fiction for describing some experimental results. In fact, it was Morgan's use of the term "gene" that helped bring it into general use in science.

To attack the issue of heredity, Morgan chose to work with the fruit fly, *Drosophila melanogaster*. This fly requires little space, breeds quickly, has many observable characteristics, and has only four chromosomes, making it an ideal model organism for genetics studies. Morgan also gathered a trio of very bright students, Hermann Muller, Alfred Sturtevant, and Calvin Bridges, and cultivated an egalitarian system of collaboration that was unknown in most other labs. The combination of the right question, the right model, the right collaborators, and some luck allowed Morgan and his group in their lab, dubbed "The Fly Room," to make their fundamental discoveries. Beginning in 1908, they proved that chromosomes do indeed carry the genes, that genes are discrete physical things arranged on chromosomes like beads on a string, that genes change places on chromosomes, that genes can be mutated and those mutations are faithfully inherited, and that mutations can be caused by exposure to high-energy radiation or other environmental phenomena.

A Lucky Discovery

This long string of seminal discoveries began with the discovery of a single male white-eyed fly among the many thousands of normal red-eyed ones. Morgan bred this mutant male with a red-eyed female. All the offspring were red-eyed, indicating the white form of the gene (called the white **allele**) was "recessive" to the dominant red allele: Flies carried the mutant allele, but its effects did not show up. When these offspring were crossed, the ratio of red to white was 3:1, just what would be expected for a classical **recessive** trait.

However, Morgan noted an unusual fact about the white-eyed flies—all of them were male. Morgan knew that the female *Drosophila* had two so-called X chromosomes, while the male had only one. Combining this fact with his discovery that only males showed the white-eye trait, he reasoned that the white-eye mutant allele must be on the X chromosome. Males show the white-eye trait because the mutant white allele is the only one they have—they don't have a second X chromosome with a normal red allele. Females rarely show the white-eye trait, because they have a normal red-eye allele on the other X chromosome.

Thomas Hunt Morgan won the 1933 Nobel Prize in physiology or medicine for his discoveries concerning the role of chromosomes in heredity.

allele a particular form of a gene

recessive requiring the presence of two alleles to control the phenotype

LINKAGE AND CHROMOSOME MAPPING

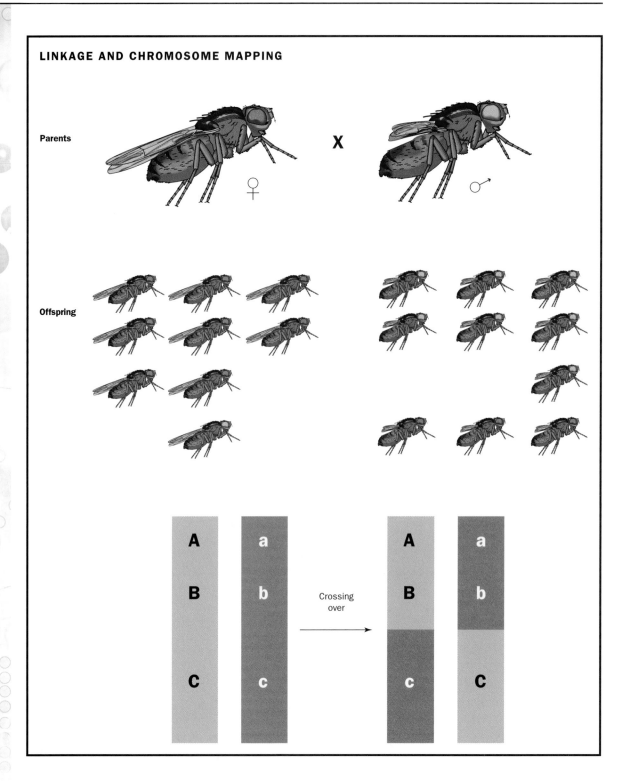

A cross between two double heterozygotes indicates linkage. Very few recombinants (long wing-purple eye or short wing-red eye) are formed if the two genes are close together.

Morgan's results showed that the white-eye allele is inherited on the X chromosome, and confirmed the discovery that the X chromosome helps determine sex, first shown in 1905 by Sutton and Nettie Stevens. In one step, his discovery proved that genes, the factors governing inheritance, are carried on chromosomes, and that specific genes are carried on specific chromosomes. This provided the crucial evidence that genes are indeed discrete physical objects.

Linkage and Chromosome Mapping

The discovery of more mutated genes allowed Morgan's group to explore how genes are arranged on the chromosome, and to discover an exception to one of Mendel's laws of inheritance. Mendel had proposed the Law of Independent Assortment, stating that the alternative forms of different traits (such as round versus wrinkled pea seeds and short versus tall plant height) separate and recombine independently of each other, so that, for instance, obtaining a wrinkled tall plant is just as likely as obtaining a wrinkled short plant.

Morgan found this was not always true. Rather, certain combinations of alleles are very unlikely to be separated from each other, a fact he attributed to co-inheritance of the two alleles on the same chromosome. While alleles on separate chromosomes assort independently, as Mendel predicted, those on the same chromosome travel together unless separated.

To explore this, Morgan crossed a red-eyed fly with normal-length wings with a purple-eyed fly with stubby wings. After two generations, Mendel's laws predicted that all possible combinations of eye color and wing length should be equally likely. Instead, Morgan found that most flies had the original trait combinations, while red-eyed, stubby-winged flies were rare, as were purple-eyed, normal-winged flies. He concluded that the genes for wing length and purple eye color were on the same chromosome. Like passengers traveling on the same ship, once the particular alleles were together, they tended to stay linked. (Note that the purple eye-color gene is not the same one as the red-white eye-color gene he discovered previously, and is not on the X chromosome.)

However, Morgan noted specific allele combinations didn't always stay together: There were a few flies whose stubby-wing allele and purple-eye allele had become separated from each other. This led Morgan to propose that chromosomes sometimes exchange segments, allowing their passengers to change vessels, so to speak. This phenomenon is known as crossing over, and was later conclusively demonstrated in maize by Barbara McClintock.

Crossing over is now known to occur only during meiosis, the chromosome division that leads to formation of eggs and sperm. During meiosis, **homologous** chromosomes originally donated from the mother and father pair up for an extended period. In this period, called synapsis, the maternal and paternal chromosomes randomly exchange several segments, resulting in a pair of chromosomes with a mix of maternally derived and paternally derived alleles. These then separate to form the eggs and sperm.

homologous carrying similar genes

Morgan's student Sturtevant reasoned that the likelihood of two alleles becoming separated during crossing over was proportional to the distance between them. In other words, the closer they are, the more likely they will stay together, and the further apart they are, the more likely they will separate. If A, B, and C are on the same chromosome, and A stays with B more often than it stays with C, then the distance from A to B is shorter than the distance from A to C. In this way, the relative distances of genes can be determined, providing a "linkage map" of the chromosomes. The unit of relative distance is called the morgan, in honor of Morgan himself. Calvin Bridges later devised a method to determine the absolute distance between genes, relying on the distinct banding patterns seen in *Drosophila* chromosomes in the larval stage.

Morgan's Legacy

In 1915 Morgan, Bridges, Sturtevant, and Muller published *The Mechanism of Mendelian Heredity*, a highly influential textbook laying out the evidence for the chromosomal theory of heredity and illustrating their methods so others could apply them in further research. In 1928 Morgan moved to the California Institute of Technology to found the Division of Biology. Sturtevant and Bridges went with him. Five years later Morgan was awarded the Nobel Prize in physiology or medicine for his work in genetics. He shared the prize money with Sturtevant and Bridges. Besides his own discoveries, Morgan's intellectual legacy includes the historically important researchers who trained with him, including Theodosius Dobzhansky, who applied the new genetics to an understanding of evolution. Another of his students was George Beadle, who discovered that mutations affect the working of proteins, and proposed the "one gene–one enzyme" definition of the gene. SEE ALSO FRUIT FLY: *DROSOPHILA*; LINKAGE AND RECOMBINATION; McCLINTOCK, BARBARA; MEIOSIS; MENDEL, GREGOR; MULLER, HERMANN.

Richard Robinson

Bibliography

Allen, Garland E. *Thomas Hunt Morgan: The Man and His Science.* Princeton, NJ: Princeton University Press, 1978.

Judson, Horace F. *The Eighth Day of Creation: The Makers of the Revolution in Biology.* New York: Simon & Schuster, 1979.

Morgan, Thomas Hunt, et al. *The Mechanism of Mendelian Heredity.* New York: Holt Rinehart & Winston, 1915. Reprint, with an introduction by Garland E. Allen, New York: Johnson Reprint Corporation, 1978.

Sturtevant, Alfred H. *A History of Genetics.* New York: Harper & Row, 1965.

Mosaicism

In 1961 Mary Lyon, an English scientist, hypothesized that one of the two X chromosomes in females becomes genetically silent early in a female embryo's development. To understand how she arrived at this idea, which has come to be known as "the Lyon Hypothesis," we need to understand what was known about the sex chromosomes.

The Sex Chromosomes

Humans have twenty-three pairs of chromosomes, including one pair of sex chromosomes and twenty-two pairs of **autosomes**. The sex chromosomes are either X or Y chromosomes. Females have two X chromosomes, and males have an X and a Y chromosome.

autosomes chromosomes that are not sex-determining (not X or Y)

In mammals, the sex of an individual is generally determined by whether the individual inherited an X or a Y chromosome from the father. The Y chromosome contains the *SRY* (Sex-determining Region Y) gene that directs male sexual development, but holds relatively few other genes. Many of the several dozen genes or gene families on the Y chromosome are necessary for the production of sperm. A handful are shared with the X chromosome, which is a medium-sized chromosome that is likely to contain more than one thousand genes.

Lyon knew that female mice that had only a single sex chromosome, the X chromosome, were normal. She also knew that mice carrying two different genes for coat color, one on each X chromosome, exhibited a mosaic, or blotchy, pattern of coat color. Some cells expressed one color gene, while others expressed the other, producing a mottled pattern.

Finally, she knew that when female cells are stained and looked at under a microscope a darkly staining region called a Barr body can be seen. She hypothesized that in female cells the Barr body is an inactive X chromosome. Thus, only one X chromosome would be active in any cell, resulting in a mottled pattern of X-linked gene expression. Furthermore, female cells lacking an X chromosome would be all right if the remaining X was the active one.

Mosaic Expression

A good example of an animal that exhibits mosaicism is a tortoiseshell cat, which has patches of black and orange fur. There is a dominant gene on the X chromosome that makes the cat's fur orange. If a female cat has this gene on only one of its two chromosomes, then the pigment-producing cells in which this chromosome is active will generate orange fur, while those that have the gene on the inactive X chromosome will make black fur.

The choice of which X chromosome to inactivate occurs very early in development, when an embryo has less than one hundred cells. While this initial choice is generally random, the same inactivation pattern is then passed on to descendant cells through subsequent cell divisions, resulting in a patch of cells with one or the other X chromosome active, and therefore producing orange or black fur in the tortoiseshell cat.

Because the single X chromosomes in males is never inactivated, male cats do not have tortoiseshell coats. XXY male cats, however, which have an extra X chromosome, can have such coats.

X Chromosome Inactivation

How does a cell manage to silence one X chromosome in a cell but not the other even though the two chromosomes are almost identical? A clue to this puzzle came from the discovery of a gene named *XIST* (X inactive specific transcripts). This is a gene that is expressed only on the inactive X chromosome. It is transcribed into an RNA that does not code for protein, unlike most genes. Instead, the RNA associates with the X chromosome from which it is made, resulting in silencing of the chromosome.

We know many of the components of this silencing process, and they are proteins that have been implicated in the silencing of other genes or regions of chromosomes as well. They are predominantly factors that influence the structure of the chromatin, which is the complex of DNA and proteins that is found in chromosomes. For example, the chromatin structure can be changed by adding methyl groups to the DNA, or by adding acetyl or methyl groups to the **histone** proteins with which the DNA interacts.

histone protein around which DNA winds in the chromosome

Effect of X Inactivation on Human Disease

Females with a mutated gene on an X chromosome have two populations of cells. One group produces the intact protein, and the other produces a protein that is affected by the mutation. Like tortoiseshell cats, these females

The tortoiseshell or calico cat is an example of mosaicism. This is caused by random X chromosome inactivation. A female has two X chromosomes. If she is heterozygous, one will carry the black allele, and one the orange allele. Random inactivation (X_i yields a population of cells with orange fur, and another population of cells with black fur. (The white colored fur seen in calico cats (that distinguishes them from tortoiseshell cats) is not associated with the X chromosome inactivation, rather, it is due to an autosomal gene.)

are mosaic. Health sometimes depends on what fraction of the cells in a tissue express the functional gene.

For reasons that are not yet well understood, some females exhibit non-random inactivation patterns. If the chromosome with the normal copy of a gene is inactivated in most of the cells in a female's body, and if the normal protein is vital for some function, the female is likely to develop a disease.

With some X-linked diseases, cells that contain a mutation on the active X chromosome proliferate less during development than cells that carry the mutation on the inactive X chromosome. In such cases, the female primarily expresses the normal gene.

Unlike females, males with an X-linked mutation will usually show signs of the disease, because they have no second functional copy. (Females will usually show symptoms if they inherit a mutated copy of the gene from each parent.) Males therefore inherit X-linked diseases, such as Duchenne muscular dystrophy, hemophilia, or colorblindness, much more commonly than females. Some X-linked disorders are almost never found in males, which may seem paradoxical until we consider that the absence of a functional gene can be so harmful that most males who inherit the disease die before being born.

Such is the case with Rett syndrome, an X-linked, dominant neurological disorder. This disorder is due to a mutation in a gene called *MECP2*. The disorder is primarily found in females, whose mosaicism gives them partial protection from its effects. Only a handful of males with Rett syndrome are known.

Other Types of Mosaicism

Since humans consist of more than ten trillion cells, it is not surprising that mutations occur in the genes in some of these cells, rendering the individ-

ual a mosaic. In some cases such changes have limited impact and are found in only a few cells. In other cases they may lead to cancer or disease.

We are all likely to have some cells in our body that have acquired mutations, and therefore everyone may be considered a mosaic at some level. Mosaicism can create differences among a person's cells. It can also result in differences between "identical" twins.

Most females are mosaics, due to X chromosome inactivation, though their mosaicism does not necessarily involve any disease gene. Chromosomal and mitochondrial mosaicism are also observed frequently.

Chromosomal Mosaicism. Every time a cell divides, the genetic material assembled in chromosomes needs to be divided too. If the chromosomes do not separate evenly, then cells are formed with missing or extra chromosomes. Such cells are called aneuploid. If this occurs early in embryonic development, a significant proportion of cells in an individual will be abnormal.

Chromosomal mosaicism may also result from the "rescue" of a fertilization that resulted from an aneuploid sperm or egg. If a fertilized egg contains three copies of a particular chromosome, a condition called trisomy, instead of the normal two, one of the extra copies can be "lost" if the chromosomes divide unevenly, restoring the normal chromosome number to the daughter cell.

Typically, in fetuses surviving the first trimester of pregnancy, the abnormal cells are found in placental but not in fetal tissues. Cells with three copies of a chromosome may be able to survive better in placental tissues, or there may be stronger selection against the growth of such cells in fetal tissues.

Trisomy is occasionally associated with pregnancy complications, such as poor fetal growth, but it may be common in placental tissues, and mosaicism confined to the placenta has been suggested to occur in up to 5 percent of births. Trisomic cells can also be found in the fetus itself, although this occurs much more rarely. Sometimes the abnormal cells will be present in only one type of tissue, such as the skin or lungs. Such variability has made it difficult to determine how often such chromosome mosaicism occurs and how it affects the health of an individual.

trisomy presence of three, instead of two, copies of a particular chromosome

Mitochondrial Mosaicism. The mitochondria are **organelles** in the cytoplasm that release energy stored in molecules for cells to use. They contain their own small chromosome. The mitochondrial chromosome contains 16,569 base pairs, compared with the nuclear chromosomes, which, together, contain three billion base pairs.

organelles membrane-bound cell compartments

Two features make mitochondria prone to mosaicism. First, their DNA is mutated more frequently than the nuclear DNA, in part because of the more dangerous cellular environment facing mitochondria and in part because mitochondria are not equipped to repair mutations as effectively as the nucleus.

Second, each mitochondrion contains numerous copies of its genome, and there are thousands of mitochondria in each cell. Thus individuals can have mutations in some of their mitochondrial genomes that are not found in their other mitochondria. This can lead to variable expression of diseases associated with mitochondrial mutations. Deletions of part of the mitochondrial genome appear to accumulate in different tissues with age and have been suggested to be a critical factor in normal human aging. SEE ALSO CHROMOSOME, EUKARYOTIC; DNA REPAIR; GENE EXPRESSION: OVERVIEW OF CONTROL; MITOCHONDRIAL DISEASES; MITOCHONDRIAL GENOME; MUSCULAR DYSTROPHY; NONDISJUNCTION; X CHROMOSOME.

Carolyn J. Brown

Bibliography

Avner, Philip, and Edith Heard. "X-Chromosome Inactivation: Counting, Choice and Initiation." *Nature Reviews: Genetics* 2 (2001): 59–67.

Lyon, Mary F. "Gene Action in the X-Chromosome of the Mouse (*Mus musculus L.*)." *Nature* 190 (1961): 372–373.

Internet Resource

Online Mendelian Inheritance in Man. Johns Hopkins University, and National Center for Biotechnology Information. <http://www.ncbi.nlm.nih.gov/omim>.

Mouse *See Rodent Models*

Muller, Hermann

Geneticist
1890–1967

Hermann Muller.

Hermann Joseph Muller was one of the founding members of the "fly lab" that was initiated by Thomas Hunt Morgan. In the early part of the twentieth century, this lab was the center of important research into the role of chromosomes in inheritance, using the fruit fly *Drosophila* as a model organism in experiments. The major members included Morgan, Alfred Henry Sturtevant, Calvin Blackman Bridges, and Muller, all working at Columbia University around 1910 to 1915, when their major contributions to classical genetics were carried out. Muller was the only one of Morgan's students to also win a Nobel Prize.

Muller's career was unusual in that he worked in several countries. He was a third-generation American, but he left the United States in 1932 to work in Germany, the Soviet Union, and Edinburgh, Scotland, before returning to his homeland. He had a productive career as an experimental and theoretical geneticist, but he also had a life-long interest in seeing genetics applied to society. This made him controversial and sometimes put his life in jeopardy.

The first phase of Muller's career was spent at Columbia, where the fly lab was established, and at Rice University, where he held his first tenured position from 1915 to 1922. During this phase he did work that contributed to the understanding of crossing over and gene mapping (discovered by Morgan and Sturtevant). Muller also clarified the meaning of genetic mutation by limiting the concept to variations in the individual gene. He proposed the gene as the basis of life, arguing that only genes had the property of replicating their errors, essential for the evolution of life. He was the first to measure mutation rates, and he designed stocks of *Drosophila* to detect them.

Muller's second phase was the decade he spent at the University of Texas, from 1922 to 1932. During this time he was heavily committed to the study of mutation, culminating in his Nobel Prize work on the induction of mutations by **radiation**. Muller rapidly followed up his initial reports and founded a new field of radiation genetics. He also used X rays as a tool to delete chromosomes, and used these small deleted chromosomes to reveal the mechanism of genetic functions such as **dosage compensation**, a phenomenon he was first to interpret. Muller showed that mutations are produced in proportion to radiation dosage and that chromosome rearrangements (such as **translocations**) were induced at higher dosages.

In his third phase, which ran from 1932 to 1940 and which he spent working in Berlin, the Soviet Union, and Scotland, Muller studied chromosome structure, gene structure, and changes in gene function when genes were moved from their normal chromosome location. He also introduced the idea that genes arise from preexisting genes, when he discovered that a fly mutation called Bar eyes arose from a physical duplication of genes. In his last phase, Muller was back in the United States (from 1940 to 1967), working mainly at Indiana University, where he worked out the mechanism of cell death from radiation exposure and calculated the amount of mutation normally occurring in humans (genetic load) each generation. He became a critic of the Cold War policies of the United States, favoring a strong nuclear defense as a protection against Stalinism but also calling for mutual treaties to limit nuclear arms. He also fought hard against the misuse of radiation by health practitioners and industry.

In addition to his fundamental work on fly genetics, Muller contributed to human genetics through studies of twins that he conducted in the 1920s. He argued that the relation of observable character traits to genes is very complex, a problem he had first studied in detail in *Drosophila*, when he investigated the verifiability of shape and size in the "truncate" and "beaded" wing mutations. Muller stressed that an observable trait such as intelligence or longevity will be influenced by many genes as well as by the environment, and that simple one-gene/one-trait relationships were the exception rather than the rule in complex organisms.

Muller felt that advocates of **eugenics** programs ignored environmental modifiers and the complex residual heredity that he called modifier genes. He denounced the American eugenics movement in 1932 at the Third (and last) International Congress of Eugenics. Yet he remained an idealist about eugenics, favoring a positive eugenics based on "germinal choice," a noncoercive way for educated people to choose the genetic character of their own children. SEE ALSO EUGENICS; FRUIT FLY: *DROSOPHILA*; EVOLUTION OF GENES; MAPPING; MORGAN, THOMAS HUNT; MUTATION; TWINS.

Elof Carlson

radiation high energy particles or waves capable of damaging DNA, including X rays and gamma rays

dosage compensation equalizing of expression level of X-chromosome genes between males and females, by silencing one X chromosome in females or amplifying expression in males

translocations movements of chromosome segments from one chromosome to another

eugenics movement to "improve" the gene pool by selective breeding

Bibliography

Carlson, Elof. *The Gene: A Critical History*. Philadelphia: Saunders Publishing Co., 1966.

Muller, Hermann J. "The Development of the Gene Theory." In *Genetics in the Twentieth Century*, L. C. Dunn, ed. New York: Macmillan, 1951.

Olby, Robert. *The Path to the Double Helix*. Seattle: University of Washington Press, 1974.

Multiple Alleles

Alleles are alternative forms of a gene, and they are responsible for differences in phenotypic expression of a given trait (e.g., brown eyes versus green eyes). A gene for which at least two alleles exist is said to be polymorphic. Instances in which a particular gene may exist in three or more allelic forms are known as multiple allele conditions. It is important to note that while multiple alleles occur and are maintained within a population, any individual possesses only two such alleles (at equivalent **loci** on **homologous** chromosomes).

loci sites on a chromosome (singular, locus)

homologous carrying similar genes

Examples of Multiple Alleles

Two human examples of multiple-allele genes are the gene of the ABO blood group system, and the human-leukocyte-associated **antigen** (HLA) genes.

antigen a foreign substance that provokes an immune response

The ABO system in humans is controlled by three alleles, usually referred to as I^A, I^B, and I^O (the "I" stands for isohaemagglutinin). I^A and I^B are codominant and produce type A and type B antigens, respectively, which migrate to the surface of red blood cells, while I^O is the recessive allele and produces no antigen. The blood groups arising from the different possible genotypes are summarized in the following table.

Genotype	Blood Group
$I^A I^A$	A
$I^A I^O$	A
$I^B I^B$	B
$I^B I^O$	B
$I^A I^B$	AB
$I^O I^O$	O

HLA genes code for protein antigens that are expressed in most human cell types and play an important role in immune responses. These antigens are also the main class of molecule responsible for organ rejections following transplantations—thus their alternative name: major histocompatibility complex (MHC) genes.

The most striking feature of HLA genes is their high degree of **polymorphism**—there may be as many as one hundred different alleles at a single locus. If one also considers that an individual possesses five or more HLA loci, it becomes clear why donor-recipient matches for organ transplantations are so rare (the fewer HLA antigens the donor and recipient have in common, the greater the chance of rejection).

polymorphism DNA sequence variant

Polymorphism in Noncoding DNA

It must be realized that although the above two are valid examples, most genes are not multiply allelic but exist only in one or two forms within a population. Most of the DNA sequence variation between individuals arises not because of differences in the genes, but because of differences in the noncoding DNA found between genes.

An example of a noncoding DNA sequence that is extremely abundant in humans is the so-called microsatellite DNA. Microsatellite sequences consist of a small number of **nucleotides** repeated up to twenty or thirty times.

nucleotides the building blocks of RNA or DNA

For instance, the microsatellite composed of the dinucleotide AC is very common, appearing about one hundred thousand times throughout the human genome.

The interesting feature about microsatellites is that they are very highly polymorphic for the number of repeat lengths. For example, one particular individual might possess the microsatellite sequence ACACACACAC at a specific locus on one chromosome, and the sequence ACACACACACA-CACACAC at the same locus on the other homologous chromosome.

Making Use of Polymorphic DNA

Multiple alleles and noncoding polymorphic DNA are of considerable importance in gene mapping—identifying the relative positions of genetic loci on chromosomes. Gene maps are constructed by using the frequency of crossing-over to estimate the distance between a pair of loci. To obtain a good estimate, one must analyze a large number of offspring from a single cross. In laboratory organisms such as the fruit fly *Drosophila*, programmed crosses can be carried out so it is possible to use gene loci to construct a reliable genetic map. In humans, this is not the case. For this reason, the more highly variable noncoding regions are of considerable importance in human genetic mapping. SEE ALSO BLOOD TYPE; IMMUNE SYSTEM GENETICS; MAPPING; POLYMORPHISMS; TRANSPLANTATION.

Andrea Bernasconi

Bibliography

Alberts, Bruce, et al. *Molecular Biology of the Cell*, 4th ed. New York: Garland, 2002.

Strachan, Tom, and Andrew P. Read. *Human Molecular Genetics*. New York: Bios Scientific Publishers, 1996.

Muscular Dystrophy

Muscular dystrophies (MDs) are a group of disorders that share three characteristics: They are inherited, they cause progressive weakness and muscle wasting, and the primary defect is localized to skeletal muscle, sparing the nerves. Although selected limb muscles develop some degree of weakness in all dystrophies, to distinguish among the different types, it is critical to know the mode of inheritance, the age of onset, and whether muscles other than limb muscles are also affected. For example, some dystrophies additionally affect eye and lip closure; another type affects eye movement ability, as well as swallowing and speech.

More than thirty types of MDs are now recognized. Three of the more prevalent forms—Duchenne, myotonic, and limb-girdle dystrophies—will be discussed from the standpoint of the presenting symptoms, age of onset, inheritance pattern, causative genes, and the availability of prenatal and presymptomatic molecular testing.

Duchenne Muscular Dystrophy

Duchenne muscular dystrophy (DMD) is an X-linked recessive disorder with a worldwide occurrence of one in four thousand newborn males, with approximately one-third of the cases arising from new mutations. DMD was

A cross-section of healthy muscle tissue.

named after the French neurologist Duchenne de Boulogne, who described the disorder in 1861. Becker muscular dystrophy (BMD), named after Peter Becker, a German geneticist who first described it in the mid-1950s, is a disorder that is very similar to DMD but has a much milder course. In 1983 these disorders were first shown to be located on the short arm of the X chromosome. The disorders are now known to be allelic, meaning an alternate form of the DMD gene causes BMD.

Because DMD is X-linked, almost all cases occur in males. Boys with DMD are normal at birth, and their early motor milestones occur at normal times. The manifestations of DMD are frequently apparent from the time they begin to walk, due to the developing weakness of the hip-girdle and upper-leg muscles. Their gait is unsteady and clumsy, resulting in frequent falls. If running is attempted, it is slow and waddling. The calf muscles often are enlarged enough to be termed "hypertrophic," implying that these children are muscular and strong. In reality calf *pseudo*hypertrophy is present: When the calves are examined microscopically, the amount of muscle tissue is markedly reduced, having been replaced by fat and fibrous tissue.

As the disorder progresses, the Achilles tendons tighten, causing toe-walking, which further compromises patients' gait and balance. Stair-climbing, rising from a fall, and even walking on level ground becomes more arduous. Even if they undergo Achilles tendon lengthening surgery or use leg braces, virtually all DMD boys require a wheelchair for mobility before the age of thirteen. Weakening of the muscles of the upper extremities and neck and of the respiratory muscles occurs in parallel to that of the lower extremities, although at a slower rate. By some time in their twenties, if not before, nearly all DMD patients will die, often due to an overwhelming respiratory infection resulting in respiratory failure, cardiac arrest, or both.

The causative gene for DMD (named *dystrophin*) and the protein product (also named dystrophin) were identified in 1986. Dystrophin is found on the inner side of the membrane that surrounds skeletal muscle fibers (the sarcolemma). It is usually absent or severely deficient in DMD boys, and this causes the sarcolemma to weaken and develop tears, allowing excess cal-

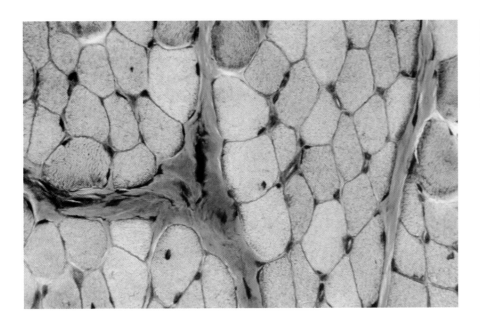

A cross-section of Duchenne muscular dystrophy tissue. Note the increase in connective tissue and the resulting decrease in healthy muscle tissue.

cium to enter the muscle fiber. This eventually leads to the death of muscle fibers, and, when a sufficient number of fibers are involved, muscle weakness results.

The *dystrophin* gene is the largest known human gene, encompassing two thousand kilobases (two million bases) of genomic DNA. In 55 percent to 65 percent of DMD or BMD cases, large deletions of the *dystrophin* gene can be found. Duplications within the gene account for about 5 percent of cases. DNA testing, available through a number of commercial laboratories in the United States, is based on the identification of these large deletions and duplications. DNA tests can confirm a diagnosis of DMD or BMD, and they can be used for accurate **carrier** or prenatal testing.

carrier a person with one copy of a gene for a recessive trait, who therefore does not express the trait

Myotonic Muscular Dystrophy

Myotonic muscular dystrophy (DM, or dystrophia myotonica) is the most common adult-onset muscular dystrophy, having a frequency of one per twenty thousand persons in the general population. Myotonia, the delayed relaxation of a voluntary muscle after it is contracted, and muscle weakness are the hallmarks of the disorder. For example, a person with DM using a hammer will not immediately be able to release his grip on the handle when finished. It is an **autosomal** dominant disorder, but there is great variability in the disorder's severity and in the number of manifestations it leads to.

autosomal describes a chromsome other than the X and Y sex-determining chromosomes

A unique feature of this dystrophy is a genetic phenomenon called pleiotropy, or multisystem involvement, despite the single genetic defect. The potential involvement includes multiple organs and organ systems other than skeletal muscle, including the cardiac, respiratory, gastrointestinal, central nervous, endocrine, and dermatologic systems, as well as bone or eyes. A congenital variety occurs in which infants are born floppy, often require respiratory assistance, have extremity deformities, and are both physically and mentally retarded.

Patients often initially complain of a loss of hand strength (they have difficulty twisting off caps from bottles, for example) or of tripping while

walking or climbing stairs, due to the weakness of muscles that extend the feet and toes. Weakness may progress to involve the shoulder and hip girdles and, in some cases, is severe enough to necessitate the use of a wheelchair. Droopy eyelids, wasting of facial and neck muscles, and frontal balding frequently occur, producing atypical facial appearance.

The gene for DM is a protein kinase gene (known as *DMPK*) and is located on the long arm of chromosome 19. The disorder arises from a repeated sequence of three nucleotides—cytosine (C), thymine (T), and guanine (G)—in the gene. Individuals without DM have C-T-G repeats that contain between 5 and 37 iterations of the triplets. By contrast, repeats that are between 40 and 170 iterations long are found in the mild phenotype, repeats between 100 and 1,000 iterations are found in the "classic phenotype," and repeats of between 500 and 3,000 are found in the congenital phenotype. A number of laboratories in the United States perform this triplet repeat assay for diagnostic, prenatal, and presymptomatic testing.

Limb-Girdle Muscular Dystrophy

Limb-girdle muscular dystrophy (LGMD) has been described both as a heterogeneous group of disorders and as a diagnosis of exclusion. Any patient who has weakness of the shoulder and hip-girdle muscles and who otherwise has been excluded from the other MDs will be diagnosed with LGMD. Using the patterns of inheritance that exist within LGMD, a classification system has been created to simplify the heterogeneity: Both autosomal dominant (LGMD1) and recessive families (LGMD2) are well recognized. The number of LGMD genes that have already been identified has further improved the classification.

The frequency of LGMD in the general population is reported to be one in twenty-five thousand. The age at onset may vary widely. In some individuals, onset is in early childhood and in others it occurs in the forties and fifties, but most commonly it occurs in the teens to early adulthood. The characteristic pattern of muscle involvement is symmetric weakness, beginning initially in the hip and shoulder girdle, but usually noticed in the hips before the shoulders. Thus slowness in running, difficulty rising from a low seat, and difficulty ascending stairs are all common complaints from affected individuals. As LGMD progresses, it will involve upper-leg and arm muscles and may eventually affect the muscles that extend the feet and wrists. Lower-extremity weakness may become severe enough to require a wheelchair.

Among the LGMD1 types, five have chromosomal linkages, but only in one is the protein product of the gene known. LGMD2 is better characterized, with nine chromosomal localizations, five with known proteins. Almost all these proteins are membrane-associated proteins (just as dystrophin is). When they are abnormal in structure or deficient in quantity, they affect the stability of the muscle membrane, resulting in the same pathological process that was described for DMD. Commercial testing in the United States is only available for some of the LGMD2 types and is performed using muscle tissue.

Treatment of the Muscular Dystrophies

As of mid-2002, gene therapy treatment of LGMD was tried in a very small number of patients. These early experiments delivered a functional gene to a very small muscle in the foot and were designed to test the long-term safety and effectiveness of the treatment. Gene therapy for DMD is much more problematic, because of the immense size of the gene and the distribution throughout the body that would be required for effective treatment. Drug treatment with prednisone or other corticosteroids is being used, although at best this provides another six to twelve months of mobility before a wheelchair becomes necessary. There are no effective treatments for myotonic dystrophy as of 2002, although research continues in many laboratories worldwide. SEE ALSO GENE THERAPY; GENETIC TESTING; INHERITANCE PATTERNS; PRENATAL DIAGNOSIS; TRIPLET REPEAT DISEASE.

Jeffrey M. Stajich

Bibliography

Emery, Alan E. H., ed. *Neuromuscular Disorders: Clinical and Molecular Genetics.* Chichester, U.K.: John Wiley & Sons, 1998.

———. *Muscular Dystrophy, The Facts.* Oxford, U.K.: Oxford University Press, 2000.

Hoffman, Eric P. "Muscular Dystrophy: Identification and Use of Genes for Diagnostics and Therapeutics." *Archives of Pathology and Laboratory Medicine* 123 (1999): 1050–1052.

Internet Resource

Muscular Dystrophy Association. <http://www.mdausa.org>.

Mutagen

A mutagen is any substance or agent that can cause a mutation, or change in the sequence or structure of DNA. Mutagens are classified on the basis of their physical nature and the types of damage they do. A mutagen is not the same as a carcinogen. Carcinogens are agents that cause cancer. While many mutagens are carcinogens as well, many others are not. The Ames test is a widely used test to screen chemicals used in foods or medications for mutagenic potential.

Chemical Mutagens

There are many hundreds of known chemical mutagens. Some resemble the bases found in normal DNA; others alter the structures of existing bases; others insert themselves in the helix between bases; while others work indirectly, creating reactive compounds that directly damage the DNA structure.

"Base analogs" are molecules whose chemical structure is similar to one of the four DNA bases (adenine, thymine, cytosine, and guanine). Because of this similarity, they can be incorporated into the helix during DNA replication. A key feature of mutagenic base analogs is that they form base pairs with more than one other base. This can cause mutations during the next round of replication, when the replication machinery tries to pair a new base with the incorporated mutagen. For instance, 5-bromodeoxyuridine (5BU) exists in two different forms. One mimics thymine and therefore pairs with adenine during replication, while the other mimics cytosine and therefore pairs with guanine. In its thymine-mimicking form, 5BU can be incorporated across from an adenine. If it then converts to its cytosine-like form, during the next round of replication, it will cause a guanine to enter the opposite strand, rather than the correct adenine.

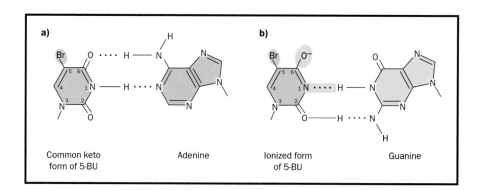

The mutagen 5 bromo-deoxyuridine exists in two forms, one of which pairs with adenine, and the other of which pairs with guanine. Adapted from <http://fig.cox.miami.edu/Faculty/Dana/baseanalog.jpg>.

organic composed of carbon, or derived from living organisms; also, a type of agriculture stressing soil fertility and avoidance of synthetic pesticides and fertilizers

"Base-altering mutagens" cause chemical changes in bases that are part of the DNA. For example, nitrite preservatives in food convert to the mutagen nitrous acid. Nitrous acid causes deamination, or loss of an –NH₂ group, of cytosine. When this occurs, cytosine becomes uracil, a base that is not normally incorporated in DNA but that is very similar to thymine. Unless repaired, this uracil will cause an adenine to enter the opposite strand instead of a guanine. Many base-altering mutagens are complex **organic** molecules. These are formed in large quantities in smoke, making up the "tar" of cigarette smoke, for example. They act as alkylating agents, combining with DNA to form bulky groups that interrupt replication.

"Intercalating agents" are flat molecules that insert themselves between adjacent bases in the double helix, distorting the shape at the point of insertion. Where this occurs, DNA polymerase may add an additional base opposite the intercalating agent. If this occurs in a gene, it induces a frameshift mutation (that is, it alters the reading of the gene transcript, changing which amino acids are added to the encoded protein). Ethidium bromide is one such agent, widely used in DNA research because its dark color allows DNA to be easily visualized. This is useful in gel electrophoresis, for instance, to find the DNA bands that have been separated in a gel.

Other damaging agents include chemicals that create "free radicals" inside a cell. Free radicals are compounds in which an atom, usually an oxygen, has an unbonded electron. Free radicals are highly reactive and can cause several types of damage to DNA.

Light and Radiation

Radiation refers to two different phenomena: light and high-energy particles. Visible light represents a small slice of the electromagnetic spectrum, which includes long-wavelength (low-energy) radio waves and short-wavelength (high-energy) ultraviolet waves, plus X rays and gamma rays. These high-energy forms can directly disrupt DNA by breaking its chemical bonds. In severe cases, this can break apart chromosomes, causing chromosome aberrations. More often, they create mutagenic free radicals in the cell. X rays were first used by Hermann Muller to induce mutations in fruit flies. They continue to be used to create mutations in model organisms to study genes and their effects.

Ultraviolet light is less energetic than X rays but causes mutations nonetheless. The higher-energy form, UV-B, is more toxic than UV-A, because of its potential to cause cross-linking between adjacent thymine or cytosine bases, creating a so-called pyrimidine dimer (cytosine and thymine are chemically classified as pyrimidines). Pyrimidine dimers interrupt replication. UV-A does not cause dimer formation but can still cause mutations by creating free radicals.

Another meaning of the term "radiation" is high-energy particles released during the breakdown of radioactive elements, such as uranium. These particles are either electrons (called beta particles) or helium nuclei (called alpha particles). Their energy is sufficient to disrupt DNA structure, or to create free radicals.

Repairing the Damage

DNA is constantly being damaged, and it is constantly being repaired as well. It is only when the damage is not repaired that a mutation can lead to cancer or cell death. The DNA repair enzymes can recognize damaged nucleotides and remove and replace them. The human liver contains a large number of enzymes whose role is to detoxify toxic compounds, mutagenic or otherwise, by chemically reacting them. However, in some cases these enzymes (called cytochrome P450s) actually create mutagens during the course of these reactions. Such "bioactivation" may be a significant source of mutagens. SEE ALSO Carcinogens; Chromosomal Aberrations; DNA Repair; Muller, Hermann; Mutagenesis; Mutation.

Richard Robinson

Bibliography

Philp, Richard B. *Ecosystems and Human Health: Toxicology and Environmental Hazards*, 2nd ed. Boca Raton, FL: Lewis Publishers, 2001.

Brusick, David. *Principles of Genetic Toxicology*, 2nd ed. New York: Plenum Press, 1987.

Mutagenesis

Mutagenesis is the process of inducing mutations. Mutations may occur due to exposure to natural **mutagens** such as ultraviolet (UV) light, to industrial or environmental mutagens such as benzene or asbestos, or by deliberate mutagenesis for purposes of genetic research. For geneticists, the study of mutagenesis is important because mutants reveal the genetic mechanisms underlying heredity and gene expression. Mutations are also important for studying protein function: Often the importance of a protein cannot be characterized unless a mutant can be made in which that protein is absent.

mutagen any substance or agent capable of causing a change in the structure of DNA

Noninduced Mutagenic Agents

Environmental agents can influence the mutation rate not only by increasing it, but also by decreasing it. For example, antioxidants, which are found commonly in fruits and vegetables, are thought by many to protect against mutagens that are generated by normal cellular respiration. In addition to protective agents, however, many plants also contain **deleterious** mutagens known as **carcinogens**. Many chemical mutagens exist both naturally in the environment and as a result of human activity. Benzo(a)pyrene, for example, is produced by any incomplete burning, whether of tobacco in a cigarette or of wood in forest fires.

deleterious harmful

carcinogens substances that cause cancer

Figure 1. First the parental males were treated and mated to females carrying the *CIB* X chromosome. Second, those female offspring with Bar eyes were mated individually so that only their offspring would be present in the bottle. Third, after the offspring hatched, the bottle was examined for the presence of males. There were no males with the *CIB* X chromosome because the *l* gene kills them. So, any bottles with no males represent one lethal mutation on the other X chromosome, which originated in the treated male.

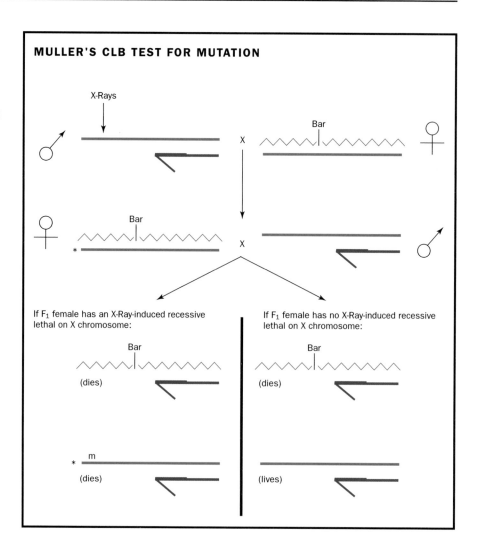

MULLER'S CLB TEST FOR MUTATION

X-Rays

Bar

X

Bar

X

If F$_1$ female has an X-Ray-induced recessive lethal on X chromosome:

Bar

(dies)

* m

(dies)

If F$_1$ female has no X-Ray-induced recessive lethal on X chromosome:

Bar

(dies)

(lives)

recessive requiring the presence of two alleles to control the phenotype

Spontaneous (noninduced) mutations are very rare, and finding them is difficult because most are **recessive**. The recessive nature of most mutations means that they will not be evident in most of the individuals who inherit them, for they will be hidden by the presence of the dominant allele. The rarity of mutations means that many individuals must be examined to find a mutant, whether they are people, other organisms, or even cells in culture.

Creating Mutations

To overcome the problem of the rarity of mutations, researchers induce mutations with a variety of agents. Hermann Muller was the first to do this when, in 1927, he used X rays on fruit flies (*Drosophila*) to increase the mutation rate by more than 100-fold. Other high-energy forms of radiation can also be used to create mutations.

The first chemical to be recognized as a mutagen was mustard gas, which had been developed during World War I, but not tested until World War II by Auerbach and Robson, at the University of Edinburgh. Since then a wide variety of chemicals have been discovered that are also mutagenic. Some induce mutations at any point in the cell cycle, by disrupting DNA structure. Others only act during DNA replication. Called base analogs, these latter chemicals have structures similar to the bases found in DNA, and are incorporated instead of the normal base.

Transposable genetic elements (also called transposons, or "jumping genes") can also induce mutations. These elements insert randomly into the genome, and may disrupt gene function if inserted into a gene or its promoter. Finding the organism with a disrupted gene is made easier if the transposon carries with it a reporter gene whose product can be identified, or a selectable marker that allows the transformed cells to live while non-transformed ones die. (The use of reporter genes and selectable markers are techniques used in genetic analysis in the laboratory.) The transposon sequence itself serves as a molecular tag. Thus, if the target gene (the gene being studied) is interrupted, finding the transposon allows the researcher to find the gene.

All of the above methods disrupt genes randomly. However, specific genes can also be targeted, for "site-directed mutagenesis," if their sequence is known. Using the known sequence, a matching DNA sequence is inserted into a single-stranded **vector**. Short, complementary, partial sequences containing the desired mutation are then synthesized. These are allowed to pair up, and DNA polymerase is then used to complete the complementary strand. Further replication **amplifies** he number of copies of the mutant. In bacteria, the mutant gene can be placed on a plasmid for transformation of the bacteria. The bacteria make the mutant protein, and the effect of the mutation can then be studied. This is a key tool in studying how amino acid sequences affect protein structure, since individual amino acids can be changed, one at a time.

In eukaryotes, the mutant gene can be inserted into the chromosome of an experimental organism by "homologous recombination," a system in which the mutant gene switches places with the normal chromosomal gene. Such techniques can "knock out" and "knock in" genes bearing the desired mutations.

The First Mutagenesis Assay

Before DNA sequencing became widespread, most mutations could only be detected by their effects on the **phenotype** of the organism. Many mutations are recessive, however, and do not affect the phenotype if present in only one allele. Hermann Muller, who pioneered the study of mutations, overcame this problem by focusing on the X chromosome in his studies of the genetics of *Drosophila*, the fruit fly. While females have two X chromosomes, males have only one, so any mutated gene carried on the X chromosome is expressed in males, even if it is recessive. Hence recessive lethal mutations on the X chromosome kill any male inheriting them, but would not kill a female. Muller's method examined all of the genes on the X chromosome that could mutate to give a recessive lethal mutation. Muller used X rays to generate mutants. X rays are a very high-energy form of radiation, and break the DNA at numerous points. The method is shown in Figure 1.

Muller treated adult males with X rays and mated them to females who carried one copy of a specially prepared X chromosome, called *ClB*. This chromosome had a gene to prevent crossing over (*C*), which kept the chromosome intact; a lethal recessive gene (*l*) to kill any males that inherit it;

Michael Smith of Canada received the 1993 Nobel Prize in physiology or medicine for invention of site-directed mutagenesis. He shared the prize with Kary Mullis, who invented the polymerase chain reaction.

vector carrier

amplifies multiplies

phenotype observable characteristics of an organism

and a dominant "bar eye" gene (*B*) that resulted in a distinctive phenotypic change, making it easy to find female flies that inherited it.

Muller mated X-ray treated males with *ClB*-carrying females. All female offspring from this crossbreeding received one treated X chromosome from the male (which might or might not have carried a lethal recessive gene). They also received one X from the female, either normal or *ClB*. He selected only the bar-eyed females for further mating. To determine which of these females carried an X-ray induced lethal recessive, he separated each female into a separate jar, and examined their offspring.

Three types of males were created in this cross, depending on what type of X chromosome they inherited. Males inheriting the *ClB* chromosome died, due to the presence of the *l* gene. Males inheriting an X-ray-treated X chromosome with a lethal recessive died. Males inheriting an X-ray-treated X chromosome without a lethal recessive lived. Therefore, any jar with live males indicated that the mother did not carry a lethal recessive. Any jars with no males indicates the mother carried a lethal recessive, originally induced in the X-ray-treated male. The analysis was rapid because an experienced person could examine a bottle of flies and see at a glance if there were males present. Subsequent studies showed that this method tested almost 1,000 genes simultaneously, thus making it practical to use when detecting rare mutations. Unfortunately the breeding takes quite a lot of time, so this assay has now fallen largely into disuse, despite its historical importance.

Detecting Mutations

Today the mutagenic potential of chemicals is considered in evaluating the mutagenic risks posed by chemical exposure. Many new methods have been developed to determine if chemicals to which people will be exposed, such as new drugs, food additives, and pesticides, are mutagens. Since mutations can occur in any organism, and because there are many different kinds of mutations, there are a correspondingly wide variety of tests to detect them. No one test detects them all.

The Ames test was the first and remains the only test to be almost universally required by regulatory agencies as a minimum standard for determining if a chemical is mutagenic. The test is conducted in *Salmonella* bacteria. Since bacteria have only one chromosome, recessive mutations can be detected readily. Rare mutations are easily detected because mutants can be selected very simply. Several variants have been added to the original test, allowing for detection of many types of mutations. In an effort to make the test more relevant to human risk, one variant uses an extract of liver to mimic the biochemical modifications of chemicals that occur in the human liver.

Assays for Chromosome Aberrations. Chromosomal aberrations can be detected by examining cells in mitosis or meiosis for changes (see Figure 2). Typically, bone marrow cells of mice or rats are examined for in vivo tests. Any cells can be used for tests of cells in culture, but Chinese hamster cells or human **fibroblasts** are most commonly used. Another test, called the micronucleus test, is also commonly used. Micronuclei are small nuclei that arise from pieces of chromosomes or whole chromosomes that have been lost during cell division. They are conveniently detected in mouse red blood

fibroblasts undifferentiated cells that normally give rise to connective tissue cells

cells, which have no normal nucleus but which often retain micronuclei. The micronucleus assay is also widely used in cultured cells.

Assays for Somatic Mutations. Recessive mutations can be detected more readily on a mammalian X chromosome than on the other chromosomes, because only one X chromosome is active. Therefore, detection of the mutagenic potential of a substance in mammals can be most efficiently performed by analyzing the X-linked mutations. A system using the X-linked gene *hprt* has been widely used because the enzyme is not essential and because the addition of the drug thioguanine kills all cells except mutants. A count of the cells that can be cultured in the presence of thioguanine is a count of *hprt* mutants. SEE ALSO AMES TEST; MUTAGEN; MUTATION; MUTATION RATES.

John Heddle

Bibliography

Griffiths, Anthony J. F., et al. *An Introduction to Genetic Analysis.* New York: W. H. Freeman, 2000.

Muller, Hermann J. "Artificial Transmutation of the Gene." *Science* 66 (1927): 84–87.

Rubin, G. M., and A. C. Spradling. "Genetic Transformation of *Drosophila* with Transposable Element Vectors." *Science* 218 (1982): 348–353.

Internet Resource

United Nations Scientific Committee on the Effects of Atomic Radiation. <http://www.unscear.org/>.

Mutation

A mutation is any heritable change in the genome of an organism. For a population, heritable mutations provide the source of genetic variation, without which evolution could not occur: If all individuals of a species were genetically identical, every subsequent generation would be identical regardless of which members of the species reproduced successfully. For an individual organism, mutations are rarely beneficial, and many cause genetic diseases, including cancer. For researchers, mutations (either spontaneous or introduced) provide important clues about gene location and function.

Phenotypic Effects and Evolution

Mutations in the **germ-line** cells are heritable and provide the raw material upon which natural selection operates to produce evolution. Mutations in **somatic** cells, which are cells that are not germ line, are not heritable but may lead to disease in the organism possessing them.

Most mutations do not cause disease and are said to be "silent" mutations. This is for at least two reasons. First, most DNA does not code for genes, so changes in the sequence do not affect the types or amounts of protein made and there is no change in the **phenotype** of the organism. Second, most sexually reproducing organisms are diploid, meaning they possess two copies of every gene. Many types of mutation simply disable one copy, leaving the other intact and functional. Therefore these mutations display a recessive inheritance pattern, with no effect on phenotype unless an individual inherits two copies of the mutation. Diploid species can accumulate a large pool of such recessive mutations, which are mostly disadvantageous and thus contribute to the burden of genetic disease.

germ-line cells giving rise to eggs or sperm

somatic nonreproductive; not an egg or sperm

phenotype observable characteristics of an organism

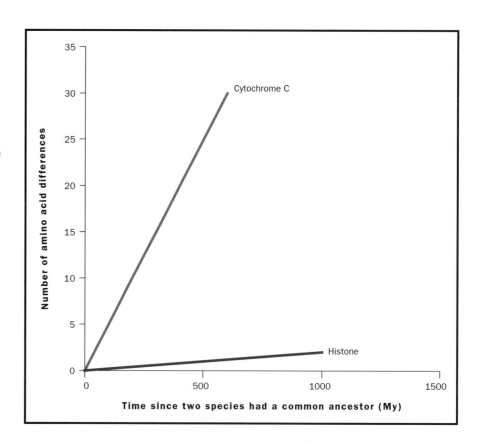

Figure 1. Since there is dated fossil evidence for the species used to generate the line, the evolutionary history of species without fossil evidence can be estimated from the number of amino acid changes (substitutions) in the protein.

allele a particular form of a gene

Some mutations lead to detrimental alterations of the normal phenotype and are, therefore, selected against. Very occasionally, the mutant phenotype is superior and provides a selective advantage, which leads to an increase in the frequency of this mutant **allele** and, thus, to evolution of the population. Alternatively, a disadvantageous mutation in one environment may become advantageous in another, again leading to increased frequency of this allele.

Molecular Basis of Mutations

DNA is composed of a double helix, each side of which is a long string of four types of nucleotides. Each nucleotide possesses identical sugar-phosphate groups that contribute to the DNA backbone but differs in the structure of the base suspended between the two backbones. The bases are adenine, thymine, cytosine, and guanine (A, T, C, G). Because of their structure, A pairs only with T across the double helix, and C only with G.

Within genes, the sequence of DNA encodes a sequence of amino acids used to build a protein. The DNA is read in triplets of bases, with each triplet coding for an amino acid. With the recognition that the genetic information lies in the sequence of bases in the DNA, it became possible to understand the chemical nature of gene mutations and how these could be as stable as the original allele of the gene.

Consideration of the genetic code linking DNA and amino acids reveals how mutations can either alter a protein, have no effect, or prevent it from being produced entirely. Mutations fall into four broad categories (point mutations, structural chromosomal aberrations, numerical chromosomal

Deletion of the T in the DNA   **TAC CAT CAT CAT CAT CAT CAT CAT ...**

becomes   **TAC CAT CAC ATC ATC ATC ATC ATC ...**

Figure 2. Frameshift mutations are deletions or insertions of bases in the DNA. Unless a multiple of three is involved in the deletion, all of the subsequent codons are altered, as illustrated by the deletion of T above. Hence, different amino acids will appear in the mutant protein, and its function will be destroyed.

aberrations, and transposon-induced mutations), each of which may be subdivided further.

Point Mutations

"Point mutations" are small changes in the sequence of DNA bases within a gene. These are what are most commonly meant by the word "mutation." Point mutations include substitutions, insertions, and deletions of one or more bases.

If one base is replaced by another, the mutation is called a base substitution. Because the DNA is double-stranded, a change on one strand is always accompanied by a change on the other (this change may occur spontaneously during DNA replication, or it can be created by errors during DNA repair. Consequently, it is often difficult to know which base of the pair was mutated and which was simply the result of repairing the mismatch at the mutation.

For example, the most common mutation in mammalian cells is the substitution of a G-C pair with an A-T pair. This could arise if G is replaced by A and subsequently the A is replicated to give T on the other strand. Alternatively, the C could be replaced by a T and the T could then be replicated to give an A on the complementary strand, the final result being the same. It is believed that the G-C to A-T conversion most commonly begins with a C-to-T mutation. This is because most of these mutations occur at DNA sequences in which C is methylated (i.e., chemically modified by the addition of a $-CH_3$ group). The methylated form of C can be converted to a base that resembles T (and thus pairs with A) by removal of an $-NH_2$ group (deamination)—a relatively common event.

Base substitution mutations are classified as transitions or transversions. Transitions are mutations in which one pyrimidine (C or T) is substituted by the other and one purine (G or A) is substituted on the complementary strand. The G-C to A-T conversion is a transition mutation, since C becomes T.

Transversions are mutations in which a purine is replaced by a pyrimidine or vice versa. Sickle cell anemia is caused by a transversion: T is substituted for A in the gene for a hemoglobin subunit. This mutation has arisen numerous times in human evolution. It causes a single amino acid change, from glutamic acid to valine, in the β subunit of hemoglobin. Sickle cell anemia was the first genetic condition for which the change in the protein was demonstrated in 1954 by Linus Pauling (a Nobel laureate from the California Institute of Technology) and subsequently shown to be a single amino acid difference by Vernon Ingram (a Nobel laureate from the Massachusetts Institute of Technology).

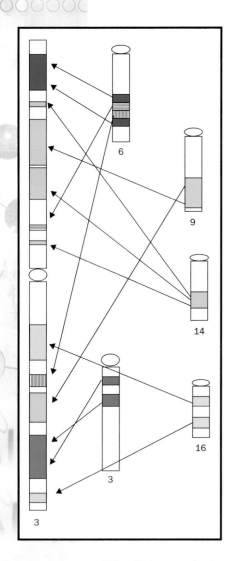

Figure 3. An example of the relationship between human and mouse chromosomes. Human chromosome 3 is on the left, and five mouse chromosomes are on the right. Although there have been numerous changes, these blocks of hundreds of genes have remained together for hundreds of millions of years in both species, and are not randomly arranged.

codon a sequence of three mRNA nucleotides coding for one amino acid

Base substitutions are sometimes silent mutations—mutations that do not change the amino acid sequence in the protein encoded by the gene. Silent mutations are possible because the original and mutated sequences can code for the same amino acid, given the redundancy of the genetic code. In the divergence between sea urchins and humans, for example, one of the histone proteins has only two amino acid substitutions, although the gene has many base pair substitutions. Histones are proteins around which DNA is wrapped in chromosomes. The very close similarity in sequence between such distantly related organisms is an indication of how critical the structure is for the function: Most mutations that change it are very disadvantageous.

One type of substitution mutation that almost always inactivates the gene is mutation to a stop **codon**. A stop codon ends the assembly of the protein, and a truncated protein is usually not active biochemically. Many recessive genetic diseases occur when a mutation converts a coding triplet to a stop codon.

Other mutations involve the insertion or deletion of one or more base pairs in the DNA. When they occur in genes, such mutations typically inactivate the encoded protein, because they change the "reading frame" of the gene. The DNA sequence is translated in groups of three nucleotides. Insertion or deletion of a nucleotide changes the sets of triplets, and thus every subsequent amino acid is altered, changing the protein completely, as shown in Figure 2. Stop codons also frequently arise from insertions or deletions.

Naturally occurring trinucleotide repeat sequences (e.g., CAGCAG CAGCAG) are hot spots for certain important human mutations that involve the insertion of more copies of the repeated sequence. For example at the locus for Huntington's disease, a sequence of 10–29 copies of CAG is normal and stable, but if there are 30–38, there is a high rate of mutation to increased numbers of copies, and if there are 39 or more copies, middle-age dementia called Huntington's disease results.

Functional Consequences and Inheritance Patterns. Mutations can be classified by their functional consequences. Mutations that inactivate the resulting protein, or prevent it from being made at all, are called loss-of-function mutations. These are usually recessive, since the organism still retains one functional copy on the other chromosome. Loss-of-function mutations may be dominant if the organism cannot compensate for the loss by using the other gene copy. Gain-of-function mutations are those in which the protein takes on a new function, or loses the ability to be regulated by other proteins. These mutations are typically dominant, since the new function may be deleterious even in the presence of a normal protein, encoded by the other gene copy.

Chromosomal Aberrations and Transposons

"Structural chromosomal aberrations," the second category of mutations, arise when DNA in chromosomes is broken. The broken ends may remain unrepaired or may be joined with those of another break, to form new combinations of genes, such as translocations. A translocation between chromosomes 8 and 21 in humans causes acute myeloid leukemia by increasing the activity of *c-myc*, a gene involved in cell replication.

Translocations often cause human infertility, because they interfere with the normal distribution of chromosomes during meiosis. Chromosomes pair up before separating, as eggs or sperm are formed, and the correct pairing depends on matching sequences between them. Structural aberrations also include inversions and duplications of pieces of chromosomes.

Most chromosomal aberrations lead to the formation of chromosomal fragments without **centromeres**. Centromeres are crucial for proper chromosomal division, during both mitosis and meiosis. Therefore a chromosomal fragment is likely to be lost from one of the daughter cells formed after cell division.

centromeres regions of the chromosome linking chromatids

Structural aberrations are nonetheless common in evolutionary history. As a result, although the chromosomes of mouse and man are quite different in appearance, most genes have the same neighbors in the two species, representing the ancestral mammalian arrangement, even if they have been moved to another chromosome as shown in Figure 3.

"Numerical chromosomal aberrations," the third category of mutations, are changes in the number of chromosomes. In some cases, the whole genome has been duplicated (called **polyploidy**) and the mutant has, for example, four of each chromosome (and is thus tetraploid) rather than the usual two (diploid, as in humans). These are much more common in the evolution of plants than animals. In other cases, only one or a few of the chromosomes are involved, which is referred to as aneuploidy. Down syndrome, in which a person has an extra chromosome 21, is an example of such a mutation. Aneuploidy may also involve the loss of a chromosome. The absence of one of the sex chromosomes, X or Y, is a mutation in humans that results in Turner's syndrome, in which there is only one X.

polyploidy presence of multiple copies of the normal chromosome set

"Transposon-induced mutations" are the fourth category of mutations. Transposable genetic elements (transposons) are pieces of DNA that can copy themselves and insert into a new location in the genome. They were first discovered by Barbara McClintock, a U.S. geneticist and Nobel laureate in 1950. When transposons jump into a new position, the insertion may disrupt a gene and thus mutate it, usually inactivating it. Sometimes the transposon jumps again, and the activity of the gene it leaves is restored. Often, however, the transposon stays in the original position, permanently disrupting the gene. Some forms of hemophilia are due to transposon insertion. Transposon mutations have been extremely common in human evolution, and such mutations are still occurring.

Mutations in Research and Medicine

Early geneticists treasured mutations in the organisms they studied, since no characteristic can be studied genetically unless heritable variants exist. If, for example, everyone had brown eyes, nothing could be learned about the inheritance of eye color, as all generations would have the same color of eyes. For this reason, geneticists collected and propagated all the mutants they could find, and methods were developed to deliberately induce mutations, a process called mutagenesis. Such techniques include exposing their experimental organisms to X rays and chemicals.

Transposons can also be deliberately used to introduce mutations in model organisms. In the plant *Arabidopsis thaliana* and in the fruit fly

Drosophila melanogaster, transposon mutagenesis is often used to induce mutations, as the mutation can be very rapidly cloned and mapped with the transposon's DNA sequence as starting point.

Comparing existing mutations can help determine the evolutionary relatedness of two organisms. During evolution, there has been a relatively constant rate of accumulation of mutations in genes for a number of proteins, so the number of changes can be used to estimate the time since two species had a common ancestor. This is called the molecular clock and is illustrated in Figure 1. Each gene has evolved at a characteristic rate—the result of mutation rates, selection, and chance changes in the gene pool.

Advances in genetics have only intensified the search for mutations, especially in complex traits such as behavior and cancer, as the key to finding the genes involved and then unraveling the underlying mechanisms. This involves mapping the mutations, cloning the genes, and studying the mutants to discover what biochemical processes are changed in the mutants.

Mutations are believed to underlie most, if not all cancers. Cancer-causing mutations found so far include genes involved in communication between cells (signal transduction) and in the control of cell division. Many of these genes have been categorized into two broad classes: oncogenes and tumor suppressor genes. The mutation that has been found most often, in a tumor suppressor gene called *p53*, usually arises as a somatic mutation but can also be inherited as Li Fraumeni syndrome.

Xeroderma pigmentosum is an autosomal recessive condition in which the ability to repair DNA damage induced by UV light is defective. Many mutations are produced, and the affected people have large numbers of skin cancers. SEE ALSO CARCINOGENS; CHROMOSOMAL ABERRATIONS; DNA REPAIR; GENE; HEMOGLOBINOPATHIES; MUTAGENESIS; MUTATION RATE; POLYPLOIDY; TRANSPOSABLE GENETIC ELEMENTS.

John Heddle

Bibliography

Drake, John W. "Spontaneous Mutation." *Annual Review of Genetics* 25 (1991): 125–46.

Hartwell, Leland H., et al. *Genetics: From Genes to Genomes.* Boston: McGraw-Hill, 2000.

Lewis, Ricki. *Human Genetics: Concepts and Applications*, 4th ed. Boston: McGraw-Hill, 2001.

Pauling, Linus, et al. "Sickle Cell Anemia, a Molecular Disease." *Science* 110 (1949): 543–548.

Internet Resource

International Agency for Research on Cancer. <http://www.iarc.fr/>.

Mutation Rate

Mutation rate refers to the frequency of new mutations per generation in an organism or a population. Mutation rates can be determined fairly precisely in experimental organisms with short generation times, such as bacteria or fruit flies. Human mutation rates are more difficult to determine accurately. Mutation rates can be used as a "molecular clock" to determine the time since two species diverged during their evolution.

Measurements of Mutation Rate

Mutation rate is often difficult to measure. The frequency of existing mutations in a population is not a good indication of the mutation rate, since a single mutation may be passed on to many offspring. In addition, there are often selective pressures that increase or decrease the frequency of a mutation in a population.

Mutation rates differ widely from one gene to another within an organism and between organisms. Generally mutation rates in bacteria are about one mutation per one hundred million genes per generation. While this sounds quite low, consider that the *Escherichia coli* bacteria in our intestines produce more than 20 billion new bacteria every day, each of which has approximately four thousand genes. This works out to about ten million new mutations in the population every day. In mice, the rate is about one mutation per ten thousand genes per generation. While this is much larger than the rate for bacteria, the mouse generation time is also much greater.

Human Mutation Rates

The appearance of rare dominant genetic diseases, such as retinoblastoma, have been used to estimate the mutation rate in the human population. Retinoblastoma is a childhood cancer of the eye and was a lethal condition until recently. Hence almost every case represented a new mutation (because individuals with the condition did not survive to reproduce and pass the genetic propensity for the disease along to their offspring), and the mutation rate could thus be readily estimated. Modern methods indicate that the mutation rate is roughly one mutation per 10,000 genes per generation. With at least 30,000 genes, this means that each person harbors about three new mutations, although this estimate may be off by a factor of ten. There are many more mutations in non-coding portions of DNA, but these are fairly difficult to study because they have no effect on the **phenotype** of the person.

phenotype observable characteristics of an organism

About 90 percent of human mutations arise in the father rather than the mother. This may be related to the difference in the number of cell divisions required to produce a sperm versus an egg; sperm are produced late in a male's development, compared to eggs, which are produced quite early in the development of a female. Older parents pass on more mutations, and these may be either mutations within a gene or chromosomal aberrations, which are deletions or rearrangements of the chromosomes and involve many genes. Human mutation rates are generally quite similar worldwide. The exception is in local populations that have been exposed to radioactivity from nuclear testing or other sources.

Factors Influencing the Mutation Rate

Within a single organism, the mutation rate of two genes can differ by a thousandfold or more, so within a species some mutations may be very rare and others quite common. Exposures to very high doses of very potent mutagens can increase the mutation rate per generation by more than a hundredfold.

Both the nature of the gene and its environment can influence the mutation rate. The size of the gene, its base composition, its position in the

genome, and whether or not it is being actively transcribed influence its mutation rate. The *dystrophin* gene, mutated in Duchenne muscular dystrophy, is thought to have a mutation rate of one in every ten thousand births, while the gene, mutated in Huntington's disease, has a mutation rate of closer to one in one million. The explanation for this difference is, at least in part, gene size: The dystrophin gene is one of the largest known. Genes whose **promoter** regions have been silenced by methylation (the addition of $-CH_3$ units to cytosine bases) are more likely to be mutated, since methylcytosine is easily converted to a base that resembles thymine.

promoter DNA sequence to which RNA polymerase binds to begin transcription

In addition, the repair capacity of the organism can be important in determining how many mutations ultimately remain in the genome. For example, Bloom syndrome, a human cancer–causing condition, causes a decreased ability to repair DNA damage and an elevated mutation rate. Exposure to environmental mutagens or to protective agents, possibly dietary, can alter the mutation rate. Since the mutation rate is partly under genetic control, it is a selected characteristic of an organism, with the burden of detrimental mutations being balanced by the benefit of rare favorable mutations that are adaptive and permit evolution of the species.

One important factor influencing observed mutation rates is the means by which mutations are detected, as some methods may detect a changes at only one or two base pairs of a specific gene, leaving others undetected. Obviously the mutation rate observed by such methods will be lower than if more altered bases could be detected.

wild-type most common form of a trait in a population

Mutations can also change a mutated gene back to the normal, **wild-type** form of the gene. Such "back" mutations are typically much rarer than "forward" mutations. This is because the number of ways to inactivate a gene is much greater than the number of ways to fix it. Imagine there are 1,000 bases that could be changed to produce a forward mutation. To reverse one of these mutations, it is necessary to change the one specific **base pair** that has mutated, and to change it back to the base it was before, rather than to a different one. Therefore, a back mutation rate of less than 1 one-thousandth of the forward rate would be expected.

base pair two nucleotides (either DNA or RNA) linked by weak bonds

The Origin of Spontaneous Mutations

The causes of most spontaneous mutations is not known, so the main factors affecting the spontaneous mutation rates are obscure. It is likely that the methylation of cytosine in the DNA is important for many spontaneous mutations, because many are found at sites in the genes where cytosine is methylated. Ionizing radiations, such as cosmic rays, probably account for less than 10 percent of spontaneous mutations. Other factors are errors made during replication of the DNA; exposure to mutagens produced by cells during their normal metabolic activity, with reactive oxygen species being the common suspect; spontaneous breakdown of DNA at body temperatures; and exposure to environmental agents. Many mutations are made when the mechanisms that repair DNA make mistakes, and many error-prone DNA repair **enzymes** are known.

enzymes proteins that control a reaction in a cell

Molecular Clocks

While it may seem unlikely, it is believed that the overall mutation rate within a species does not vary much over long periods of time. This means that the mutation rate serves as a "molecular clock." The clock can be used to determine the time since the evolutionary divergence of two species. Two organisms with very few DNA sequence differences between them diverged more recently than two that display more accumulated differences. The absolute amount of time for these divergences can be determined if the clock is calibrated, that is, if a known number of sequence differences can be correlated with a known time since divergence. This is done by comparing sequence data with data from the fossil record.

Early work in this field concentrated on a small handful of genes and gave conflicting results. Because mutation rates vary among genes, the best results come from analyzing changes in many genes. A 1998 study of the evolution of mammals analyzed 658 nuclear genes from 207 vertebrate species. It showed that the ancestors of most contemporary mammals arose more than 80 to 110 million years ago, long before the extinction of dinosaurs, and demonstrated that the fossil record from that time has some very large gaps in it. SEE ALSO CARCINOGENS; DNA REPAIR; GENE; METHYLATION; MOLECULAR ANTHROPOLOGY; MUSCULAR DYSTROPHY; MUTAGEN; MUTATION.

John Heddle

Bibliography

Crow, J. F. "The High Spontaneous Mutation Rate: Is It a Health Risk?" *Proceedings of the National Academy of Science* 94 (1997): 8380–8386.

Griffiths, Anthony J. F., et al. *An Introduction to Genetic Analysis.* New York: W. H. Freeman, 2000.

Kumar, Sudhir, and S. Blair Hedges. "A Molecular Timescale for Vertebrate Evolution." *Nature* 392 (1998): 917–920.

Nature of the Gene, History

Although Wilhelm Johannsen coined the term "gene" in 1909, our understanding of the nature of the gene has changed significantly over the course of the twentieth century. Gregor Mendel's elements of inheritance were given a material basis in the chromosome theory of the early twentieth century. Attempts to understand the nature of gene action and mutation spurred interest in the biochemical role and molecular basis of the gene, culminating in the discovery of the structure of DNA.

From Elements to Genes

In 1865, when Mendel articulated the laws of inheritance that now bear his name, he did not use the terms "gene" or "**allele**." He referred instead to "elements" and "characters." Mendel described the patterns of inheritance he observed in terms of character pairs. These pairs segregate to form the next generation of character pairs and remain independent of the behavior of other character pairs. The external characters of the pea plants he described corresponded to elements within the **germ cells** of the same plants. Whether Mendel thought that pairs of characters were expressions of pairs of cellular elements is not clear.

allele a particular form of a gene

germ cells cells creating eggs or sperm

Edward Tatum. He and his colleague George W. Beadle developed the "one gene-one enzyme" model of gene function.

phenotype observable characteristics of an organism

By the time of Mendel's rediscovery in 1900 by Hugo de Vries, Carl Correns, and Erich Tschermak, however, visible characters were understood to be expressions of hereditary particles within each cell. Just as characters occurred in pairs, Mendelians interpreted hereditary particles as occurring in pairs. William Bateson called these pairs of hereditary particles "allelomorphs," a term that would eventually be shortened to "alleles." The idea that something within the gametes (sperm and egg cells) specified the characteristics of the organism was captured by Johannsen's term "gene."

It was not at all clear in the first decade of the twentieth century that the segregation and assortment of alleles could explain patterns of inheritance. While Bateson forcefully advocated Mendelian principles, many of his contemporaries, such as Karl Pearson and Walter F. R. Weldon, explained patterns of heredity in terms of continuous characters, instead of discrete Mendelian character pairs. For Pearson, Weldon, and other biometricians, as they called themselves, traits were expressed as continuous distributions from tall to short, for instance. The mating of a tall parent with a short parent would yield offspring with a range of heights, as the parental traits were blended together.

Because Mendelians, like Bateson, strongly identified characters with alleles, they insisted that both were discrete. The apparent blending of parental traits was a stumbling block until the distinction between genotype and **phenotype** was consistently applied and single continuous traits such as height were seen to be the expression of many individual Mendelian factors. Thus, multifactor inheritance allowed geneticists to explain a continuously distributed phenotype as an expression of many discreet Mendelian factors or genes.

The Chromosome Theory

Of the many difficulties facing early genetics, one of the most important was figuring out what a gene was actually made of. As early as 1902, Walter Sutton and Theodore Boveri had observed that, during meiosis, chromosomes separated just as Mendelian particles were proposed to separate. The discovery of sex chromosomes a few years later suggested that chromosomes might play a genetic role, but the association of a specific gene with a chromosome would not occur until 1910, when Thomas Hunt Morgan demonstrated the sex-limited inheritance of the white-eye mutation in *Drosophila*.

Using the wealth of new information provided by their experiments with *Drosophila*, Morgan and his colleagues articulated the chromosome theory of inheritance, which treated genes as indivisible particles arranged like beads on a string to form a chromosome. Their patterns of association provided the clues to map their linear order on chromosomes and to understand processes of chromosomal recombination and rearrangement.

Gene Action and Mutation

The tremendous successes of the Morgan group often overshadow a simultaneous tradition of exploring the nature of gene action. The problem of how genes produce their effects was the domain of physiological genetics. As early as 1911, A. L. and A. C. Hagedoorn had proposed that genes acted as chemical catalysts. In 1916 Richard Goldschmidt interpreted genes as

enzymes, while Sewall Wright explained coat-color patterns in terms of genetic regulation of enzymes in pigment-formation pathways. Many different geneticists sought to understand the action of genes in terms of their regulation of the rates of chemical reaction, the production of specific chemical products, and the induction of developmental processes. The incredible biochemical complexity of developmental and physiological processes, however, meant that physiological genetics made relatively slow progress.

The gap between genetics and biochemistry was narrowed in 1941, when George Beadle and Edward Tatum began to use the microorganism *Neurospora* to dissect biochemical processes. By growing *Neurospora* on media with different chemical compositions, Beadle and Tatum were able to devise a system for detecting specific changes in the biochemical abilities of their organism. Careful study of biochemical mutants led the two researchers to propose the "one gene–one enzyme" theory, linking genes to specific enzymes and the chemical reactions they catalyzed. Although this association was not new, Beadle and Tatum's work invigorated the field and encouraged the creation of biochemical genetics as a field of study.

Later it was discovered that many proteins are not enzymes, but instead may be signaling molecules or receptors, or may play structural roles. Thus "one gene–one enzyme" was modified to "one gene–one protein." It was also discovered that many functional proteins are composed of several distinct amino acid chains (polypeptides) whose corresponding DNA sequences were not necessarily close together or even on the same chromosome, leading to the "one gene–one polypeptide" formulation of the gene definition.

Like the study of gene action, the nature of the gene itself was understood as a biochemical problem early in the twentieth century. Because chromosomes were known to be composed of proteins and nucleic acids, many geneticists proposed specific molecular mechanisms to explain genetic changes or mutations. In 1919 Carl Correns had proposed that the gene was a large molecule with a number of side chains. Mutations were caused by changes in these side chains. Hermann Muller's pioneering work on the ability of X rays to induce mutations led Nikolay Timofeeff-Ressovsky, Karl Zimmer, and Max Delbrück to investigate the relationship between dose and mutability. The resulting model of mutation, published in 1935, provided a quantum mechanical mechanism for the molecular effects of the ionizing energy of X rays. As the nature and causes of mutation became a more prominent part of genetics, the importance of understanding the molecular basis of the gene also became widely recognized.

The Molecular Gene

Genetic material was understood in the mid-twentieth century to have two key chemical properties: the ability to **catalyze** reactions to make more genetic material and the ability to catalyze reactions to make a wide array of chemical products found in organisms of all sorts. For most biologists, proteins were the only molecules that seemed to have the ability to play so many specific roles. Researchers naturally focused their attention on the protein component of chromosomes.

The realization that the nucleic acid (DNA) portion of the chromosome was actually the hereditary material was the consequence of two sets of experiments. In 1941 Oswald Avery, Colin MacLeod, and Maclyn McCarty

George W. Beadle, colleague of Edward Tatum.

enzymes proteins that control a reaction in a cell

catalyze aid in the reaction of

The Hershey-Chase
experiment established
that DNA, not protein,
entered bacteria, and
thus was presumably the
genetic material. Adapted
from <http://www
.accessexcellence.com/
AB/GG/HERSHEY.gif>.

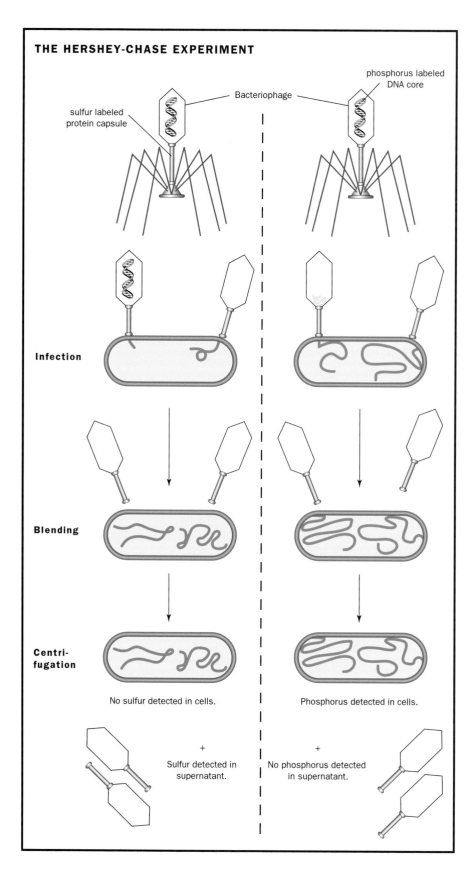

THE HERSHEY-CHASE EXPERIMENT

Bacteriophage

sulfur labeled
protein capsule

phosphorus labeled
DNA core

Infection

Blending

**Centri-
fugation**

No sulfur detected in cells.

Phosphorus detected in cells.

+

Sulfur detected in
supernatant.

+

No phosphorus detected
in supernatant.

extended work by Fred Griffith on the transformation of nonvirulent bacteria into virulent bacteria. Working from the premise that some hereditary chemical component of the virulent bacteria was transforming the nonvirulent bacteria, these researchers isolated DNA and proteins from the virulent bacteria in order to determine which was the "transforming principle." The surprising result that DNA caused transformation contributed to growing interest in DNA, but DNA was not widely accepted as the genetic material until much later.

In 1952 Alfred Hershey and Martha Chase used radioactive labels to follow DNA and proteins. Hershey and Chase worked with bacteriophages—viruses that infect bacteria. Bacteriophages are composed of proteins and DNA. To determine which was the genetic material, Hershey and Chase created DNA-specific labels and protein-specific labels using radioactivity. They were then able to determine that only DNA was injected into the bacteria to provide the genetic blueprint for the next generation of viruses. This elegant experiment was soon followed by the discovery of the structure of DNA by James Watson and Francis Crick. The double helix structure for DNA immediately suggested a mechanism for its own replication. Thus, by 1953 DNA was identified as having the key catalytic properties required of the genetic material.

At about the same time, the physicist-turned-geneticist Seymour Benzer was using **bacteriophages** to show that genes were not indivisible units; rather, they could break and recombine within their structures. This focused even more attention on the molecular nature of the gene. An understanding of recombination led slowly to a more dynamic view of genes and chromosomes, exemplified by Barbara McClintock's discovery that some genetic elements are mobile, moving from place to place around the chromosomes. In the early 1960s, Francois Jacob showed that bacterial gene expression is controlled by several noncoding DNA segments. Jacob developed the concept of the operon, a set of coding genes controlled by a common set of regulatory regions.

bacteriophages viruses that infect bacteria

Identifying DNA as the molecular basis of the genetic material sparked interest in cracking the DNA code and determining how it specifies its products. Reconciling the structure of the DNA sequence with its function became a central preoccupation of molecular genetics.

Ever since Morgan, the gene as a hereditary unit had been a unit of structure and function. Morgan's particulate gene theory, however, had begun to dissolve in the late 1930s, as it became clear that rearrangements in the chromosome could alter genetic function. Various units of structure and function were suggested (**enzymes, polypeptides,** etc.) in the wake of the particulate gene, but the discovery of the genetic code suggested that the molecular gene could be identified as a continuous coding sequence of DNA.

enzymes proteins that control a reaction in a cell

polypeptides chains of amino acids

While most coding sequences lead to the formation of a temporary RNA intermediate (messenger RNA) that is then translated into protein, some sequences code for RNA molecules that are not translated but are functional themselves (ribosomal RNA, transfer RNA, and a host of small nuclear RNAs). The discovery of noncoding sequences (introns) within coding regions (exons) further complicated any simple formulation of the structure-function relationship, as did the growing understanding of regulatory

regions, which may or may not be located near the coding regions. Finally, recent discoveries indicate that exons can be joined in different ways in different tissues and that this alternative splicing allows a single set of exons to code for a group of related protein products. By the end of the twentieth century genes could be seen as sequences of DNA (that may be interrupted by noncoding introns) that code for RNA products, many of which are translated into proteins (or a group of related proteins). SEE ALSO ALTERNATIVE SPLICING; CHROMOSOMAL THEORY OF INHERITANCE, HISTORY; DELBRÜCK, MAX; DNA STRUCTURE AND FUNCTION, HISTORY; GENE; INHERITANCE PATTERNS; QUANTITATIVE TRAITS; MEIOSIS; MENDEL, GREGOR; MORGAN, THOMAS HUNT; MULLER, HERMANN.

Michael Dietrich

Bibliography

Carlson, Elof. *The Gene: A Critical History.* Philadelphia: W. B. Saunders, 1966.

Portin, Petter. "The Concept of the Gene: Short History and Present Status." *The Quarterly Review of Biology* 69 (1993): 173–223.

Sturtevant, Alfred H. *A History of Genetics.* New York: Cold Spring Harbor Press, 2001.

Internet Resource

Access Excellence. The National Health Museum. <http://www.accessexcellence.com/AB/GG/HERSHEY.gif>.

Nomenclature

Like any other field in science, genetics has its own language. However, genetics is also a multidisciplinary field that encompasses expertise, and hence terminology, from diverse areas of science, including molecular biology, statistics, clinical medicine, and, most recently, **bioinformatics**. Despite all of the new and changing language in the field, two of the most frequently used terms in genetics are still "chromosomes" and "genes."

Humans have twenty-three pairs of chromosomes. One member of each pair is inherited from the person's mother, and the other from the father. Of the pairs, twenty-two are known as **autosomes**. The remaining pair consists of the sex chromosomes, which determine a person's gender. Females have two X chromosomes, and males have one X chromosome and one Y chromosome.

Chromosomes are located in the nucleus of a cell. During cell division, which is known as mitosis, the chromosomes' long strands coil up tightly, to the point where they can be seen as individual units under the microscope. At this stage, each chromosome is composed of two identical strands, called chromatids (each of which further consists of two strands of nucleotides). Chromatids are attached at a constriction point called the centromere.

Chromosomes can be distinguished by their size and by their "banding pattern." Researchers use a chemical staining process in the laboratory to create the banding pattern, allowing them to see the chromosomes more easily. Each chromosome is divided into two sections, or "arms," with one arm on each side of the centromere. The short arm is called the p arm, and

bioinformatics use of information technology to analyze biological data

autosomes chromosomes that are not sex-determining (not X or Y)

the long arm is the q arm. The bands on each arm are numbered. As new and better staining techniques are developed, the numbering system is also refined, so that band 32, for example, would be subdivided into 32.1, then 32.15, and so on. One particular position on the long arm of chromosome 5 would be referred to as 5q32.15.

Almost every human chromosome contains more than a thousand genes. Therefore, even a small extra piece or missing piece of a chromosome results in hundreds of genes being added or deleted from an individual. When researchers study a person's chromosomes, they try to determine if there are any missing or extra chromosomes or chromosome pieces. The addition or deletion of genes sometimes causes a recognizable genetic disorder. Down syndrome, for example, results when there are three copies of chromosome 21 rather than the normal two.

Genes are very small structures that lie on chromosomes. They are the instructions, or blueprints, for producing proteins, which are the building blocks that our bodies use to grow, develop, and function. Humans have an estimated thirty thousand to forty thousand genes.

As happens with other types of scientific discovery, the person who discovers a gene names it, and a scientist can name a gene anything he or she wants. This has led to some confusion, as different naming schemes are used by different groups. To bring order to the situation, several international working groups are trying to standardize the naming of genes. There are separate working groups that focus on naming genes from humans, mice, fruit flies, plants, and other organisms.

Some scientists choose names based on the clinical disorder that is thought to be associated with changes in the gene. For example, one gene was named *CFTR* because changes in its sequence are associated with the disease cystic fibrosis. Geneticists studying fruit flies traditionally use single-word names, such as wingless, hunchback, and sevenless, that refer to the effect of a mutation in the gene. "Sevenless" refers to the absence of the R7 protein.

For human genes, abbreviations are commonly used. Abbreviated names are especially useful for genes with long names. *WNT2*, for example, stands for "wingless-type MMTV integration site family member 2." Although the word "wingless" seems unnecessary (humans, of course, don't have wings!), *WNT2* is named after similar genes in fruit flies. Genes are often named after genes they resemble in other organisms.

Sometimes the gene name is actually a variation of the name of the protein it makes. For example, the *RELN* gene in the human encodes the "reelin" protein. The "reelin" protein was named for the reeling motion exhibited by mice that lack a functional copy of the protein.

Other genes are classified based on what their proteins do. For example, HOX genes (short for homeobox) are genes involved in development. Individual HOX genes are named with additional letters and numbers, such as *HOXA1* or *HOXD9.* The consistent naming system lets scientists know that any gene with the name *HOX* is likely to play a specific role in development.

There are even playful gene names that have nothing to do with a disorder or protein. An example is the *SHH* gene, which is involved in the

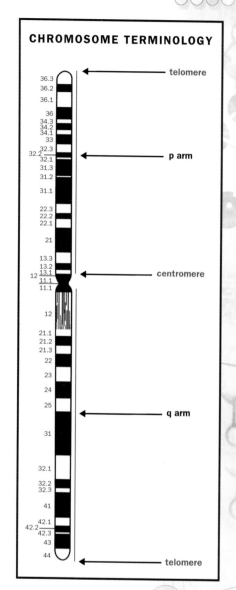

CHROMOSOME TERMINOLOGY

A chromosome, with banding patterns, band numbers, and major physical landmarks.

This is a photograph of a high-resolution chromosome analysis commonly referred to as a "karyotype." This analysis shows 46, XX which means that this patient has the normal number of chromosomes and is a female since she has two X chromosomes.

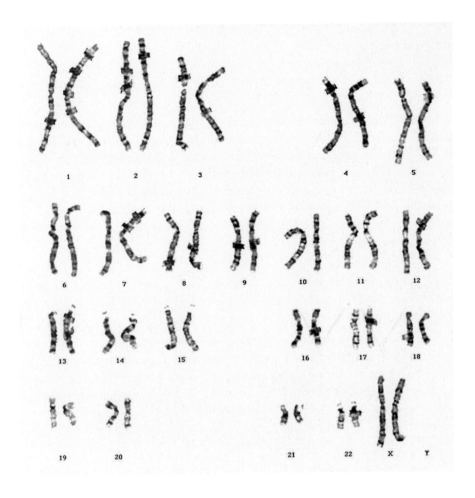

development of the brain, spinal cord, and limbs. The *SHH* gene is named after the cartoon character Sonic the Hedgehog! SEE ALSO BIOINFORMATICS; CHROMOSOMAL BANDING; CHROMOSOME, EUKARYOTIC; FRUIT FLY: *DROSOPHILA*.

Chantelle Wolpert

Bibliography

Internet Resource

HUGO Gene Nomenclature Committee. <http://www.gene.ucl.ac.uk/nomenclature>.

Nondisjunction

Nondisjunction is the failure of two members of a homologous pair of chromosomes to separate during meiosis. It gives rise to gametes with a chromosomal content that is different from the norm. The consequences of this are usually quite severe, and a number of clinical conditions are the result of this type of chromosome mutation.

The Mechanism of Nondisjunction

Homologous chromosomes are virtually identical chromosomes that occur in pairs, one member inherited from each parent. Humans have

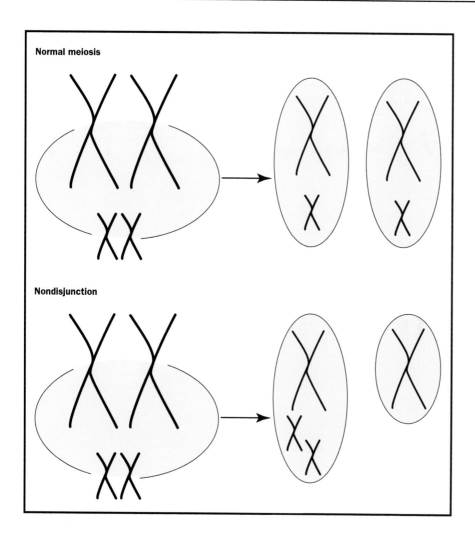

Normal meiosis

Nondisjunction

Nondisjunction occurs when homologous chromosomes fail to separate, creating some cells with too many, and others with too few, chromosomes.

forty-six chromosomes, or twenty-three homologous pairs. In normal meiosis, homologous chromosomes pair up and, by attachment to the spindle fibres, become aligned at the cell equator. Prior to the first meiotic division, the members of each homologous pair migrate to opposite poles of the cell by means of the pulling action of the spindle fibers. This ensures that, upon completion of meiosis, each gamete will contain one copy of every chromosome.

However, the segregation process is not error-free, and every so often it happens that two **homologous** chromosomes fail to separate (disjoin) and thus both migrate to the same pole. This gives rise to two types of gamete. One type possesses two copies of the chromosome, whereas the other type lacks that chromosome altogether. This condition, involving the loss or gain of a single chromosome, is referred to as aneuploidy. Fusion of an aneuploid **gamete** with a normal gamete gives rise to a **zygote** with an odd number of chromosomes.

A zygote which has one less than the normal diploid number of chromosomes $(2n-1)$ is said to be monosomic, and such zygotes do not usually develop to term. Zygotes containing an extra chromosome $(2n+1)$ are trisomic for the chromosome of interest, and these may develop, though usually with severe abnormalities.

homologous carrying similar genes

gamete reproductive cell, such as sperm or egg

zygote fertilized egg

Various complements of sex chromosomes that yield different phenotypes: two XX, normal female; one X, a Turner syndrome female; one X and one Y, normal male; two X and one Y, a Klinefelter male.

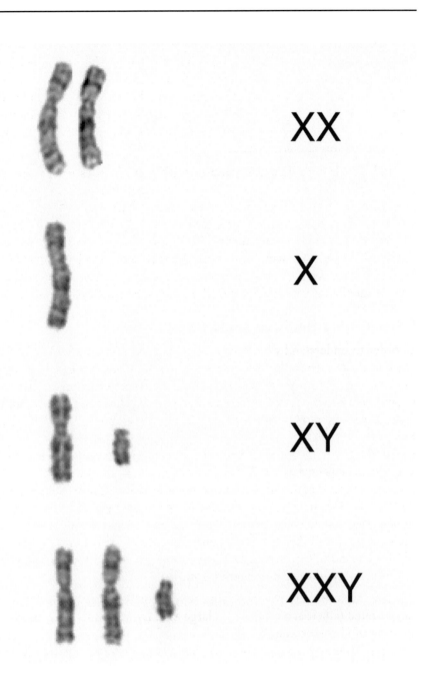

XX

X

XY

XXY

Non-Fatal Human Aneuploid Conditions

The most common example of non-fatal trisomy in humans is that of Down syndrome, caused by the presence of an extra copy of chromosome 21. Affected individuals suffer from mental retardation, congenital heart disease, and increased suceptibility to infection. Physical characteristics include a short, stocky body, flattened facial features, and almond-shaped eyes.

Down syndrome is an example of an autosomal trisomy as it involves one of the autosomes ("autosome" is the term that designates all the chromosomes other than the X and Y, or sex, chromosomes). There are many human conditions that are caused by nondisjunction of the sex chromosomes, and these usually affect the individual's secondary sexual characteristics and fertility.

For example, the fusion of an XY sperm with a normal X egg, or the fusion of a Y sperm with an XX egg gives rise to an XXY individual (with normal autosomes). This condition is known as Klinefelter's syndrome. Individuals affected by this disorder usually have below-average intelligence. They are are phenotypically male, but present some female secondary sexual characteristics. They may develop breasts, and they have little facial hair, very small testes and are sterile.

Individuals with Turner's syndrome (XO) are females with a single X chromosome. They are sterile and have underdeveloped secondary sexual characteristics, and they are shorter than normal. Females with genetic constitution XXX, on the other hand, have a normal appearance and are fertile, but suffer from a mild mental handicap. Similarly, XYY males have relatively few clinical symptoms and appear phenotypically normal. They are taller than average and may show aggressive behavior and a below-average intelligence. Both XXX and XYY conditions usually pass undiagnosed.

Fatal versus Nonfatal Conditions

In order to understand why some aneuploid conditions are fatal and others (such as those mentioned above) are not, one must understand the concept of gene dosage and its importance in development. A normal human possesses twenty-two pairs of autosomes and two sex chromosomes (XY in the case of males and XX in the case of females). Such an individual develops normally because there is a situation of genetic balance: Each gene is present in the correct amount (or dose), such that its contribution towards development is appropriate and ideal. However, if a chromosome is either removed from or added to the normal set, a situation of imbalance is immediately established: The contribution (or gene dosage) of each gene contained within that chromosome is altered and as a result development is compromised. While the duplication or silencing of an individual gene is not usually fatal, the wholesale addition or loss of a chromosome, which contains a thousand or more genes, almost always is.

It is obvious from this reasoning that a small change is more likely to be tolerated (albeit at a cost) than a large one. Down syndrome is caused by trisomy of chromosome 21, which is one of the smallest human chromosomes (containing a relatively small number of genes). This provides an explanation as to why this condition is not fatal, while a trisomy involving another, larger autosome would most likely be fatal.

With the sex chromosomes, a lot more flexibility is allowed: Although the X chromosome is very large, only one is used in development (in every female cell one of the two X chromosomes is inactivated at random). The Y chromosome, on the other hand, contains very few genes and is not necessary for normal female development. It is only required for male development. With a knowledge of these facts it is relatively easy to understand why aneuploidies involving the sex chromosomes tend not to be fatal. Note, however, that the YO condition is fatal due to the lack of the essential X chromosome.

The Causes of Nondisjunction and Its Frequency in Humans

Meiosis is a very tightly regulated process, and a whole series of control mechanisms (constituting a number of "checkpoints") exist to ensure that

everything proceeds in the correct manner. If an error should occur during the process, it is usually corrected. Nondisjunction is the result of a mistake at the level of chromosome segregation, which involves the spindle fibers. In normal meiosis, there is a mechanism that monitors the correct formation of the spindle fibers, the correct attachment of the chromosomes to the spindle fibers, and the correct segregation of chromosomes. Collectively, this is referred to as the spindle checkpoint. Failure of this checkpoint to function correctly results in nondisjunction of chromosomes.

Nondisjunction is known to occur more frequently in the cells of older individuals. This is illustrated by the fact that older women are more likely to give birth to children affected by an aneuploid condition than are younger women. For instance, the risk of a twenty-year-old mother giving birth to a child with Down syndrome is about one in two thousand, compared to an approximate one in thirty risk in the case of a woman of age forty-five. The precise reason for this is not entirely certain, but a simple explanation could be that the older a cell is, the more loosely controlled are the processes occurring within that cell. This would mean that an older cell undergoing meiosis would be more likely than a younger one to ignore the constraints of the spindle checkpoint and hence give rise to aneuploid cells. SEE ALSO CHROMOSOMAL ABERRATIONS; CROSSING OVER; DOWN SYNDROME; MEIOSIS.

Andrea Bernasconi

Bibliography

Alberts, Bruce, et al. *Molecular Biology of the Cell*, 3rd ed. New York: Garland Publishing, 1994.

Lewin, Benjamin. *Genes VII*. New York: Oxford University Press, 2000.

Strachan, Tom, and Andrew P. Read. *Human Molecular Genetics*. Oxford, U.K.: Bios Scientific Publishers, 1996.

Nucleases

polymers molecules composed of many similar parts

nucleotides the building blocks of RNA or DNA

phosphodiester bonds the links between two nucleotides in DNA or RNA

endonucleases enzymes that cut DNA or RNA within the chain

DNA and RNA are **polymers** made by linking together smaller units called **nucleotides**. Nucleases are enzymes that break the chemical bonds, called **phosphodiester bonds**, that hold the nucleotides of DNA or RNA polymers together. Enzymes that cleave the phosphodiester bonds of DNA are called deoxyribonucleases, and enzymes that cleave the phosphodiester bonds of RNA are called ribonucleases.

Nucleases can be divided into two classes, exonucleases and endonucleases, based on the positions of the cleaved bonds within the DNA or RNA polymers. The exonucleases are involved in trimming the ends of RNA and DNA polymers, cleaving the last phosphodiester bond in a chain. This cleavage results in the removal of a single nucleotide from the polymer. If the enzyme removes nucleotides from the 3′ ("three prime") end, it is referred to as a 3′ exonuclease. If cleavage is at the 5′ end, the enzyme is called a 5′ exonuclease. The endonucleases cleave phosphodiester bonds of DNA or RNA at positions other than at the end of the polymer. The cutting reactions of the **endonucleases** produce fragments of DNA or RNA.

Individual nucleases frequently show preferences for various structures of DNA and RNA. Some nucleases prefer single-stranded polymers, while

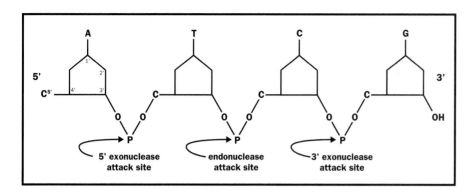

Nucleases are distinguished by their sites of attack. Exonucleases attack the end of a chain, while endonucleases attack intern postions. In reality, the nucleotide polymer could range in length from dozens to thousands of units.

others prefer double-stranded polymers. Some nucleases cleave at specific nucleotide sequences, and others cleave at positions in the polymers independent of nucleotide sequence. Exonucleases can show preference for DNA ends that are correctly base-paired or for ends that are mispaired. Some endonucleases involved in DNA repair recognize damaged nucleotides and **cleave** phosphodiester bonds at these sites. The preferences exhibited by the nucleases reflect the wide biological functions for these **enzymes**.

cleave split

enzymes proteins that control a reaction in a cell

The Nuclease Mechanism

There are two different mechanisms used by various nucleases to cleave the chemical bonds of the DNA or RNA polymer. The most common mechanism is one in which a water molecule is used to break the **phosphodiester bond**. This is called a hydrolysis reaction. Under most conditions the P–O (phosphorous and oxygen) bond of the DNA or RNA polymer is very stable, and the H_2O molecule is not usually very reactive. However, nucleases that use the hydrolysis mechanism make the H_2O reactive by removing one of the hydrogens to generate a highly reactive OH^- (hydroxyl). The negatively charged OH^- can then attack the P–O bond to cleave the polymer. An alternative mechanism used by some DNA repair endonucleases involves the initial cleavage of a C–O (carbon and oxygen) bond and subsequent P–O bond cleavage. This is called a lyase reaction and does not involve water.

phosphodiester bond the link between two nucleotides in DNA or RNA

Deoxyribonucleases in DNA Replication and Repair

During DNA synthesis the 3′ and 5′ exonucleases function to remove unwanted nucleotides from the DNA. Occasionally, a DNA polymerase will add an incorrect nucleotide to the growing DNA polymer. A 3′ exonuclease removes nucleotides that have been incorrectly polymerized into DNA chains. These exonucleases are referred to as "proofreading" exonucleases. The proofreading exonucleases work in close association with the DNA polymerases to increase the overall accuracy of DNA synthesis.

In many cases the exonuclease activity is contained in the same protein as the DNA polymerase activity. For example, the *Escherichia coli* DNA polymerase I is a single polypeptide with three separate domains, or regions of function. Each of these three domains contains an enzymatic activity. The DNA polymerase activity is in one domain, and the two other domains contain 3′ and 5′ exonuclease activities. The 3′ exonuclease proofreads for the DNA polymerase, and the 5′ exonuclease removes unwanted nucleotides in advance of the DNA polymerase. In contrast, the proofreading exonuclease

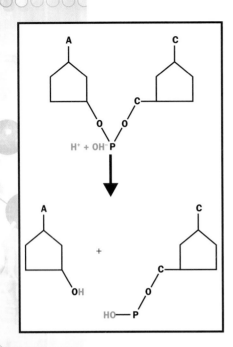

Hydrolysis reaction.

topological describes spatial relations, or the study of these relations

Inhalation of recombinant human DNA endonuclease breaks up sticky airway secretions in cystic fibrosis.

of *E. coli* DNA polymerase III is located in a separate protein called the ϵ subunit, while the polymerase activity is contained on the α subunit. These two separate proteins, encoded by different genes, associate and interact in a complex to assure a high level of accuracy during DNA replication.

Multiple DNA repair pathways also use nucleases to restore the correct nucleotide base-pairing if it becomes altered during the life of the cell. Reactive molecules originating from inside the cell during normal metabolism or brought into the cell during exposure to external sources can damage the nucleotides of DNA. A damaged nucleotide opposite a normal nucleotide creates a distortion in the shape of the DNA double helix that is recognized by DNA repair proteins. A DNA helix distortion is also generated when normal but mismatched nucleotides are generated during DNA replication—for example, if a nucleotide is paired with C rather than A. Mismatches occur when DNA polymerases misinsert nucleotides and fail to proofread the misinserted base. These DNA helix distortions are repaired to minimize introduction of mutations into the genome. The steps in these DNA repair pathways include recognition of the distorted DNA, incision of the DNA by endonucleases on the 5′ or 3′ side of the damage, excision (removal) of nucleotides by exonucleases from the damaged region, and synthesis of a new DNA strand by a DNA polymerase. Some of the genes encoding the repair endonucleases and exonucleases have been identified in *E. coli* and in human cells, and the precise functions of these enzymes in cells are an active area of research.

The topoisomerases are a specialized class of nucleases functioning in cells to alter the **topological** structure of DNA. During replication the DNA becomes twisted, creating a barrier to progression of the DNA replication apparatus. The topoisomerases recognize these twisted regions of DNA and restore them to their untwisted state. This is accomplished by incising the DNA, removing the topological strain by unwinding, then resealing the DNA to regenerate the intact polymer. The Type I topoisomerases function by cutting one of the DNA strands. The Type II topoisomerases cut both DNA strands. The incision of DNA is transient, and both classes of topoisomerases reseal the DNA strands.

The restriction endonucleases are involved in the DNA restriction-modification systems of bacteria, which protect these cells from invading viruses. These enzymes have become powerful tools for DNA manipulation by molecular biologists. They recognize specific sequences in DNA and cut the DNA at these sites. The recognition sequences are usually between four and six nucleotides in length in duplex DNA. Each restriction enzyme has a different recognition sequence, making it possible to cut DNA in a variety of very predictable patterns.

Ribonucleases in RNA Maturation and Degradation

The expression of genes into protein products requires the generation of a messenger RNA (mRNA) by transcription and the subsequent translation of the mRNA into protein. In bacteria, the mRNA is transcribed, translated, and then degraded by ribonucleases in rapid succession. Thus, the ribonucleases are primarily responsible for mRNA degradation in bacteria. In animal cells, RNA molecules are transcribed as precursors that require processing by ribonucleases to generate functional RNAs. This RNA maturation process requires cleavage by endonucleases and trimming by exonucleases. After the mRNA is translated into protein it is degraded by additional ribonucleases. SEE ALSO CARCINOGENS; DNA POLYMERASES; DNA REPAIR; MUTATION; NUCLEOTIDE; RESTRICTION ENZYMES.

Fred Perrino

Bibliography

Gerlt, John A. "Mechanistic Principles of Enzyme-Catalyzed Cleavage of Phosphodiester Bonds." In *Nucleases*, 2nd ed., Stuart M. Linn, R. Stephen Lloyd, and Richard J. Roberts, eds. Cold Spring Harbor, NY: Cold Spring Harbor Laboratory Press, 1993.

Nucleotide

Nucleotides are the building blocks of deoxyribonucleic acid (DNA) and ribonucleic acid (RNA). Individual nucleotide monomers (single units) are linked together to form polymers, or long chains. DNA chains store genetic information, while RNA chains perform a variety of roles integral to protein synthesis. Individual nucleotides also play important roles in cell metabolism.

Structure

The nucleotide molecule contains three functional groups: a base, a sugar, and a phosphate (see diagram). It may seem puzzling that a nucleic acid should contain a base. While the base portion does have weakly basic properties, the nucleotide as a whole acts as an acid, due to the phosphate group.

The names DNA and RNA are generated from the deoxyribose and ribose sugars found in these two polymers. Both are five-carbon sugars, whose carbons are numbered around the ring from 1′ to 5′ ("one prime" to "five prime"). The prime distinguishes the carbons on the sugar from the carbons on the base. The sugar in RNA nucleotides is ribose. The sugar in DNA is 2′-deoxyribose, which lacks an –OH group at the 2′ position. This small difference has some important consequences: The extra oxygen in RNA interferes with double helix formation between RNA chains (though it does not completely prevent it), and makes RNA more susceptible than DNA to base-catalyzed cleavage (breakdown into individual monomers).

A base attaches to the sugar at the sugar's 1′ position. Because of their nitrogen content, the bases are called nitrogenous bases, and are further classified as either purines or pyrimidines. Purine structures have two rings, while pyrimidines have one. The two purine bases found in both DNA and in RNA are guanine (G) and adenine (A). The two pyrimidine bases found in DNA are cytosine (C) and thymine (T), and the two pyrimidine bases found in RNA are cytosine and uracil (U). The only difference between thymine and uracil is the presence of a methyl group in thymine that is lacking in uracil. A base plus a sugar is called a nucleoside.

The phosphate groups are linked to the sugars at the 5′ position. The addition of one to three phosphate groups generates a nucleotide, also known as a nucleoside monophosphate, nucleoside diphosphate, or nucleoside triphosphate. For instance, guanosine triphosphate (GTP) is an RNA

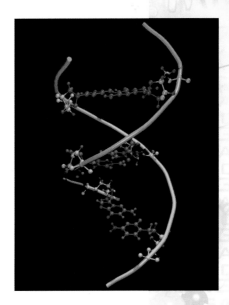

DNA double helix molecular model. The ball-and-stick models linking the backbones represent the nucleotides A, T, C, and G.

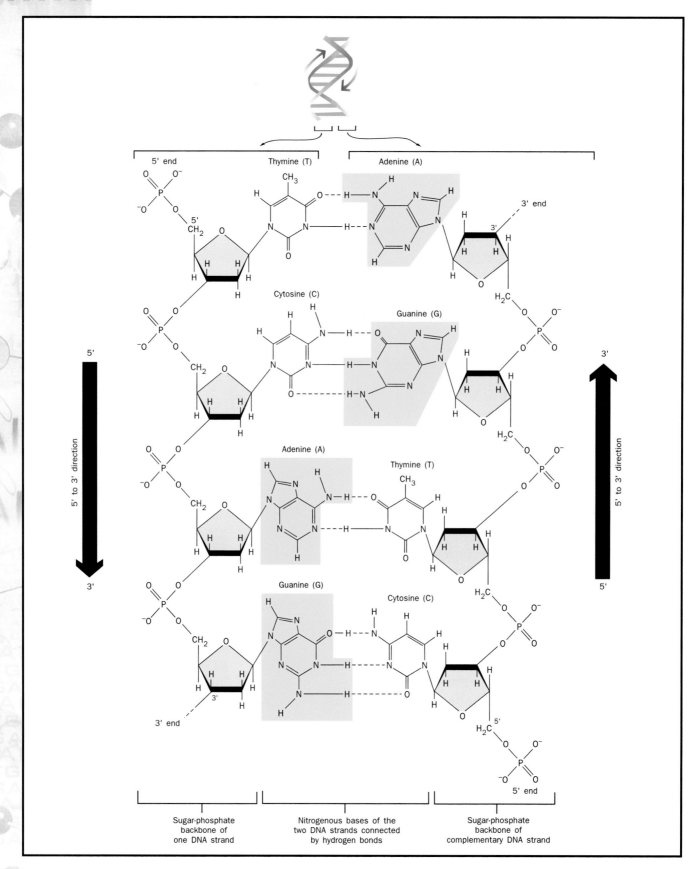

DNA nucleotides hydrogen bond across the double helix. A pairs with T, and C with G, due to the specific positioning of the hydrogen bonds (dashes) they form. Each side of the helix has a directionality imposed by the differing characters of the 5′ and the 3′ ends of the nucleotides. The two sides are oriented in opposite, "anti-parallel" directions.

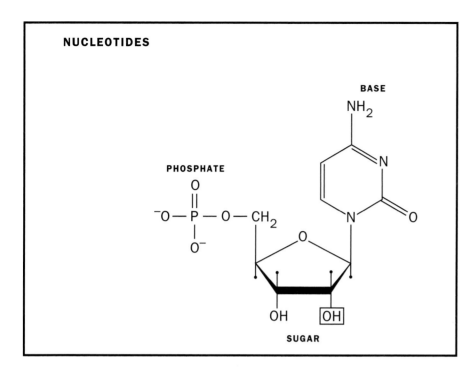

NUCLEOTIDES

PHOSPHATE

BASE

SUGAR

A nucleotide consists of a nitrogen-containing base, a 5-carbon sugar, and one or more phosphate groups. Carbons occupy four vertices of the pentagon, and are numbered counterclockwise from 1′ ("one-prime"), the point of attachment of the base, to 5′, the point of attachment of the phosphate. The sugar depicted is ribose; Deoxyribose has an H instead of an OH at the 2′ position, indicated by the box.

nucleotide with three phosphates attached. Deoxycytosine monophosphate (dCMP) is a DNA nucleotide with one phosphate attached.

Adenosine triphosphate, ATP, is the universal energy currency of cells. The breakdown of energy-rich nutrients is coupled to ATP synthesis, allowing temporary energy storage and transfer. When the ATP is later broken back down to ADP or AMP (adenosine diphosphate or monophosphate), it provides energy to power cell reactions such as protein synthesis or cell movement.

Polymer Formation

DNA and RNA polymers are constructed by forming **phosphodiester bonds** between nucleotides. In this arrangement, a phosphate group acts as a bridge between the 5′ position of one sugar and the 3′ position of the next. This arrangement is called the "sugar-phosphate backbone" of DNA or RNA; the bases hang off to the side.

In the cell, DNA or RNA polymers are synthesized using nucleoside triphosphate monomers as precursors. During polymer synthesis, two of the phosphate groups of the incoming nucleoside triphosphate are cleaved off, and this provides the energy needed to power the reaction. The remaining phosphate takes its place in the sugar-phosphate backbone of the growing nucleic acid chain. A pyrophosphate molecule (two linked phosphates) is released.

Just as an arrow has a tip and a tail, DNA or RNA chains have directionality, due to the structure of the sugar. At one end of a chain, a 5′ carbon will be left free. This is known as the 5′ end of the chain. At the other end, the 3′ carbon will be free; this is the 3′ end of the chain. Segments of DNA that are not free at their ends can also be discussed in terms of their 5′ and 3′ ends. This directionality has important consequences. When DNA

phosphodiester bonds the links between two nucleotides in DNA or RNA

RNA nucleotides join by linking the 5′ phosphate of one to the 3′ ribose carbon of the next. This linkage is called a phosphodiester bond.

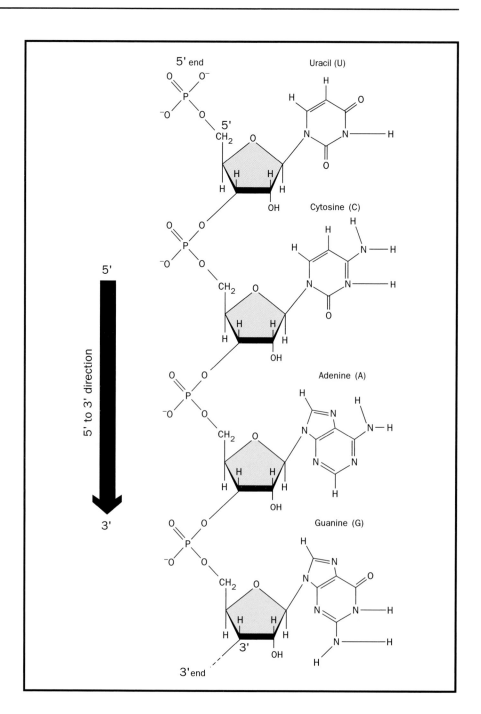

replication occurs, it always moves from the 5′ end to the 3′ end, and the incoming triphosphate joins the 3′ end of the chain. Transcription (RNA synthesis from a DNA gene) also moves in this 5′-to-3′ direction. The 5′ end is considered the "upstream" end of the gene, and is the end on which the gene promoter (the transcription initiator) is located.

The Double Helix of DNA

In the double helix of DNA, guanine nucleotides are base-paired opposite cytosine nucleotides. Adenine nucleotides are base-paired opposite thymine nucleotides. This pairing is due to the **complementary** natures of the structures involved. Note that G is a two-ringed purine, while its partner C is a

complementary matching opposite, like hand and glove

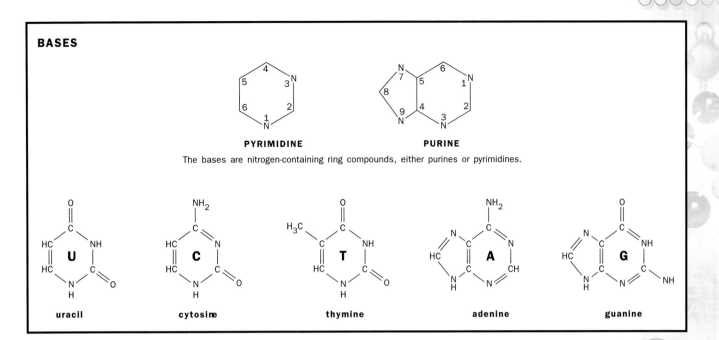

BASES

PYRIMIDINE   PURINE

The bases are nitrogen-containing ring compounds, either purines or pyrimidines.

uracil   cytosine   thymine   adenine   guanine

one-ringed pyrimidine. Similarly, A is a purine and T is a pyrimidine. These pairings give the interior of the helix a fixed diameter, without bulges or gaps. Just as importantly, the arrangement of atoms in the rings allows the partners to form sets of weak attractions, called **hydrogen bonds**, across the interior of the helix. The hydrogen bonds contribute greatly to the stability of the double helix, and the specificity of the G-C, A-T pairing is the structural basis of faithful replication of DNA. SEE ALSO DNA; DNA POLYMERASE; REPLICATION; RNA; RNA POLYMERASES.

Fred Perrino

The molecular structures of the five nitrogenous bases.

hydrogen bonds weak bonds between the H of one molecule or group and a nitrogen or oxygen of another

Bibliography

Watson, James. *The Double Helix: A Personal Account of the Discovery of the Structure of DNA.* New York: Atheneum, 1968.

Stryer, Lubert. *Biochemistry*, 4th ed. New York: W. H. Freeman, 1995.

Nucleus

The largest of the membrane-bound organelles, the nucleus first was described in 1710 by Antoni van Leeuwenhoek using a simple microscope. In 1831 the Scottish botanist Robert Brown characterized the organelle in detail, calling it the "nucleus," from the Latin word for "little nut." The nucleus is the site of gene expression and gene regulation.

Distinctive Features

A distinguishing characteristic of eukaryotes, the nucleus contains the genetic information (**genome**) of the cell in the form of its chromosomes. It is within the nucleus that the DNA in the chromosomes is duplicated prior to cell division and where the RNAs are synthesized. Ribosomes are partially assembled around the newly synthesized ribosomal RNAs (rRNA) while still in the nucleus and then transported into the cytoplasm to continue

genome the total genetic material in a cell or organism

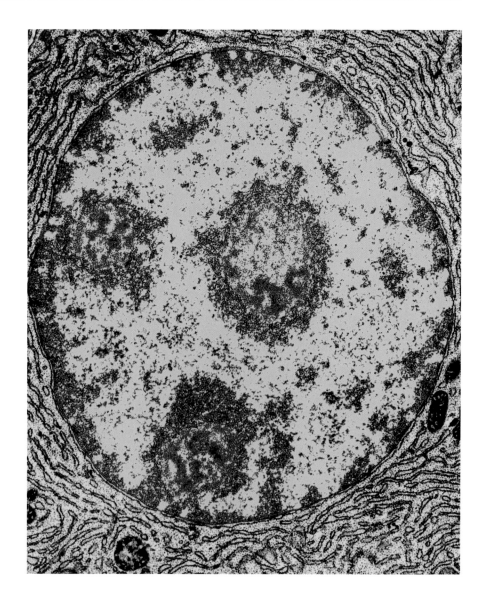

Figure 1. Pancreatic cell nucleus magnified nearly 3,000 times by an electron microscope. Stains help to differentiate the various compartments of the nucleus. Note the membrane around the outside, and the extensive endoplasmic reticulum beyond. The dark, stained material within the membrane is chromatin.

their final assembly. Similarly, messenger RNAs (mRNA) are synthesized, packaged, and subsequently transported to the cytoplasmic ribosomes, where they are translated into protein.

Typically spherical in shape and taking up 10 percent of the volume of a cell, the nucleus is bounded by a double membrane called the nuclear envelope (Figures 1 and 2). Most material passes in and out of the nuclear envelope through large openings called the nuclear pores. The outside surface of the envelope is directly connected to the endoplasmic reticulum of the cytoplasm and is surrounded by a network of cytoplasmic intermediate filaments. The inside surface of the nuclear envelope is lined with the nuclear lamina. Internally, the nucleus contains several structures: the chromosomes themselves, which together constitute the chromatin; the interchromatin compartment; the large nucleolus; and a variety of different granules collectively called the subnuclear bodies, which include Cajal (coiled) bodies, gems, PML bodies, and speckles. Every time a cell divides, the nuclear envelope must break down to release the recently duplicated chromosomes. After the chromosomes have segregated to the new daughter cells, the nucleus and its components must be rebuilt.

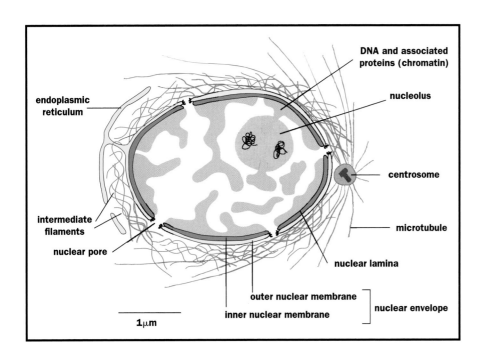

Figure 2. The nucleus holds the cell's DNA. It is surrounded by the nuclear envelope, which is continuous with the endoplasmic reticulum.

If the DNA of each cell were stretched out linearly, it would be over six feet in length. Although the chromosomes of a nucleus appear as a diffuse network in the electron microscope, they are highly compacted into nucleosomal units. Because of nucleosomal folding, the six feet of DNA yields an organelle tightly packed with chromosomal material. Consequently, it was thought that the nucleus in nondividing cells was a fairly static structure, with its various substructures locked into place. Since the 1980s, however, technological advances have permitted investigators to "paint" chromosomes, parts of chromosomes, genes, proteins, RNAs, or subnuclear bodies with genetically defined fluorescent tags. Combined with new techniques that permit these procedures in living cells, and coupled with time-lapse photography and computer simulation, an entirely different image of the cell nucleus is emerging. The nucleus is now understood to be a dynamic organelle composed of a highly ordered architecture that permits a great deal of structural flexibility and movement of molecules and particles between its various subcompartments.

Chromosomal Territories

Each chromosome is specifically anchored through its **telomeres** to a discrete place on the nuclear envelope by the proteins of the nuclear lamina. Thus each occupies a geographically distinct nuclear space called a chromosomal territory (Figure 2). The homologous chromosomal pairs (matching chromosomes derived from mother and father) do not necessarily lie next to each other.

telomeres chromosome tips

Chromosomal territories are separated by channels of open **nucleoplasm** called the interchromatin compartment. Within each territory, DNA can be highly condensed (heterochromatin) or less condensed (euchromatin). Heterochromatin, defined as DNA that is not currently undergoing active transcription, can contain important chromosomal elements such as **centromeres**. Euchromatin are those chromosomal areas more likely to be active in gene

nucleoplasm material in the nucleus

centromeres regions of the chromosome linking chromatids

Figure 3. Each chromosome in the nucleus is localized into a territory, separated by more open domains. Early-replicating regions tend to be closer to the center than late-replicating regions.

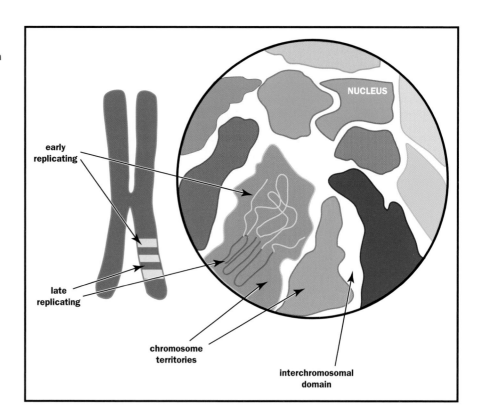

transcription. The heterochromatin of any given chromosome is found within its territory close to the nuclear envelope (Figure 1), but can often project into the interior of the nucleus as patches and/or surround the nucleolus. The euchromatin of each territory extends into the center of the nucleus. In addition, those specific areas of euchromatin undergoing active RNA transcription (gene expression) are typically found on the very periphery of the chromosomal territory, at its juncture with the interchromatin channels.

Chromosomal territories contain at least one other known functional subdomain. Those portions of the DNA that replicate late are found near the nuclear envelope, while earlier-replicating DNA is found in the interior of each territory, projecting into the center of the nucleus. Thus each chromosome not only occupies a discrete place in the nucleus, but each is additionally highly organized into different functional subcompartments. The DNA in each chromosome is highly contorted, looping back and forth within its territory. Chromosomes appear capable of shuffling segments to the correct spot within their territories to carry out gene expression or DNA replication. Indeed, painting of chromosomal segments, including specific genes, with fluorescent tags clearly indicates that chromosomes are constantly shifting around within their territories. Thus the architecture of the chromosomal territories, although highly organized, has a considerable degree of flexibility that is closely tied to both gene expression and DNA replication.

Interchromatin Compartment

The interchromatin (interchromosomal) compartment is best viewed as a series of channels in and around the individual chromosomal territories that are in direct connection with the nuclear pores of the nuclear envelope. It is filled with nucleoplasm containing subnuclear bodies, nuclear proteins,

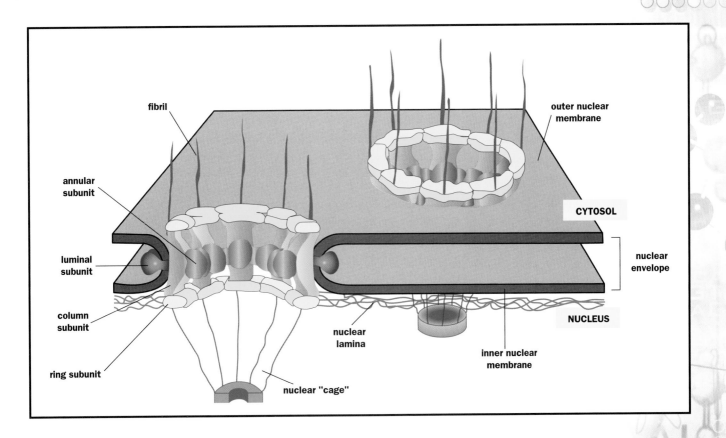

Figure 4. The double-layered nuclear envelope is studded with nuclear pores, composed of numerous sub-units and related structures.

and RNAs, which move rapidly through its channels. It is thought that as RNA is transcribed from genes along the periphery of the chromosomal territory, it drops into the interchromatin compartment for processing, packaging, and transport out of the nucleus through the nuclear pores.

Hormone receptors, **histones**, and DNA repair enzymes are all known to move actively through these channels, seeking their nuclear targets. Trafficking of molecules is highly efficient; it takes only seconds for a newly synthesized RNA particle to exit through a nuclear pore. Thus the nucleus is a very busy place, with a rapid and continuous exchange of proteins involved in nuclear function and genomic expression occurring both between nuclear compartments and deep within individual compartments, through which access is guaranteed by transportation through the interchromatin compartment.

histones proteins around which DNA winds in the chromosome

Nucleolus

The most prominent nuclear feature, the nucleolus is a ribosomal factory. To make the large number of ribosomes needed, eukaryotic genomes carry multiple rRNA gene copies. The human genome contains 180 rRNA genes located on the tips of five different chromosomes (chromosomes 13, 14, 15, 21, and 22). Anchored by the opposite end to the nuclear envelope, each in their own chromosomal territory, the tip of these five chromosomal pairs (ten chromosomes in a diploid cell) extend into the center of the nucleus and come together, and the rRNA genes align to form what is called the nucleolar organizer. Transcription of the rRNA genes 28S, 18S, and 5S occurs rapidly. The transcripts are immediately processed and sequentially packaged through multiple stages into ribosomal subunits. The processing

is complicated, requiring many cytoplasmic proteins and enzymes that are transported through the nuclear pores, diffusing through the interchromatin compartment until they reach the nucleolus, where they bind and remain. The nucleolus itself is composed of three subdomains: the nucleolar organizer; rRNA in the process of being transcribed, which is seen as dense fibrils; and granules, which are ribosomes very early in the assembly process.

Proteomic analysis indicates that human nucleoli contain at least 271 different proteins of a diverse array of known functions, with 31 percent encoded by unknown genes. This has raised the distinct possibility that the nucleolus performs other functions besides ribosome synthesis. Corroborating data suggest that the nucleolus entraps specific cell-cycle regulatory proteins (such as CDC14), inhibiting their activity until needed. When released from the nucleolus, they regain activity. Nucleoli may also synthesize and/or transport other ribonucleoprotein particles besides the ribosome, and may play a role in the processing and transport of mRNA or tRNA. Because nucleoli are often seen associating with other subnuclear bodies such as Cajal bodies, additional functions are likely.

Subnuclear Bodies

Although the function of many of the subnuclear bodies remains elusive, they are indeed true nuclear structures. They are seen both by light and electron microscopy and can be studied in living cells through the use of fluorescent tags. They are known to contain complexes of proteins with or without RNA. Like nucleoli, they are not surrounded by a membrane. They often move through interchromatin channels and are thought to represent dynamic complexes that may form and re-form with each other and other nuclear components to process and transport nuclear components.

Cajal (Coiled) Bodies. By electron microscopy, Cajal bodies are seen as tangled balls of thread. They number one to ten per nucleus, with more seen in growing cells. They are often found in association with nucleoli or specific chromosomal territories. Although their true function remains unknown, their ability to associate regularly with nucleoli has led to speculation that they are somehow involved in processing either mRNA or rRNA.

Gems. More tightly coiled, smaller versions of Cajal bodies, gems are frequently seen interacting with Cajal bodies and are distinct structures. They are known to contain a protein called SMN (which stands for "survival of motor neurons") that, when mutated, is responsible for a severe inherited form of a human muscular wasting disease called spinal muscular atrophy. Based upon the known function of the normal SMN protein, it is speculated that gems are involved in trafficking mRNA spliceosome subunits through the nucleus and may indirectly help remove mRNA introns.

PML Bodies. Nuclei typically have ten to twenty PML bodies (also known as PODs, Kremer bodies, or ND10) that take the shape of dense rings. They contain proteins that, when mutated, have been identified with such disease processes as retinoblastoma and Bloom's syndrome. Their normal pattern is altered in the nuclei of human acute promyelocytic leukemia. When cells are infected with herpes simplex virus type 1, adenovirus, or human cytomegalovirus, PML bodies are disrupted. Although their function

proteomic derived from the study of the full range of proteins expressed by a living cell

remains unknown, the fact that they are altered in diseased or malignant cells suggests that they play an important role in the normal cell, including growth control and **apoptosis**.

Speckles (Interchromatin Granules). Speckles are clusters of dense structures seen by electron microscopy that, when stained with fluorescent tags specific to small nuclear ribonucleoproteins (snRNP), give rise to a "speckled" nucleus. Small nuclear ribonucleoproteins are RNA-protein complexes that are subunits of the spliceosome involved in mRNA intron removal. The twenty to fifty speckles per nuclei are typically found in the interchromatin compartment, where mRNA undergoes processing prior to transport through the nuclear pore and into the **cytoplasm**.

Nuclear Envelope

Completely surrounding the nucleus, the nuclear envelope sequesters the genomic information of the cell, probably protecting it from the various enzymes and processes that occur within the cytoplasm. It is composed of two concentric membranes, each of which has a distinct protein composition: the outer membrane, which faces the cytoplasm; and the inner membrane, facing the nuclear interior. The inner and outer membranes are separated by the perinuclear space. Both the outer membrane and the perinuclear space are continuous with the endoplasmic reticulum and studded with ribosomes. Any proteins made on the nuclear outer membrane-bound ribosomes drop into the perinuclear space and are transported through the inner membrane into the nucleus. The major transport pathway in and out of the nucleus, however, is thought to be through nuclear pores.

The inner membrane is coated with a mesh-like network of intermediate filaments called the nuclear lamina. Various nuclear structures, including the chromosomes, attach directly to the lamina, which is essential for maintaining the overall architecture and function of the nucleus. Mutations in the lamina proteins, lamin and emerin, can cause the chromosomes to dissociate from the nuclear envelope and disrupt the organization and properties of the nuclear pores, both of which result in embryonic death. In humans, other lamin mutations cause several rare, inherited diseases, including Emery-Dreifuss muscular dystrophy, an inherited form of muscular dystrophy, or Dunnigan-type lipodystrophy, a disease that results in loss of adipose tissue and late-onset, insulin-resistant diabetes beginning at puberty. How lamina protein mutations cause these two diseases is unknown.

Nuclear Pores

Perhaps the most startling feature of the nuclear envelope are the very large, basket-like transport structures called the nuclear pores (figure 4). These structures have a molecular weight of 125 million daltons, making them thirty times larger than a ribosome. Composed of 100 to 200 different proteins collectively called nucleoporins, each nuclear pore pierces through both membranes of the nuclear envelope and probably opens into the interchromatin space of the nucleus. Some nucleoporins are structural components of the nuclear pore; others facilitate transport. Each mammalian cell nucleus contains 3,000 to 5,000 of these pores. The large number is needed to transport the tremendous quantity of proteins, enzymes, RNAs,

apoptosis programmed cell death

cytoplasm the material in a cell, excluding the nucleus

125

factors, and complexes in and out of the nucleus to maintain its function and integrity. Small molecules, ions, and proteins up to 45,000 daltons passively diffuse through the pores. However, the vast majority of material transported is through a highly controlled process called "gating," which is responsible for keeping complexes such as the ribosomes in the cytoplasm from entering the nucleus.

Some proteins require multiple crossings through the nuclear pore. Ribosomal proteins are first made in the cytoplasm, transported into the nucleus, assembled into ribosome subunits by the nucleolus, and then transported back out into the cytoplasm. Viruses infect nuclei by taking advantage of the presence of nuclear pores. Some can be transported intact, while others "dock" on the cytoplasmic side of the pore and inject their DNA into the nucleus through the pore's opening. Each nuclear pore can both import and export material in one of two ways.

Any protein transported in or out of the nucleus must contain a nuclear localization signal, which is a specific sequence of four to eight amino acids that triggers either nuclear import or export. Each nuclear pore contains nucleoporins that recognize either the import or export signal, called importins or exportins, respectively. Importins, located on the cytoplasmic side of the nuclear pore, bind their import "cargo" and flip or slide it to the inside of the nucleus. They then move back into their original position, ready to "transport" their next "cargo." The opposite happens on the side of the nuclear pore facing the interior of the nucleus. Here, exportins bind proteins within the nucleus carrying the export signal and flip or slide them through the pore and into the cytoplasm. RNA molecules and complexes can also move through the pores, but only if the importins and or exportins recognize them as cargo. SEE ALSO CELL CYCLE; CHROMOSOME, EUKARYOTIC; MUSCULAR DYSTROPHY; PROTEIN TARGETING; RIBOSOME; TELOMERE.

Diane C. Rein

Bibliography

Alberts, Bruce, et al. "The Cell Nucleus." In *Molecular Biology of the Cell*, 3rd ed. New York: Garland, 1995.

Dundr, Miroslav, and Tom Misteli. "Functional Architecture in the Cell Nucleus." *Biochemical Journal* 356 (2001): 297–310.

Lamond, Angus I., and William C. Earnshaw. "Structure and Function in the Nucleus." *Science* 280 (1998): 547–553.

Lewis, Joe D., and David Tollervey. "Like Attracts Like: Getting RNA Processing Together in the Nucleus." *Science* 288 (2000): 1385–1389.

Olson, Mark O. J., Miroslav Dundr, and Attila Szebeni. "The Nucleolus: An Old Factory with Unexpected Capabilities." *Trends in Cell Biology* 10 (2000): 189–196.

Pederson, Thoru. "Protein Mobility within the Nucleus: What Are the Right Moves?" *Cell* 104 (2001): 635–638.

Wilson, Katherine L. "The Nuclear Envelope, Muscular Dystrophy and Gene Expression." *Trends in Cell Biology* 10 (2000): 125–129.

Wilson, Katherine, et al. "Lamins and Disease: Insights into Nuclear Infrastructure." *Cell* 104 (2001): 647–650.

Wolffe, Alan P., and Jeffrey C. Hansen. "Nuclear Visions: Functional Flexibility from Structural Instability." *Cell* (2001) 104: 631–634.

Oncogenes

An oncogene is a gene that causes cancer. Oncogenes arise from normal cellular genes, often ones that help regulate cell division.

Early Oncogene Research

The first clues that cancer has a genetic basis came from several independent observations. In 1914 the German cell biologist Theodor Boveri viewed cancer cells through a microscope and noted that they often carried abnormal chromosomes. However, recognition that a specific chromosomal abnormality was routinely associated with a particular type of cancer did not come until 1973, when Janet Rowley showed that chronic myelogenous leukemia (CML) cells carried a chromosomal translocation in which the ends of chromosomes nine and twenty-two are exchanged. Several other studies showed that certain types of cancer can run in families, suggesting that cancer risk can be inherited. Then, in 1981 the laboratories of Robert Weinberg, Michael Wigler, Geoff Cooper, and Mariano Barbacid showed that DNA from a human bladder cancer cell line could cause nonmalignant cells in tissue culture to become cancerous.

Since the Weinberg and Wigler observations, dozens of oncogenes have been identified and characterized. It is clear that oncogenes represent certain normal cellular genes that are aberrantly expressed or functionally abnormal. Such normal cellular genes, or "proto-oncogenes," can be altered to become oncogenes through a variety of different molecular mechanisms.

RNA Tumor Viruses and Proto-Oncogenes

In 1911 Peyton Rous reported that a class of RNA viruses can cause tumors in animals. These RNA tumor viruses, called "retroviruses," carry an RNA genome that, once inside a cell, is copied into DNA, which then is inserted randomly into the **genome** of a host cell. Some retroviruses are slow to cause tumors. After infection and spread to a large number of cells, a DNA copy of the viral genome, by chance, integrates into a host cell's DNA next to a normal gene that plays an important role in cell growth. If this viral integration disrupts the expression or structure of the normal cellular gene, it induces abnormal growth signals that can lead to cancer.

genome the total genetic material in a cell or organism

Other retroviruses cause tumors to appear very quickly. In the process of copying viral RNA into DNA, RNA that is expressed from cellular genes can be mistakenly copied into the viral genome. The **progeny** of the virus transfer the cellular gene to many other cells. If this "captured" cellular RNA is from a gene that stimulates cell growth, it then causes abnormal growth stimulation, leading to cancer. This process is termed "gene capture."

progeny offspring

Through molecular cloning, the genes that are activated or captured by retroviruses have been identified and characterized. Almost three dozen such retroviral oncogenes and their related cellular proto-oncogenes are now known.

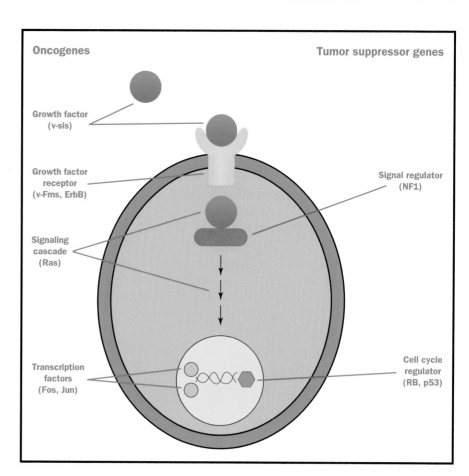

Oncogenes and tumor suppressor genes collaborate to "short circuit" normal growth regulatory pathways in cells. A generic growth-signaling pathway is shown where a growth factor binds to its receptor, which stimulates a cascade of signals inside the cell. The signals culminate in the nucleus, where they induce expression of genes involved in stimulating cell proliferation. On the left side are different steps in the signaling pathway that can be affected by activation of oncogenes. On the right side are growth inhibitory steps that can be lost by inactivation of tumor suppressor genes. In parentheses are examples of oncogenes and tumor suppressor genes that are mentioned in the text.

Proto-Oncogene Activation without Retroviruses

The first human oncogene, called *Ras*, was identified in the Weinberg and Wigler experiments. The protein product of the *Ras* gene serves as a switch that turns growth signals on and off. Normally, the activity of this *Ras* switch is tightly regulated. However, single mutations (called point mutations) in critical sites of the *Ras* gene cause the *Ras* growth switch to remain constantly turned on, which contributes to cancer. Thus, some proto-oncogenes can become oncogenes by genetic point mutation. Such mutations are not dependent upon the presence of retroviruses. Instead, they can occur during normal cell division and can be caused by environmental factors such as chemicals, ultraviolet rays from the sun, and X rays. DNA repair mechanisms usually, but not always, correct such mutations.

Oncogenes also can be activated by structural changes, called "amplifications" or "translocations," that occur in the chromosomes. DNA amplification increases by several-fold a specific region of a chromosome. This can produce many copies of any genes that lie in the amplified region. If one of the genes in this region is important in driving cell growth, its overexpression due to amplification leads to uncontrolled proliferation.

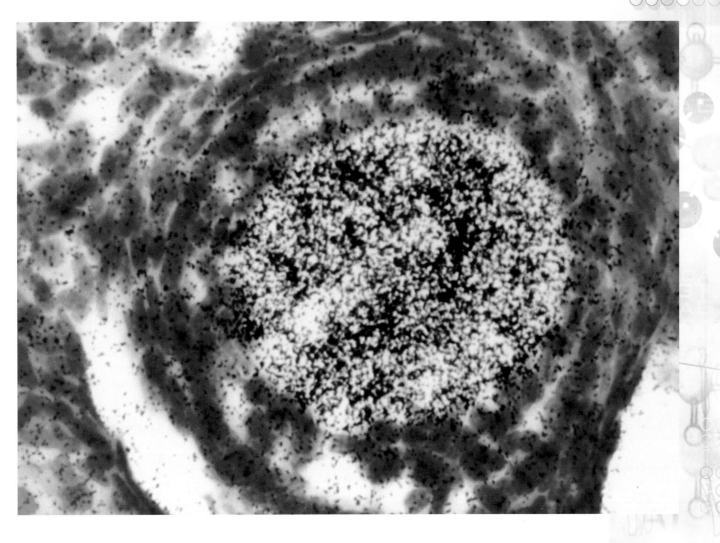

In the case of chromosome translocations, a proto-oncogene on one chromosome might be moved to another chromosome, resulting in the gene's structural alteration and/or aberrant expression. For example, in the translocation between chromosomes 9 and 22 that is found in CML, a proto-oncogene on chromosome 9, called c-*Abl*, is moved to chromosome 22, where it is fused to another gene called *Bcr*. Normally, c-*Abl* is a nuclear enzyme called "tyrosine kinase," which adds a phosphate molecule to proteins at an amino acid called tyrosine. **Phosphorylation** regulates the function of certain proteins that play important roles in stimulating cell proliferation. The fusion of *Bcr* and c-*Abl* genes creates an oncogene, called *Bcr/Abl*, which makes a highly overactive tyrosine kinase variant that is found in the cytoplasm instead of the nucleus. These changes in the activity and cellular location of the c-*Abl* proto-oncogene lead to chronic myelogenous leukemia.

In situ hybridization reveals that the product of a known oncogene, inserted into a mouse egg cell, hybridizes to multiple copies of RNA (black spots).

phosphorylation addition of the phosphate group PO_4^{3-}

Short-Circuiting Normal Cell Growth Mechanisms

Normal cell growth is controlled by the availability of growth factors, which are hormone-like molecules that bind to specific receptors embedded in the surface membrane of cells. When this happens, the receptor stimulates a signaling cascade inside cells that ultimately tells the cells to divide. Many gene products in this signaling pathway are proto-oncogenes that can become oncogenes when activated by the different mechanisms described

above. When a signaling proto-oncogene is activated, the signaling cascade becomes "short-circuited" and cells behave as if they are continually stimulated by their growth factor.

For example, the v-*sis* oncogene from a monkey cancer virus known as simian sarcoma retrovirus (SSV) comes from a gene that encodes platelet-derived growth factor, which stimulates growth of different cell types. Cells infected with SSV are, therefore, constantly bathed in the v-*sis* growth factor and stimulated to proliferate. Other oncogenes are mutated growth factor receptors where mutation leaves the receptor in the "on" status even in the absence of the growth factor. Two examples of mutated receptor oncogenes include v-*erbB*, found in a bird retrovirus that causes various cancers, and v-*fms*, which is carried by a mouse retrovirus that causes leukemia.

Inside the cell, components of the signaling cascade that connect cell surface growth receptors to the nucleus also can cause cancer when their activity is altered by mutation or overexpression. The *Ras* proto-oncogene is an example of a signal-transmitting molecule inside cells that can mutate into an oncogene. In the nucleus, these normal growth signals trigger other proteins, called transcription factors, that regulate gene expression needed for cell growth. Many transcription factors are proto-oncogenes. Two examples of proto-oncogene transcription factors are c-*Fos* and c-*Jun*, both of which were first identified as retroviral oncogenes.

Tumor Suppressor Genes

In 1983 Raymond White and Webster Cavanee, using a technique called chromosome mapping, learned that a loss of a small segment of human chromosome 13 was a recurring feature in retinoblastoma, a rare childhood cancer of the retina that can run in families. In this deleted region they discovered a gene called *RB* (for retinoblastoma), both copies of which are inactivated either by DNA deletion or by a mutation within the gene that destroys its function. Such inactivation of both copies of the *RB* gene occurs in about 40 percent of human cancers. The product of the normal *RB* gene functions as a brake to cell division, so that loss of this brake can lead to unregulated cell growth.

Another gene associated with cancer when both copies are affected by mutation is the *p53* gene, which acts as a "guardian" of the genome. Normally, this gene product induces a cell suicide program called apoptosis in cells with damaged DNA. Loss of *p53* activity allows cells with damaged DNA to grow and pass DNA mutations to their daughter cells. A third type of gene that plays a role in cancer when it is inactivated is *NF1*. This gene encodes a protein that turns off the *Ras* growth signal mentioned above. In this case, loss of *NF1* function is another way that the *Ras* signal can be left constantly on.

Since the discovery of *RB*, researchers have identified several additional genes in which both copies are inactivated due to mutation or chromosomal deletion. These genes normally block cell growth; hence they are called tumor suppressor genes. Since both copies of these genes need to be inactivated in order to release cancer cells from growth inhibition, tumor suppressor genes act recessively. This contrasts to the oncogenes described above, where only one copy needs to be activated in order to promote cancer. Oncogenes, therefore, act in a dominant fashion.

Multiple Genetic "Hits"

In 1971 Alfred Knudson Jr. proposed that retinoblastoma resulted from at least two separate genetic defects. In families with a high risk of retinoblastoma, the first defect is inherited and the second occurs sometime during childhood. This came to be known as Knudson's "two-hit theory." Subsequent research has shown that most, if not all, cancer arises from multiple genetic events, or "hits."

In many cancers, more than two hits are required. Bert Vogelstein and coworkers first showed this in colon cancer in the late 1980s. Colon cancer begins with a precancerous stage, called a benign polyp. Left untreated, this will progress through successively more cancerous stages until it becomes an aggressive carcinoma. Vogelstein's group found that progression of colon cancer through these different stages was associated with the acquisition of genetic changes in oncogenes such as *Ras*, as well as in a number of different tumor suppressor genes, including *p53*. Together, sequential activation of different oncogenes along with inactivation of various tumor suppressor genes drive the step-wise progression of precancerous cells to highly malignant tumors. SEE ALSO APOPTOSIS; BREAST CANCER; CANCER; CARCINOGENS; CELL CYCLE; COLON CANCER; INHERITANCE PATTERNS; MUTATION; RETROVIRUS; SIGNAL TRANSDUCTION; TUMOR SUPPRESSOR GENES; TRANSCRIPTION FACTORS.

Steven S. Clark

Bibliography

Bishop, J. Michael. "Oncogenes." *Scientific American* 246 (1983): 80–92.

Cavenee, Webster K., and Raymond L. White. "The Genetic Basis of Cancer." *Scientific American* 272 (1995): 72–79.

Croce, Carlo M., and G. Klein. "Chromosome Translocations and Human Cancer." *Scientific American* 252 (1985): 54–60.

Varmus, Harold. "Retroviruses." *Science* 240 (1988): 1427–1435.

Weinberg, Robert A. "A Molecular Basis of Cancer." *Scientific American* 249 (1983): 126–142.

———. "Tumor Suppressor Genes." *Science* 254 (1991): 1138–1146.

Operon

An operon is a genetic regulatory system found in prokaryotes and the bacterial viruses (**bacteriophages**) that attack bacteria. It is a cluster of genes that share regulatory elements and are usually functionally related.

bacteriophages viruses that infect bacteria

The Discovery of Operons

French scientists Jacques Monod and François Jacob first coined the term "operon" in a short paper published in 1960 in the *Proceedings of the French Academy of Sciences*. They elaborated the concept of the operon in several papers that appeared in 1961, based on their studies on the *lac* genes (genes for the **metabolism** of lactose sugar) of the bacterium *Escherichia coli* and the genes of bacteriophage lambda. Monod and Jacob received the Nobel Prize in 1985 for this work.

metabolism chemical reactions within a cell

Lac repressor

Promoter

Operator

ß-Galactosidase

Lactose permease

Thiogalactoside acetyltransferase

Typical Features of Operons

The genes of an operon are usually functionally related. Genes are the basic unit of biological information, and consist of specific segments of deoxyribonucleic acid (DNA). The segments of DNA that constitute a gene consist of distinctive sets of nucleotide pairs located in a discrete region of a chromosome that encodes a particular protein. Within an operon, the genes encode proteins that execute related functions. For example, the five genes of the tryptophan (*trp*) operon in *E. coli* each encode one of the enzymes necessary for the biosynthesis of the amino acid tryptophan from a metabolite called chorismate. This condition is mimicked in many bacteria.

Exceptions do occur; the genes of some operons may lack an obvious functional relationship. For example one operon in *E. coli* contains one gene that encodes a ribosomal protein S21 (*rpsU*), another that encodes DNA primase (*dnaG*), and one that encodes the sigma subunit of RNA polymerase (*rpoD*). The protein products of these genes are all involved in starting up the **synthesis** of **macromolecules**, but beyond that they have no obvious functional relationships to one another. Nonetheless, the clustering of these genes and their common regulation qualify them to be treated as elements of a single operon.

Another common feature of operons is that their genes are clustered on the bacterial chromosome. This chromosome is a large circular molecule of DNA. The genes of an operon are arranged in a consecutive and linear fashion at a specific location on the bacterial chromosome.

In the case of the lactose utilization (lac) operon of *E. coli*, three genes necessary for the successful utilization of the **disaccharide** lactose, a common sugar found in milk, are arranged in a linear fashion on the chromosome (see Figure 1). The *lacZ* gene, which encodes the lactose-degrading enzyme, β-galactosidase, is directly followed by the *lacY* gene, which encodes a membrane protein, called lactose permease, that allows the entry of lactose into the cell. The *lacY* gene is immediately followed by the *lacA* gene, which encodes the thiogalactoside acetyltransferase enzyme that detoxifies lactose-related compounds that might be toxic to the cell. The linear arrangement of these functionally related genes is a hallmark of an operon. The arrangement is significant because the proteins made from these genes will all be easily turned on in concert, so that lactose metabolism proceeds rapidly and efficiently.

Other Typical Characteristics

The clustered genes of the operon typically share a common promoter and a common regulatory region, called an operator. Gene expression requires

synthesis creation

macromolecules large molecules such as proteins, carbohydrates, and nucleic acids

disaccharide two sugar molecules linked together

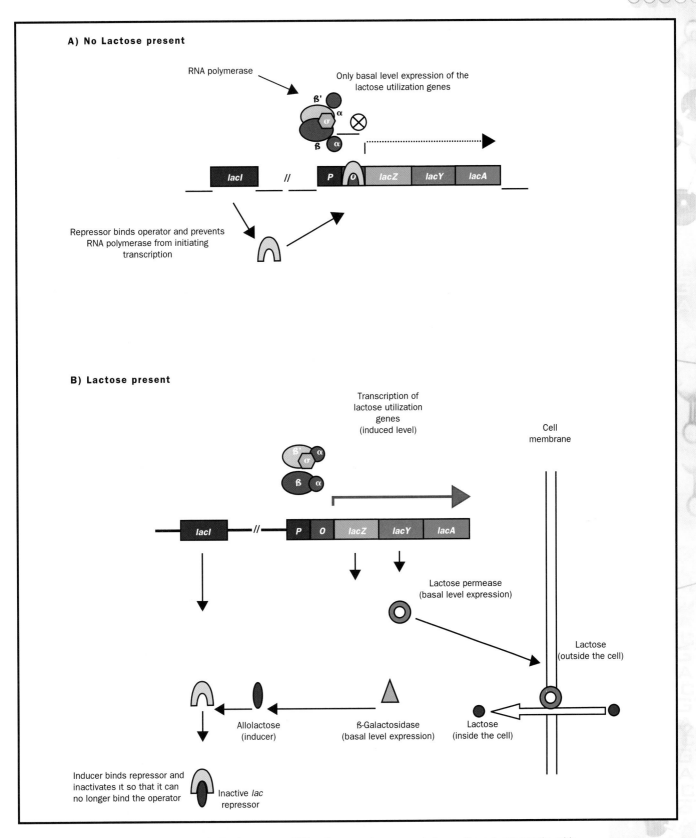

Figure 2. Regulation of gene expression in the *lac* operon. RNA polymerase is represented as a five-subunit protein, with two α subunits, one β subunit, one β' subunit and one σ subunit. The α subunit binds the promoter and dissociates from the polymerase once elongation of the RNA molecule begins. When lactose is present, some is converted to allolactose, the inducer which stimulates transcription of the *lac* operon. Creation of lactose permease increases the cell's ability to take up lactose.

RNA polymerase
enzyme complex that
creates RNA from DNA
template

promoter DNA
sequence to which RNA
polymerase binds to
begin transcription

attenuation weaken or
dilute

allolactose "other lactose"; a modified form
of lactose

the enzyme **RNA polymerase** to transcribe (synthesize an RNA copy of) the gene. This RNA copy is called a messenger RNA (mRNA), which is translated by ribosomes to produce the protein encoded by the gene. In all genes, RNA polymerase begins transcription at a specific site or sequence called the **promoter** (designated "P" in Figure 1). The genes in an operon usually share a common promoter from which the genes of the operon are transcribed.

Operons almost always contain a common promoter region, but not all operons contain only a single promoter. For example the *E. coli* operon for galactose utilization (*gal*) contains two promoters. One of these promoters is active in the presence of glucose, and the other is not (both glucose and galactose are sugars). Some operons, like the *trp* and isoleucine-valine (*ilv*) operons, both from *E. coli*, also have internal promoters that allow the expression of some but not all of the genes in the operon. (Isoleucine and valine are amino acids.)

Operons also have one or more control regions, called operators, that mediate the expression of the genes in the operon (the operator is designated "O" in Figure 1). Like a promoter, an operator is a site on the DNA, but it does not bind with RNA polymerase. Operators function in one of two ways. They can contain DNA sequences that specifically bind particular proteins. Once bound onto DNA, these proteins can prevent the expression of the operon by interfering with the action of RNA polymerase, as in the case of the *lac* repressor. Other proteins bound on other operons can greatly enhance the expression of the operon, as in the case of the AraC protein. Operators can thus prevent or facilitate gene expression.

Instead of acting as target sites for DNA-binding proteins, operators also act as the sites of regulation by **attenuation**. Amino acid biosynthesis operons such as *trp* are usually regulated by attenuation. In such operons the operator provides both a start site for transcription and a ribosome-binding site for the synthesis of a short leader peptide. Through a clever mechanism, the presence of sufficient amino acid in the cell causes the ribosome to disrupt transcription. When the supply of the amino acid is low, transcription of the operon continues without interruption. In this way, if the proteins coded for by the operon genes are needed to synthesize amino acids, then early transcriptional termination does not occur. If they are not needed, because the amino acid is already present, then early termination ensues. This prevents the wasteful production of unnecessary proteins.

The genes of an operon also show a common mode of regulation. The clustering of the genes of an operon and the related functions of these genes requires a mode of regulation that equally affects all the genes of the operon. In the case of the *lac* operon of *E. coli*, the product of the *lacI* gene is a DNA-binding protein that specifically binds to the *lac* operator and prevents RNA polymerase from initiating transcription of the lactose utilization genes from the promoter. Therefore, in the absence of lactose, the lactose utilization genes are only expressed at a very low basal level (see Figure 2A). This low level of expression allows synthesis of a few lactose permease molecules, which permit the entry of lactose into the cell when lactose is present, and a few β-galactosidase molecules, which metabolize lactose or convert it to **allolactose**.

Allolactose is the inducer of the lac operon, acting as a signal that lactose is present. Allolactose binds to the repressor protein, changing its shape in such a way that the repressor can no longer bind to the operator. This allows RNA polymerase to effectively initiate transcription from the *lac* promoter (see Figure 2B).

Transcription of an operon generates an mRNA transcript of all the genes contained within the operon. Ribosomes can translate this single mRNA to generate several distinct proteins. In the case of the lac operon, transcription produces an mRNA molecule that is translated by ribosomes to generate β-galactosidase, lactose permease, and thiogalactoside acetyltransferase. Messenger RNA molecules that encode more than one gene are called polycistronic mRNAs. The common regulation mechanism determines when each polycistronic mRNA is synthesized. This is the main means by which operons commonly regulate the expression of one or more functionally regulated genes. SEE ALSO DNA; *ESCHERICHIA COLI* (*E. COLI* BACTERIUM); GENE; GENE EXPRESSION: OVERVIEW OF CONTROL; PROTEINS.

Michael Buratovich

Bibliography

Hartwell, Leland, et al. *Genetics: From Genes to Genomes.* Berkeley: McGraw-Hill, 2000.

Miller, Jeffrey H., and Reznikoff, William S., eds. *Operon,* 2nd ed. Cold Spring Harbor, NY: Cold Spring Harbor Press, 1980.

Overlapping Genes

Overlapping genes are defined as a pair of adjacent genes whose coding regions are partially overlapping. In other words, a single stretch of DNA codes for portions of two separate proteins. Such an arrangement of genetic code is ubiquitous. Many overlapping genes have been identified in the genomes of **prokaryotes**, **eukaryotes**, **mitochondria**, and viruses.

For two genes to overlap, the signal to begin transcription for one must reside inside the second gene, whose transcriptional start site is further "upstream." In addition, the "stop" signal for the second gene must not be read by the ribosome during **translation**, using the RNA copy of the gene. This is possible because RNA is read in triplets, meaning that it can contain three separate sequences that can be "read" by the cell's protein-making machinery. Such sequences of nucleotide triplets are called reading frames, and they are different in the RNA transcripts of the overlapping genes.

Overlapping genes enable the production of more proteins from a given region of DNA than is possible if the genes were arranged sequentially. Indeed, for the bacteriophage PhiX174, overlapping of genes is necessary. The amount of DNA present in the circular, single-stranded DNA genome of this virus would not be sufficient to encode the eleven bacteriophage proteins if transcription occurred in a linear fashion, one gene after another.

The genome economy afforded by overlapping genes extends to the human genome. The recently completed sequencing of the human genome has revealed between 30,000 and 70,000 genes. Yet evidence suggests that the human genome encodes 100,000 to 200,000 proteins. At least part of

prokaryotes single-celled organisms without a nucleus

eukaryotes organisms with cells possessing a nucleus

mitochondria energy-producing cell organelle

translation synthesis of protein using mRNA code

the information for the extra proteins may come from the presence of hitherto undiscovered overlapping genes, although more may come from alternative splicing of exons in a single gene.

In algae called *Guillardia*, a structure called a nucleomorph contains only about 500,000 base pairs of DNA, a very small genome, yet produces almost 500 proteins. Part of the efficient packaging of the genome is due to 44 overlapping genes. A nucleomorph is a remnant of a nucleus from an ancient eukaryotic organism that became incorporated into the algae.

One consequence of overlapping genes is to reduce the tolerance for mutation. Virtual experiments conducted within the past several years using a software system called Avida have indicated that overlapping reduces the probability of accumulating so-called neutral mutations in a gene (mutations that have no effect). Neutral mutations are unlikely with overlapping genes, because the mutation must have no effect on two genes with different reading frames.

The evolutionary origin of overlapping genes is not yet clear. Recent research indicates that they may have arisen due to several mutational events. These may include the loss of a signal to stop the transcription process in a gene and a shift in the reading sequence of the genetic components. SEE ALSO ALTERNATIVE SPLICING; DNA REPAIR; GENE; GENETIC CODE; MUTATION; TRANSCRIPTION.

Brian Hoyle

Bibliography

Douglas, S., et al. "The Highly Reduced Genome of an Enslaved Algal Nucleus." *Nature* 410 (April 2001): 1091–1096.

Lewis, R., and B. A. Palevitz. "Genome Economy." *The Scientist* 15 (June 2001): 21.

Internet Resources

"-1 Programmed Frameshift in the Regulation of the PhiX174 Lysis Gene." Carnegie Mellon University (2000). <http://info.bio.cmu.edu/Courses/03441/TermPapers/2000TermPapers/group1/phix.html>.

"The Evolution of Genetic Codes." Michigan State University (1999). <http://www.cse.msu.edu/~ofria/home/pubs/abstracts/THESIS.html>.

Patenting Genes

A patent is a legal right granted by the government that gives the patentholder the exclusive right to manufacture and profit from an invention. While naturally occurring substances in their natural form are not patentable, a very wide range of biological materials have been the subject of patents. In 1980 the U.S. Supreme Court decision in *Diamond* v. *Chakrabarty* indicated that "anything under the sun made by man" is patentable. Under certain conditions, patent protection is available for genetic information, plants, non-human animals, bacteria, and other organisms.

For naturally occurring substances, patent protection can only be obtained if someone has changed the substance so that it is no longer the same as it is found in nature. For genes, this means that the particular gene of interest must be isolated from other genetic material, such as a chromosome. Although the *Chakrabarty* decision involved bacteria that were mod-

ified using recombinant DNA technology, this case has been viewed as mandating the patentability of other living organisms; nonnaturally occurring, nonhuman multicellular organisms such as transgenic animals, genetic materials, and purified biologically produced compounds such as enzymes. Other cases, such as the 1977 *In re Application of Bergy* and 1979's *In re Kratz*, have provided further support for the patentability of various biological materials, including cells, proteins, and organisms.

Patent Requirements

Although isolating the gene from other genetic material renders the gene "made by man," various other requirements must be met in order to gain patent protection. For example, in order to be patentable the gene must have a substantial and credible use. Thus, a patent on a gene would not be allowed if the only use described in the patent application was for the use of the gene in some area that was totally unrelated to the function of genetic sequence.

Patents may be obtained for specific DNA sequence information as well as for RNA and amino acid sequences, and for the use of these sequences in various methods. For example, some patents are directed to the use of genetic information in tests to diagnose disease or in test compounds that might be useful to fight disease. Other "methods" patents include the use of the genetic information for tests to identify people with a predisposition to acquiring a certain disease. Other patents are directed toward gene therapy to replace defective genes.

Patent laws apply to gene patents in the same way that they apply to mechanical inventions. In the United States, there is a "first to invent" standard that must be met. This standard means that whoever first invents, discovers, purifies, or isolates a gene is entitled to all patent rights arising from that invention. The law specifies that the invention must have been made by the individual submitting the application. It also requires that the genetic information be "novel." This means that it cannot have been described in a printed publication more than one year prior to the filing of the patent application. Further, the genetic information cannot have been known or used by someone other than the inventor more than one year prior to the invention by the person who has filed the patent application. If the gene is newly described and has not been publicly disclosed, the novelty requirement is usually easily met. Unlike the United States, however, most countries do not have a one-year grace period between the disclosure of the gene and the filing of a patent application. In those countries it is therefore much more difficult to meet patent law requirements.

To qualify as patentable, genetic information must also be "unobvious." This means that the gene cannot be an obvious modification of something that is already known. It is usually easy to meet this requirement when DNA, RNA, or amino acid sequences are involved. However, complications sometimes arise when variants of the genetic information are already known.

The patent application must also provide a description of the "best mode" contemplated for making and using the gene. In addition, there are "written description" and "enablement" requirements. These requirements mean that the application must include a description of the gene, its function, and use that is sufficiently thorough to enable someone "skilled in the art" to reproduce the invention. In some cases, meeting the enablement

requirement requires that the inventor provide a deposit of the biological material that is being patented. For example, an applicant seeking to patent genes produced using recombinant DNA technology would have to provide a sample of the genes to a biological materials patent depository, and would have to guarantee that the material will be available for the life of the patent.

The Patent Process

Once the patent application is filed, the application is assigned to a patent examiner, who then has the responsibility of reviewing the application and determining whether the invention is entitled to patent protection. Often, the examiner initially rejects the claims that indicate what the inventor wants to protect. This occurs when the examiner finds these claims to be too broad or unfounded. This leads to a procedure known as the "prosecution" of the application, in which the examiner and the patent attorney or patent agent attempt to resolve the problems. If they are successful, a patent will be issued; otherwise, the examiner will refuse allowance of the patent. If the patent is denied, the inventor can then either appeal the decision or simply abandon the efforts to gain patent protection.

There are various other considerations in filing for patent protection in the United States and other countries. For example, in the United States there is a "duty of candor." This means that anyone filing a patent application must provide truthful information to the examiner, along with copies of all publications and other materials that might be relevant to the examination of the application. Failure to meet these honesty requirements can result in the invalidation of patents, the disbarment of the patent attorney or patent agent involved, or both.

In addition to patent protection in the United States, many inventors apply for protection in other countries. Most countries, including Japan, Mexico, Canada, Australia, and the nations of Western Europe, recognize genes as being patentable subject matter. However, these other countries have different standards for patentability than those followed in the United States, and many do not allow patents for medical devices. Although there are some patent attorneys who are licensed to practice both in the United States and abroad, most inventors employ the services of foreign patent attorneys who are familiar with the patent laws and rules in the foreign country of interest. SEE ALSO Attorney; Biotechnology Entrepreneur; Legal Issues; Transgenic Animals.

Kamrin T. MacKnight

Bibliography

Diamond v. *Chakrabarty*, 447 U.S. 303, 206 USPQ 193 (1980).

In re Application of Bergy, 563 F.2d 1031, 195 USPQ 344 (CCPA 1977).

In re Kratz, 592 F.2d 1169, 201 USPQ 71 (CCPA 1979).

U.S. Patent and Trademark Office. *Manual of Patent Examining Procedure.* Washington, DC: Superintendent of Documents, U.S. Government Printing Office, 2000.

Pedigree

One of the most important tools used by genetic professionals is the pedigree, a pictorial description of a family tree. A complete pedigree provides

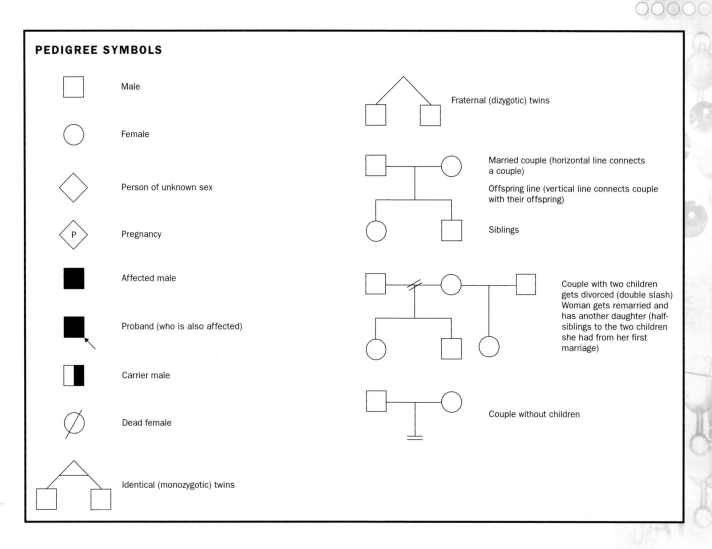

PEDIGREE SYMBOLS

Male

Female

Person of unknown sex

Pregnancy

Affected male

Proband (who is also affected)

Carrier male

Dead female

Identical (monozygotic) twins

Fraternal (dizygotic) twins

Married couple (horizontal line connects a couple)

Offspring line (vertical line connects couple with their offspring)

Siblings

Couple with two children gets divorced (double slash) Woman gets remarried and has another daughter (half-siblings to the two children she had from her first marriage)

Couple without children

Pedigree symbols used to indicate familial relationships.

information about the biological relationships of individuals in the family, their medical history, the pattern of inheritance of a genetic disorder in the family, variable expression of the disorder, which family members are at risk, fertility of individuals (including pregnancies, miscarriages, and stillbirths), and family members who are dead. Physicians sometimes refer to a pedigree as a "genogram." However, genograms usually contain more social information about family relationships than a traditional pedigree used by geneticists and genetic counselors. For example, a genogram can show a teenage child who has a poor relationship with a parent or an individual estranged from the family.

Use of Pedigrees

Pedigrees can be used in the clinical setting, such as genetic counseling sessions or genetic evaluations, or in genetic research. By analyzing how many family members have a genetic disorder, how these individuals are related, and the sex of the affected individuals, it is often possible to determine the inheritance pattern of the genetic disorder in the family. Together, the inheritance pattern and an accurate diagnosis help the genetic professional provide accurate risk information to the family. This includes risk information for future pregnancies or relatives who are currently unaffected, but who are at risk for developing the disorder based

In this sample pedigree, Sue is the proband, or person first coming to the attention of the geneticist. Birth dates and relevant medical history is also included.

SAMPLE PEDIGREE

Bill
8/18/39

Betty
5/5/40

Bob
7/3/40

Jane
11/19/41
asthma

Emily
6/2/65
diabetes

Steve
9/1/68

Sue
5/3/67
onset 13 years old

Mary
4/11/85
allergies

Jack
10/30/87
onset 15 years old

on family history information. Genetic testing options, if available, can then be offered to those at risk.

The pedigree is also a standard tool used by researchers. For example, in studies aimed at identifying genes that cause human genetic disorders, researchers must collect detailed information on relatives participating in the study, particularly those relatives who are affected with the disorder. Researchers compare the genes of affected individuals with the genes of those who did not inherit the disorder to identify the specific genes responsible. In other studies the disease-causing gene is known, and researchers study the gene **mutation**(s). A pedigree can help identify which family members should be included in mutation analysis, as only those family members who are affected or are at-risk could carry a mutation. Researchers can also pictorially show laboratory data, such as **genotypes** or **haplotypes**, on the pedigree.

mutation change in DNA sequence

genotypes sets of genes present

haplotypes sets of alleles or markers on a short chromosome segment

Terminology

Standard symbols are used to draw pedigrees. For example, males are represented by squares and females by circles. The individual who brought the family to the attention of the medical professional or researcher is called a "proband" and is identified on the pedigree by an arrow pointing towards the symbol for that individual. This individual usually has the disorder of interest to the researcher or physician.

The informant is the individual who was interviewed to obtain the pedigree. The informant may or may not be the same person as the proband. The name of the informant, date the pedigree is drawn, and interviewer are also noted on the pedigree. So, too, is the founder of the pedigree, who is the first family member known to have had the disorder.

Drawing and Recording Pedigrees

Pedigrees are hand-drawn or created using special computer software. The standard pedigree typically includes at least three generations, with each gen-

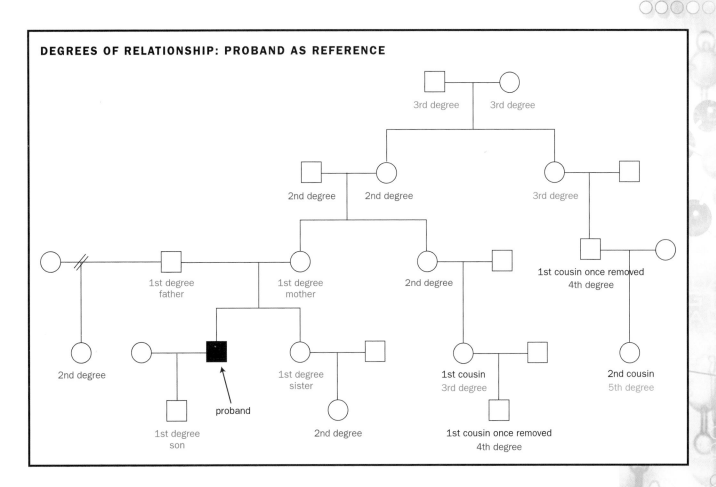

DEGREES OF RELATIONSHIP: PROBAND AS REFERENCE

3rd degree   3rd degree

2nd degree   2nd degree   3rd degree

1st degree father   1st degree mother   2nd degree   1st cousin once removed 4th degree

2nd degree   proband   1st degree sister   1st cousin 3rd degree   2nd cousin 5th degree

1st degree son   2nd degree   1st cousin once removed 4th degree

Degrees of relationship shown within a simplified pedigree. The proband is the individual first coming to the attention of the geneticist.

eration arranged horizontally and connected to the other generations by lines. Family members who have the genetic disorder in question are colored-in or shaded. Unique symbols represent carriers, miscarriages, people of unknown sex, twins, and other categories of individuals. Furthermore, different patterns within the pedigree symbol may represent variable expression (variation in symptoms of individuals with the same disorder). For example, in a pedigree showing myotonic dystrophy, a shaded upper-left quadrant of the male or female symbol may represent cataracts, but a shaded right-lower quadrant may represent heart problems.

The process of collecting the family history used to draw the pedigree includes interviewing the informant and asking questions in three general categories. The detailed information about family members is recorded under the symbol for that individual. First, the interviewer asks standard questions about the family members, resulting in a skeleton picture of the family. These questions collect identifying information about each family member, such as their names, names of their parents, siblings, and children, dates of birth, dates of death, cause of death, and pregnancies. Detailed information is also included about general health problems, such as cancer and heart disease, and their specific symptoms, onset, and age at diagnosis. The interviewer may also ask about the ethnic background of family members and any possible consanguineous relationships (relationships between individuals who are related to each other by blood, such as first cousins).

Second, the interviewer may ask general questions to identify common genetic syndromes in the family, such as birth defects, mental retardation,

vision and hearing loss, and common genetic disorders found more prevalently in certain ethnic backgrounds. For example, cystic fibrosis is relatively more common in people of Western European decent, whereas sickle cell disease is more common in those of African decent. If any common genetic syndromes are identified, they are then discussed during the genetic counseling appointment.

Lastly, the interviewer may collect targeted information about the specific genetic disease for which the family was referred. Information is gathered on the symptoms of the disease, age of onset, and age at diagnosis. These targeted questions help genetic counselors and researchers identify the typical course of disease in the family, disease severity, possible variable expression, reduced penetrance, or genetic anticipation. Because accurate diagnoses are essential for accurate genetic counseling and research studies, all diagnoses should be confirmed by medical records. Further, pedigree updates can be obtained over time during follow-up genetic counseling sessions or research interviews.

Confidentiality

Pedigrees contain very personal and identifying information about families. It is therefore essential that pedigrees be kept confidential. This is important not only in the genetic counseling clinic or research group, but also within the family. It is unethical to share information given about family members with other family members, insurance companies, or other professionals not involved in the genetic counseling sessions or research study. Information about individuals should only be released with proper written consent by that individual. SEE ALSO DISEASE, GENETICS OF; GENETIC COUNSELING; HUMAN DISEASE GENES, IDENTIFICATION OF; INHERITANCE PATTERNS.

Elizabeth C. Melvin

Bibliography

Bennett, Robin L., et al. "Recommendations for Standardized Human Pedigree Nomenclature." *American Journal of Human Genetics* 56 (1995): 743–752.

Stroop, Jennifer B. "The Family History as a Screening Tool." *Pediatric Annals* 29 (2000): 279–282.

Internet Resource

March of Dimes. "Taking a Family Health and Social History." <http://www.fullcirc.com/mod/outlines/family/history.htm>.

Pharmaceutical Scientist

The role of the pharmaceutical scientist in drug discovery and development is highly varied. Duties range from the synthesis of novel compounds designed to alter disease processes, to the formulation of these compounds into a tablet or capsule, to the development of assays (tests) to measure the drug and its **metabolites** in the body, to the testing of compounds for their effects in animals and humans. In addition, scientists in this industry pursue more basic questions, such as which genes or processes are critical in disease, as well as better ways to diagnose disease and predict clinical outcomes of treatment strategies. While the pharmaceutical industry employs thousands of chemists and biologists, the skills needed to work in the phar-

metabolites molecules involved in a metabolic pathway

maceutical industry are much broader and are being substantially changed by the infusion of genetics and genomics into the drug development process. Therefore those interested in a scientific career in the pharmaceutical industry should seriously consider training in the important areas of genetics and **genomics**.

genomics the study of gene sequences

The drug development processes can be divided into three major sections: research, where compounds are synthesized and tested against potential drug targets and for activity in animal models of disease; preclinical safety, where the compounds are analyzed for their potential toxicity in the laboratory and in animal model studies; and manufacturing of clinical-grade material and testing in human clinical trials. Individuals with a range of skills in chemistry, biology, manufacturing, and clinical sciences are very important. However, it is not limited to these areas, since the pharmaceutical industry employs most types of scientists. The increasing amount of genome sequence available at the present time has generated a need for individuals trained in bioinformatics. These scientists use computational methods to answer biological questions, particularly methods involving massive amounts of data produced by the field of genomics. In addition, scientific expertise is needed in many of the support areas of the drug development process, including the business, legal, and regulatory aspects. As a consequence, the training, skills, and qualifications needed for work in the industry are very broad and varied.

Beginning at the technical level with a college degree, there is a variety of entry-level positions in all areas of drug development. Opportunities also exist for individuals to work on postgraduate degrees within a pharmaceutical company, and there are many options for conducting postdoctoral research throughout the industry. Naturally, the work environment varies as much as the types of positions. Pharmaceutical scientists are typically based in the laboratory, manufacturing facility, or office. Those working on clinical trial design will typically work from corporate offices and then implement the patient treatment with investigators at universities or clinics, rather than actually conducting the patient research themselves.

Due to the wide variety and types of positions in this global industry, salary ranges are very broad. Historically, pharmaceutical scientists receive competitive salaries and they may also receive cash or stock bonuses. The pharmaceutical industry is oriented toward extremely high-quality research that leads to new treatments for people in need. Successful scientific work leads to useful drugs that benefit patients as well as the company that develops them. Therefore, the success of pharmaceutical scientists is tied to the degree to which their work benefits patients and to the degree of financial success achieved by the company as a whole.

A career as a pharmaceutical scientist can be exceptionally rewarding. It provides the professional with an opportunity to participate in a team that seeks to discover useful new drugs. There is satisfaction in knowing that, when approved and sold on the market, such a discovery can help millions of people for years to come. In addition, pharmaceutical scientists have the opportunity to work in cutting edge areas, using new methods for studying genetics in clinical trials. One such field of study, called pharmacogenetics, examines the reasons that individuals have different responses to the same drug. This area of study is expected to greatly improve the understanding of how drugs work and enable physicians to prescribe them to those who

are most likely to benefit, while minimizing the risk of adverse reactions. SEE ALSO BIOINFORMATICS; PHARMACOGENETICS AND PHARMACOGENOMICS; PHYSICIAN SCIENTIST.

Kenneth W. Culver and Mark A. Labow

Pharmacogenetics and Pharmacogenomics

genome the total genetic material in a cell or organism

The complete sequencing of the human **genome** in 2000, along with new technologies, such as DNA microarrays, for analyzing human genes on a genome-wide scale, provides scientists with the tools to study the molecular basis of diseases on a level and scale that previously had not been possible. Pharmacogenomics is a biomedical science that aims to use this knowledge to tailor drug therapies based on patients' individual genetic makeup. Doctors hope to use pharmacogenomics to develop safer and more effective medical treatments. For some diseases, this promise has already been realized.

Molecular Interactions and Drug Effectiveness

heritable genetic

Pharmacogenomics is a branch of pharmacogenetics, a science that deals with the **heritable** traits responsible for the individual differences in the ways people respond to drugs. It is remarkable, considering the myriad of medications we have today, how little we understand about how most of them actually work. There are many factors that influence the effectiveness of a particular drug, including how the drug enters the body's cells, how rapidly it is degraded by metabolic **enzymes**, and how it interacts with target molecules in the body, such as drug receptors.

enzymes proteins that control a reaction in a cell

Consider, for example, a common general anesthetic drug, called succinylcholine. In the 1950s doctors noticed that some patients suffered prolonged respiratory apnea (difficulty breathing) after being treated with succinylcholine. This syndrome was found to be caused by mutations in a gene for the enzyme butyrylcholinesterase. Normally butyrylcholinesterase in the blood degrades succinylcholine, and the anesthetic effects of the drug wear off with time. But in patients with mutations that inactivate or weaken butyrylcholinesterase, the anesthetic persists in the body, causing the dangerous side effect.

alleles particular forms of genes

Cytochrome P450 is a member of a large family of enzymes that inactivate more than half of all drugs. There are many different **alleles** of cytochrome P450. Some alleles are very inefficient at inactivating drugs. In individuals with these alleles, termed "poor metabolizers," drugs can accumulate in the body to levels that produce toxic effects. In contrast, some people have extra copies of cytochrome P450 genes and produce excessive levels of the enzyme. In these individuals ("ultrarapid metabolizers"), drugs become inactivated so rapidly that they may not accumulate to the concentrations needed to be effective.

genotype set of genes present

Adverse drug reactions such as the examples just discussed account for over one hundred thousand deaths each year in the United States. If a physician can determine the **genotype** of patients with respect to the genes that

Stephen Liggett uses this cell harvester to perform research in his laboratory at the University of Cincinnati. Dr. Liggett works towards understanding better how genes and drugs interact.

encode these proteins, then he or she may be able to prescribe the most appropriate dosage for a particular drug, or a better-suited drug.

A major obstacle, however, is that so many different proteins affect each drug. Some are well characterized, but most are completely unknown. This is where pharmacogenomics promises to revolutionize medicine. With the sequence of the entire human genome having been determined, and with the development of modern gene analysis technologies, scientists may now be able to pinpoint every gene that influences the effectiveness of any drug. Physicians would then be able to genotype their patients. For example, they could determine whether a patient has P450 alleles that make him a poor metabolizer or an ultrarapid metabolizer.

Not only can pharmacogenomics provide information about the best drug therapy for patients, but it can also be used to predict whether a person is predisposed to contracting a heritable disease. Mutant alleles of many genes have been shown to predispose people to diseases such as breast cancer, Alzheimer's disease, and Huntington's disease. If doctors can identify such mutant alleles in patients long before any sign of disease becomes apparent, they may be able to treat the disease better when it first appears or even prevent it before it strikes.

Of course, this powerful technology carries with it many ethical questions: If you carried a gene that gave you a moderate probability of eventually contracting a fatal disease, should you be told? What if there were no treatments for the disease? Who should have access to a patient's genetic information? If a health insurance company finds out that a person has a set of genes that predispose her to a disease that is costly to treat, should it be allowed to deny her insurance coverage?

Genetic Diagnoses That Can Improve Treatments

In addition to the complex interactions that drug compounds can have with molecules in the body, there are other reasons why some patients experience different responses to drugs. Two people that appear to have the same disease may actually have different diseases that may not respond to the same drug treatment. Modern genomic analysis methods, such as microarray gene expression profiling, can distinguish two diseases that, by all other clinical and diagnostic methods, appear to be identical.

An example is diffuse large B-cell lymphoma (DLBCL), the most common form of non-Hodgkin's lymphoma, a cancer of the white blood cells. About 40 percent of patients can be cured by the standard chemotherapy for DLBCL, but 60 percent respond poorly and eventually die. Using microarray technology to gather and compare the gene expression patterns of cancer cells from many different DLBCL patients, molecular biologists have discovered that DLBCL actually comprises two distinct disease forms. Patients with one form are treatable with the standard chemotherapy, while those with the other form are not. Using this kind of genomic information, doctors can now diagnose patients more accurately and put them on the most appropriate drug-therapy regime. Similar genomic diagnoses based on gene expression profiles are being developed for important diseases such as breast cancer and Alzheimer's disease.

Using SNPs to Identify Disease Genes

nucleotides the building blocks of RNA or DNA

Of the approximately 3 billion **nucleotides** in each of the two sets of chromosomes in human cells, the vast majority are identical from person to person. On average, though, the genomes of individuals differ from one another by about 1 million nucleotides. This is largely what accounts for the enormous diversity of humans. A nucleotide in a gene from one person that is different from the nucleotide found at the same position of the gene in most other people is called a single nucleotide **polymorphism**, or SNP (pronounced "snip").

polymorphisms DNA sequence variant

Almost 1.5 million human SNPs have been identified and catalogued as part of the human genome sequencing project. Many of these polymorphisms are within the protein-coding or control regions of genes and may

contribute to particular diseases, to a predisposition to a disease, or to adverse drug reactions. By comparing the SNP patterns of many different people, geneticists can infer whether a particular SNP (and therefore the gene it is in) is correlated with a disease or adverse drug reaction. Once a correlation is found, doctors can determine if their patients have the SNP in their genomes, to test for the likelihood of contracting the disease or experiencing the adverse reaction. SEE ALSO DNA MICROARRAYS; GENOMIC MEDICINE; POLYMORPHISMS.

Paul J. Muhlrad

Bibliography

Alberts, Bruce, et al. *Molecular Biology of the Cell*, 4th ed. New York: Garland Science, 2002.

Lodish, Harvey, et al. *Molecular Cell Biology*, 4th ed. New York: W. H. Freeman, 2000.

Internet Resource

The SNP Consortium. <http://snp.cshl.org/>.

Physician Scientist

According to Webster's II New College Dictionary, a physician is "a natural philosopher, a person skilled in physic, or the art of healing; one duly authorized to prescribe remedies for, and treat, diseases; a doctor of medicine." A scientist is one who is "learned in the observation, identification, description, experimental investigation, and theoretical explanation of phenomena." The combination of these definitions precisely describes a physician scientist.

A physician scientist typically has two very different aspects to his or her career. All of the traditional duties of a physician are performed, including caring for patients. The usual responsibilities of conducting rounds in the hospital and treating patients in clinic are also carried out. The physician may be a generalist who sees all types of illnesses, or a specialist whose practice focuses on a certain organ system. An example would be a nephrologist or renal doctor, who cares for people with kidney diseases. In addition to these conventional medical duties, however, the physician scientist typically maintains a laboratory, designs experiments, and supervises a staff of technicians who carry them out.

There are commonalities between both careers. They both characteristically take a great deal of self-motivation, and both require a commitment to continuing education in order to stay abreast of advances in science and medicine. And both the physician and the scientist are responsible for other individuals. Patients rely on the physician for honesty and competent care; technicians rely on the scientist for supervision, support, and employment. In either capacity, the physician scientist may also be required to teach students. The physician uses the knowledge of his or her particular field to enhance his science career and vice versa.

To pursue a career as physician scientist, a candidate completes college, medical school, and a residency program, as is required to earn a medical degree. He or she may then subspecialize in a field by completing a fellowship program. From college to the conclusion of such a fellowship takes

at least thirteen to fifteen years. The physician scientist may or may not have a Doctorate of Philosophy (Ph.D.) specifically as a scientist. Such training characteristically includes approximately five years of graduate school and usually some postdoctoral training as well. Occasionally, physician scientists may replace such formal education with "on-the-job" training in the laboratory, but in such cases they supplement the scientific skills and knowledge they developed in medical school with courses and mentorship from a senior scientist.

The physician scientist may work in a number of different environments, including hospitals, universities, the government, or industry. Because physician scientists do not usually see as many patients in a year as a typical doctor, they are usually salaried, and their salaries are usually covered by grants. Thus, for physician scientists who are employed in a hospital affiliated with an academic institution, part of their income may be based on the number of patients in their care, or how many months of the year they perform rounds in the hospital. To secure grants, physician scientists submit applications to a number of organizations, including the government, charities, and private companies. Their applications provide a description of the proposed research, the questions they want to ask and answer with their research, and how they plan to find these answers (their experimental plan). Alternatively, a physician scientist may work for a private research company such as a pharmaceutical firm. The salary range for physician scientists is from approximately $70,000 to more than $200,000, depending on experience and productivity. Productivity is usually measured by how much grant funding has been obtained, how many scientific papers have been published, and recognition by peers. In the industrial sector, productivity will also be measured by how much the research has contributed to successfully bringing products to market.

The combination of physician and scientist results in a very exciting career choice. The physician scientist has the pleasure of taking care of patients and potentially having a direct impact on their future health by the discovery of an important clue to their illness. This is sometimes referred to as conducting science "from-bedside-to-bench-to-bedside." It also provides variety, because the physician scientist is not only able to practice medicine in a very concrete clinical way, but can also think through a variety of more abstract issues, extending his or her intellect in very different directions. SEE ALSO GENETICIST; PHARMACEUTICAL SCIENTIST.

Michelle P. Winn

Bibliography

Coombs, Robert H., D. Scott May, and Gary W. Small, eds. *Inside Doctoring: Stages and Outcomes in the Professional Development of Physicians.* Westport, CT: Greenwood Publishing Group, 1986.

National Academy Press. *Careers in Science and Engineering : A Student Planning Guide to Grad School and Beyond.* Washington, DC: National Academy Press, 1996.

Internet Resource

The National Academy of Sciences. <http://www4.nationalacademies.org/nas/nashome.nsf>.

Plant Genetic Engineer

"Plant genetic engineer" is a popular term that describes scientists working in any of several fields who manipulate DNA or **organelles** such as **chloroplasts** and **mitochondria** in plant cells. The specific titles of such a scientist can include plant physiologist, plant pathologist, weed scientist, cell biologist, botanist, molecular biologist, plant geneticist, and biochemist. The typical career path is to earn a doctorate in any of these fields, and then to go to work in industry, for the government, or in academia.

Jobs that involve genetic manipulation of plants are also available at the technician level. This requires a bachelor's or master's degree in any of the fields listed above. As of 2001, the median salary for an academic life scientist is $42,000, and for a scientist in industry, $70,000. Technician salaries begin at about $25,000. Government salaries are similar to academic compensation.

Scientists work with plant genes in basic research as well as in developing new crops. In basic research, a plant may serve as a model system, enabling researchers to study a fundamental structure or function at the cellular or molecular level. The mustard plant *Arabidopsis thaliana* is the most popular plant model system. Most work on this plant is not to alter or develop a product from it, but to study its basic biology and then extrapolate what is learned to other plant species. Similarly, *Zea mays* (corn) was used extensively in the mid-twentieth century to reveal how chromosomes interact as a cell divides. Today, entire genomes of plants and other types of organisms are being sequenced, which is enabling researchers to work with several genes at a time. The genomes of *Arabidopsis* and several major crop plants are being sequenced.

Some plant scientists manipulate genes to create new variants. Adding a gene from another species forms a transgenic organism. For example, "bt corn" plants harbor a gene from a bacterium that enables plant cells to produce a toxin that kills certain insect larvae that feed on the leaves. "Golden rice" produces beta carotene, a precursor to vitamin A, which is not naturally abundant in the grain portion of this plant. The ability to manufacture beta carotene comes from three genes, taken from daffodils and a bacterium, that specify enzymes that interact in a biochemical pathway, along with genetic instructions to express the genes in the grain, so that it can serve as a vitamin-enriched food.

A plant genetic engineer must be familiar with the characteristics that distinguish plants from other types of organisms. Unlike animal cells, plant cells have tough cell walls, which must be penetrated to reach the DNA. Also, some genetic material resides in the organelles called **plastids**, the largest of which is the chloroplast. The engineer must also be able to regenerate an altered plant cell into a plant, test that plant in a greenhouse, and, finally, see how well it flourishes in a field environment. For example, tomato plants can be given a gene from *A. thaliana* that enables them to grow in very salty water. Developers of such a crop must analyze how the plant that can now grow in **brackish** or salty water will affect other types of plants that normally grow in that environment. Thus, in addition to understanding genetics, molecular biology, and biochemistry, a plant genetic engineer working on an agricultural variant must also have expertise in plant development and

organelles membrane-bound cell compartments

chloroplasts the photosynthetic organelles of plants and algae

mitochondria energy-producing cell organelle

plastids plant cell organelles, including the chloroplast

brackish a mix of salt water and fresh water

The tissue removed from this sunflower plant will be used in genetic research at the Sungene Technologies Laboratory in Palo Alto, California.

ecology. SEE ALSO AGRICULTURAL BIOTECHNOLOGY; *ARABIDOPSIS THALIANA*; BIOPESTICIDES; BIOTECHNOLOGY ENTREPRENEUR; COLLEGE PROFESSOR; GENETICALLY MODIFIED FOODS; GENETICIST; LABORATORY TECHNICIAN; MAIZE; MOLECULAR BIOLOGIST.

Ricki Lewis

Bibliography

Palevitz, Barry A. "Corn Goes Pop, Then Kaboom." *Scientist* 16 (2002): 18–19.

Stone, Richard. "Biologist Gets Under the Skin of Plants—and Peers." *Science* 296 (2002): 1597–1599.

Plasmid

Plasmids are naturally occurring, stable genetic elements found in bacteria, fungi, and even in the mitochondria of some plants. They may be composed of DNA or RNA, double-stranded or single-stranded, linear or circular.

Plasmids almost always exist and replicate independently of the chromosome of the cell in which they are found.

Types of Plasmids

Plasmids are not usually required by their host cell for its survival. Instead, they carry genes that confer a selective advantage on their host, such as resistance to heavy metals or resistance to naturally made antibiotics carried by other organisms. Alternatively, they may produce antibiotics (toxins) that help the host to compete for food or space. For instance, antibiotic resistance genes produced by a plasmid will allow its host bacteria to grow even in the presence of competing bacteria or fungi that produce these antibiotics.

Plasmids are subgrouped into five main types based on phenotypic function. R plasmids carry genes encoding resistance to antibiotics. Col plasmids confer on their host the ability to produce antibacterial polypeptides called bacteriocins that are often lethal to closely related or other bacteria. The col proteins of *E. coli* are encoded by plasmids such as ColE1. F plasmids contain the F or fertility system required for conjugation (the transfer of genetic information between two cells). These are also known as episomes because, under some circumstances, they can integrate into the host chromosome and thereby promote the transfer of chromosomal DNA between bacterial cells. Degradative or catabolic plasmids allow a host bacterium to metabolize normally undegradable or difficult compounds such as various pesticides. Finally, virulence plasmids confer **pathogenicity** on a host organism by the production of toxins or other virulence factors.

Replication

One common feature of all plasmids is a specific sequence of nucleotides termed an origin of replication (ori). This sequence, together with other regulatory sequences, is referred to as a replicon. The replicon allows a plasmid to replicate within a host cell independently of the host cell's own replication cycle. If the plasmid makes many copies of itself per cell, it is termed a "relaxed" plasmid. If it maintains itself in fewer numbers within the cell it is termed a "stringent" plasmid. Two different plasmids can coexist in the same cell only if they share the same replication elements. If they do not, they will be unable to be propagated stably in the same cell line, and are termed incompatible.

In nature, plasmid inheritance can occur through a variety of mechanisms. During conjugation between two bacterial strains, plasmids can be transferred along with the bacterial DNA, and this activity is controlled by a set of transfer (*tra*) genes that are located on the plasmid and not on the bacterial chromosome. The proteins produced by these transfer genes bind to the DNA at the ori site to form a DNA-protein complex known as a relaxosome. This complex makes a nick, or break, in one of the two strands of the double-stranded plasmid DNA molecule. The place where this break occurs is called the "nic" site, and the nicked DNA is said to be "relaxed" because the DNA unwinds as a result of the nick in one of the strands. The single-stranded DNA that is generated by the nick is thought to be unwound and transferred through the pilus, or mating bridge, that connects the two bacteria entering the recipient bacteria. The other strand is left in the donor

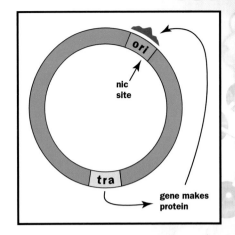

The transfer gene (tra) makes a protein that binds to the origin of replication site (ori). The protein nicks the DNA, relaxing it and allowing it to be transferred to another cell.

pathogenicity ability to cause disease

bacteria. It acts as a template for the synthesis of a new complementary DNA strand forming a double-stranded plasmid DNA molecule.

Some nonconjugative plasmids can also be transferred into bacteria by means of a process called mobilization, as long as they carry the necessary (*mob*) genes. Others are taken up by bacterial cells during the process known as transformation. Finally, plasmids that exist in a host cell that undergoes fission (cell division) are simply divided between the resultant two daughter cells.

Use in Research and Technology

Because of their ability to move genes from cell to cell, plasmids have become versatile tools for both research and biotechnology. In the laboratory, researchers use plasmids to carry marker genes, allowing them to trace the plasmid's inheritance across host cells. Transferred or "cloned" genes are used to produce a variety of important medical, agricultural, or environmental products that can be economically used by humans.

Researchers have also engineered plasmids to be extremely efficient cloning **vectors**. To be used in this way, the plasmid must contain at least one origin of replication, a multiple cloning site (called a polylinker) where a variety of **restriction enzymes** can cut so that foreign DNA can be inserted, a selectable genetic marker, and **transcription** and **translation** signals recognized by the host cell, so that the expression of a cloned gene can be easily identified.

The foreign DNA is often inserted in such a way that the expression of the foreign gene is tied to the expression of a marker gene. For example, one of the most popular methods to show that a foreign DNA has been inserted and expressed in the host is by the insertional inactivation of the *lac Z* gene. In this case, the foreign DNA is inserted in the middle of the *lac Z* gene so that the gene becomes defective and the enzyme it codes for no longer works. The damaged enzyme therefore cannot **cleave** the artificial substrate Xgal to produce a blue color or blue colony, as it normally would, and white colonies of bacteria are produced. Therefore, the white colonies indicate that artificial DNA has been successfully cloned or recombined into the plasmid in the *lac Z* gene, whereas nonrecombinant colonies are blue. The white colonies can thus be easily isolated for further expansion and experimentation.

Under certain circumstances, recombinant DNA experiments using plasmids are considered to be hazardous, and the ease with which plasmids are acquired by bacteria has led them to be classed as biohazards. They are therefore subject to guidelines and may require registration and approval. A publication produced by the National Institutes of Health, titled *Guidelines for Research Involving Recombinant DNA Molecules*, is the definitive reference for recombinant DNA research in the United States and should be consulted when considering research, particularly biomedical research, involving plasmids. SEE ALSO ANTIBIOTIC RESISTANCE; CLONING GENES; CONJUGATION; INHERITANCE, EXTRANUCLEAR; MARKER SYSTEMS; RESTRICTION ENZYMES; TRANSFORMATION.

Linnea Fletcher

vectors carriers

restriction enzymes enzymes that cut DNA at a particular sequence

transcription messenger RNA formation from a DNA sequence

translation synthesis of protein using mRNA code

cleave split

Bibliography

Alberts, Bruce, et al. *Molecular Biology of the Cell*, 4th ed. New York: Garland Science, 2002.

Bloom, Mark V., Greg A. Freyer, and David A. Micklos. *Laboratory DNA Science: An Introduction to Recombinant DNA Techniques and Methods of Genome Analysis.* Menlo Park, CA: Addison-Wesley, 1996.

Alcamo, I. Edward. *DNA Technology: The Awesome Skill*, 2nd ed. Burlington, MA: Harcourt Press, 2000.

Lodish, Harvey, et al. *Molecular Cell Biology*, 4th ed. New York: W. H. Freeman, 1999.

Pleiotropy

Pleiotropy is the phenomenon whereby a single gene has multiple consequences in numerous tissues. Pleiotropic effects stem from both normal and mutated genes, but those caused by mutations are often more noticeable and easier to study. Pleiotropy is actually more common than its opposite, since in a complex organism, a protein from a single gene is likely to be expressed in more than one tissue, and the cascade of problems caused by a mutation is likely to lead to numerous complications throughout the organism. Single-gene defects with effects in only one tissue are more common for nonessential features such as hair texture or eye color.

Sickle cell disease is a classic example of pleiotropy. This disease develops in persons carrying two defective alleles for a blood protein, beta-hemoglobin. Mutant beta-hemoglobins are misaligned inside a blood cell and cause misshapen red blood cells at low oxygen concentrations. Deformed blood cells impair circulation. Impaired circulation damages kidneys and bone. In this case, the gene defect itself only affects one tissue, the blood. The consequences of that defect are found in other tissues and organs.

One baby in three thousand to four thousand births is born with neurofibromatosis, an **autosomal dominant** disease caused by mutation in a tumor suppressor gene that helps regulate cell division and cell-cell contacts. A truncated version of the tumor suppression protein, neurofibromin (NF I) is implicated in the disease. This mutant protein can come from missense or nonsense mutations, or from reading-frame shifts after a repetitive element called Alu is inserted upstream of the NF I reading frame (a reading frame is the DNA that codes for proteins). Because the mutant protein is unable to regulate cell division, tumors grow on the nerves throughout the body. The tumors produce collateral damage: low blood sugar, intestinal bleeding, café-au-lait spots on the skin, mental retardation, heart problems, high blood pressure, fractures, spinal cord lesions, blindness, aneurysms, arthritis, and respiratory distress.

autosomal dominant pattern of inheritance in which inheritance of a single allele from either parent results in expression of the trait

Signaling Pathways

Many pleiotropic conditions arise from genes whose products are involved in signaling and regulation pathways. Because these proteins coordinate daily life in numerous tissues, defects in them have numerous consequences, as one breakdown leads to another.

Myotonic dystrophy is another autosomal dominant disorder. A gene for a protein—a **kinase**, involved in signalling and communication within the cell—is burdened with up to three thousand extra pieces of DNA. The extra DNA comes from trinucleotides $(CTG)_n$ that are added by mistake during DNA duplication, both in germ line cells and during early cell divi-

kinase an enzyme that adds a phosphate group to another molecule, usually a protein

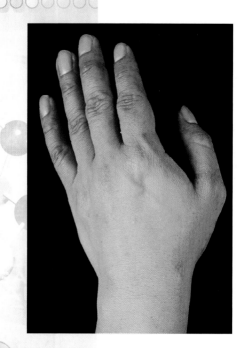

The café-au-lait spots typical of neurofibromatosis are evident on this patient's hand.

sions in the embryo. During transcription and translation, the kinase is not put together right, and the kinase's work in muscles goes badly. Muscles contract but cannot relax quickly. Young persons may have heart attacks, generalized muscle weakness, and loss of bulk. Swallowing and speech is hard, due to weak muscles in the tongue and neck. Other pleiotropic effects include baldness, cataracts, and changes in intelligence.

Pleiotropic outcomes are common with hormones. Hormones are signals that create multiple responses in tissues that carry receptors for them. The receptor binds to the hormone and triggers a cascade of reactions inside the cell. A defective receptor loses or misinterprets the signal. When the hormone insulin meets defective insulin receptors on an individual's cells, the person is more likely to develop type II diabetes. Cells do not open their gateways to let sugar in from the bloodstream, and the cells almost starve to death in the midst of plenty. Meanwhile, sugar accumulates in the blood and causes all sorts of ramifications for blood circulation, and it damages capillaries in all areas, from kidneys, to eyes, to feet. Gangrene, mental disturbances, kidney failure, and blindness can and do occur. Diminished give-and-take of sugar molecules across cell membranes leads to the multifaceted disease diabetes. SEE ALSO DIABETES; DISEASE, GENETICS OF; HEMOGLO-BINOPATHIES; MUSCULAR DYSTROPHY; SIGNAL TRANSDUCTION.

Susanne D. Dyby

Bibliography

Solomon, Eldra Pearl, and Linda R. Berg. *The World of Biology*, 5th ed. Philadelphia: Saunders College Publishing, 1995.

Internet Resources

National Center for Biotechnology Information. <www.ncbi.nlm.nih.gov>.

United States National Library of Medicine. <www.nlm.nih.gov>.

Polygenic Trait *See Complex Traits*

Polymerase Chain Reaction

The polymerase chain reaction (PCR) is a laboratory technique for "amplifying" a specific DNA sequence. PCR is extremely efficient and sensitive; it can make millions or billions of copies of any specific sequence of DNA, even when the sequence is in a complex mixture. Because of this power, researchers can use it to **amplify** sequences even if they only have a minute amount of DNA. A single hair root, or a microscopic blood stain left at a crime scene, for example, contains ample DNA for PCR.

amplify produce many copies of

PCR has revolutionized the field of molecular biology. It has enabled researchers to perform experiments easily that previously had been unthinkable. Before the mid-1980s, when PCR was developed, molecular biologists had to use laborious and time-consuming methods to identify, clone, and purify DNA sequences they wanted to study. Kary Mullis was awarded the 1993 Nobel Prize in Chemistry for inventing PCR.

PCR is based on the way cells replicate their DNA. During DNA replication, the two strands of each DNA molecule separate, and DNA polymerase, an enzyme, assembles nucleotides to form two new partner strands for each of the original strands. The original strands serve as **templates** for the new strands. The new strands are assembled such that each nucleotide in the new strand is determined by the corresponding nucleotide in the template strand. The nucleotides adenine (A) and thymine (T) always lie opposite each other, as do cytosine (C) and guanine (G). Because of this base-pairing specificity, each newly synthesized partner strand has the same sequence as the original partner strand, and replication produces two identical copies of the original double-stranded DNA molecule.

In PCR, a DNA sequence that a researcher wants to amplify, called the "target" sequence, undergoes about thirty rounds of replication in a small reaction tube. During each replication cycle, the number of molecules of the target sequence doubles, because the products and templates of one round of replication all become the templates for the next round. After n rounds of replication, 2^n copies of the target sequence are theoretically produced. After thirty cycles, PCR can produce 2^{30} or more than ten billion copies of a single target DNA sequence. This is called a polymerase chain reaction because DNA polymerase catalyzes a chain reaction of replication.

Designing Primers

To replicate DNA, DNA polymerases require not only a template, but also a primer. A primer is a sequence of single-stranded DNA that "anneals," or binds, to the template by specific base-pairing. An automated apparatus called an oligonucleotide synthesizer, sometimes nickname a "gene machine," can produce primers of any chosen sequence.

Primers for PCR are typically short sequences, around twenty nucleotides long. It is the primers' sequences that are responsible for PCR's enormous specificity. Researchers design primers so they are likely to bind to sequences on either side of the target DNA. They do so by making the primers complementary to the appropriate sequences and by making them long enough that they are unlikely to bind elsewhere.

The longer the primer, the more likely it is that it will be complementary only to the target sequence. Because any single position in a DNA sequence can be occupied by either an A, T, C, or G, there is a one in four chance that any position will contain an A, for example. (This is an approximation, because the nucleotides are not distributed equally or randomly in DNA.) The odds that any specific DNA sequence that is n nucleotides long would be present at a given spot in a DNA sequence is therefore 1 in 4. The chance that a particular twenty-nucleotide sequence (a typical length for a PCR primer) would occur in a given spot at random is less than one in one trillion (10^{-12}). The human genome has only about three billion (3×10^9) nucleotide pairs, so any twenty-nucleotide-long sequence is very unlikely to occur more than once by chance in the human genome.

Researchers design two primers that will bind to opposite strands of the DNA on either side of the target sequence. They design them to "point" the right way, so that the section of DNA between, not outside of them, is copied. Designing the primers to "point" in the right direction simply requires building them so that their 3' ends lie toward the target DNA and their 5' ends lie away from it. The ends of any segment of DNA, including the complete strand, are chemically different.

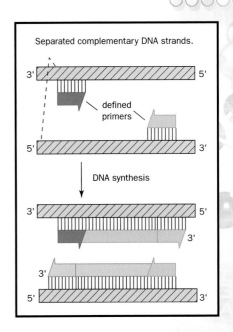

Separated complementary DNA strands.

defined primers

DNA synthesis

PCR begins by separating complementary DNA strands, and ends by adding short primers that match opposite ends of complementary strands. Adapted from Alberts, 1995.

templates master copies

155

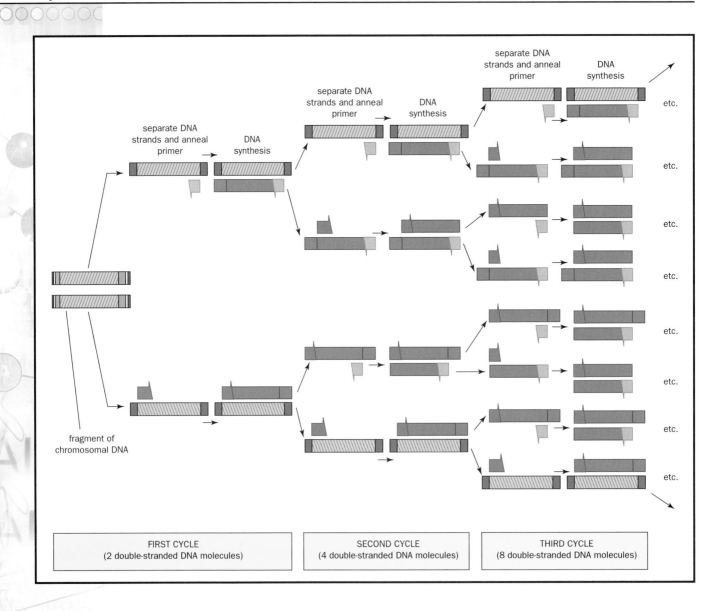

separate DNA strands and anneal primer → DNA synthesis

separate DNA strands and anneal primer → DNA synthesis

separate DNA strands and anneal primer → DNA synthesis

etc.

etc.

etc.

etc.

etc.

etc.

etc.

etc.

fragment of chromosomal DNA

| FIRST CYCLE (2 double-stranded DNA molecules) | SECOND CYCLE (4 double-stranded DNA molecules) | THIRD CYCLE (8 double-stranded DNA molecules) |

Multiple rounds of separation and copying can quickly produce billions of copies of the original DNA sequence. Adapted from Alberts, 1995.

microliters one thousandth of a milliliter

catalyze aid in the reaction of

buffers substances that counteract rapid or wide pH changes in a solution

One end is called the 5′ (pronounced "5-prime") end. The other is called the 3′ end. In DNA replication, nucleotides are always added to the 3′ end of a growing strand of DNA. DNA synthesis is said to proceed in a 5′ to 3′ direction. The two complementary strands of DNA are anti-parallel, which means that they run in opposite directions. The 5′ end of one strand lies next to the 3′ end of the other, as shown in the diagram.

A Typical PCR Reaction

A typical PCR reaction consists of the following components, mixed together in a solution with a total volume of between 25 and 100 **microliters**. The solution must include the template DNA, the primers, nucleotides to serve as building blocks for the newly forming DNA, DNA polymerase to **catalyze** the synthesis, and **buffers** and salts, usually including magnesium, that are required for optimal activity of the DNA polymerase. The template can be an unpurified mixture of DNA, such as DNA extracted from a swab of cheek cells from a patient or crime suspect.

To perform the PCR reaction, the tube containing the solution is placed into a machine called a DNA thermal cycler. Thermal cyclers are basically programmable heating blocks. They usually contain a thick aluminum block with holes in which PCR reaction tubes can fit snugly. The block can be rapidly cooled or heated to specific temperatures, for specific lengths of time, under programmable computer control. Each cycle in a PCR reaction is controlled by changing the temperature of the block and, therefore, of the reaction mixture.

The first step in PCR is to heat the mixture to a high temperature, usually 94 to 95 °C, for about five minutes. The hydrogen bonds that hold together the two strands of a double helix are broken at these temperatures, and the DNA separates into single strands. This process is termed denaturation.

In the second step, the PCR mixture is cooled to a lower temperature, typically between about 50 °C and 65 °C. This allows the primers to **anneal** to their specific complementary sequences in the template DNA. The temperature for this step is chosen carefully to be just low enough to allow the primers to bind, but no cooler. A lower annealing temperature might allow the primers to bind to regions in the template DNA that are not perfect complements, which could lead to the amplification of non-specific sequences.

anneal join together

The optimal annealing temperature for a set of primers can be determined by a formula that is based on the nucleotide composition of the primers, but it is often a matter of trial and error to find the best annealing temperature. The annealing step usually takes about fifteen to thirty seconds, an amazingly short time considering that the primers must "scan" through the template DNA to find their proper binding sites.

In the third step, the reaction is heated again, usually to about 72 °C, the temperature at which the DNA polymerase is most active. Most enzymes are destroyed at 72 °C. In the early days of PCR, scientists used a DNA polymerase that was derived from the bacterium *Escherichia coli*, which itself is most active at human body temperature, 37 °C. But the *E. coli* polymerase was destroyed at the high temperatures required for the denaturation and annealing steps, and the polymerase therefore had to be added anew to the reaction, during each PCR cycle.

To solve this problem, scientists purified DNA polymerases from microorganisms that live in hot springs or in deep-sea thermal vents. These organisms' enzymes are most active at high temperatures. The most commonly used enzyme for PCR is called Taq DNA polymerase, which was originally purified from the hot-spring bacterium *Thermus aquaticus*. (Most commercially available preparations today are recombinant versions, produced in engineered *E. coli* strains.)

At 72 °C, Taq DNA polymerase adds nucleotides to the 3′ ends of annealed primers at the rate of about two thousand nucleotides per minute. Therefore, to amplify a sequence that is one thousand nucleotides long, the primer extension step must last about thirty seconds at 72 °C. By the end of this step, each template strand has a new complementary strand. This completes the first cycle of the PCR reaction.

The cycle can be repeated, at that point, by restarting the denaturation step. In the next cycle, the original two DNA strands will serve again as

templates, as will the two newly synthesized strands. In this way, the number of templates has doubled, and it will double again with each successive cycle.

At the end of the reaction, the tube contains DNA fragments that are almost solely copies of the target DNA. The original template DNA mixture is still present, but for the purpose of most applications (with the exception of subsequent PCR experiments), it is present in negligible amounts compared to the PCR product. The amplified DNA can be analyzed by gel electrophoresis, **ligated** into a cloning vector, labeled for use as a hybridization probe, or used in numerous other experimental procedures.

ligated joined together

Contamination in PCR Reactions

The extreme sensitivity of PCR for amplifying rare DNA sequences is a mixed blessing. Just as PCR can easily amplify any sequence that a researcher wants to amplify, it can also amplify other sequences.

Amplifying a minute amount of DNA isolated from an ancient mosquito preserved in amber, for instance, could be extremely difficult. DNA from other sources could contaminate the sample during every step, including during the recovery of the amber, while researchers are drilling into it, and while the needle is prepared to remove the mosquito tissue. Contamination of the sample by even a single cell from another source can lead to amplification of that DNA, along with or instead of the mosquito's DNA. Especially if there are segments of contaminating DNA that are similar to the target DNA, primers may bind to the wrong segments.

allele a particular form of a gene

Or consider a human geneticist who has designed a PCR assay to detect a particular genetic disease. Imagine that a positive result, amplification of the disease **allele** from a patient's DNA sample, would indicate the patient is a carrier for the disease. If even a trace of DNA from a disease carrier contaminates any of the PCR reagents, then assays performed on samples from non-disease carriers are likely to produce the diagnostic PCR amplification product. It isn't difficult to imagine that a lab that routinely performs this PCR assay might have lots of the tell-tale DNA contaminating benches, pipettes, and even lab coats.

Two approaches address the contamination problem. First, laboratory practices for PCR aim to ensure the utmost cleanliness. Whole new industries have been created to produce contamination-resistant supplies, including micropipette tips. The second solution, as in all carefully planned experiments, is to use controls. Negative controls, including mixtures that have not had any template added or that contain a template known to lack the target sequence, are particularly important for PCR experiments.

PCR Applications and Variations

PCR is such a powerful, easy, and relatively inexpensive technique that it seems that molecular biologists are always looking for ways to use PCR in their research. Every month, scientific journals describe modifications to tailor the basic PCR approach to new applications.

One variation that has proved very fruitful in gene identification is the use of "degenerate primers." Many genes tend to be highly conserved among different species. Homologues, which are genes from different organisms whose protein products have similar functions, tend to have very similar, but not necessarily identical, sequences. The differences in sequence make it challenging to design standard PCR primers to search for homologues.

However, by comparing the DNA sequences of the gene as it occurs in many different species and finding portions of the sequence that are the same in all the species, a researcher can make an educated guess regarding which nucleotides in an unidentified homologue are likely to be identical to those in a known homologue.

The researcher can design a set of "degenerate" PCR primers, which are primers whose nucleotide sequence is fixed only in those positions where the nucleotides are presumed to be known. In the other positions, nucleotides are allowed to incorporate at random. This makes it likely that at least one of the primers will amplify the unknown target. By conducting PCR with degenerate primer sets, and by using primer annealing temperatures that are lower than normal to allow for less-than-perfect base-pairing, a researcher can often amplify a gene in a single experiment, thus isolating the new homologue and allowing it to be sequenced and studied.

Another important variation on PCR is reverse-transcription PCR. This technique involves first copying RNA into DNA molecules, using the enzyme **reverse transcriptase**, and subsequently using the standard PCR technique to amplify this complementary DNA (cDNA). Because the messenger RNA content of a cell or tissue represents only the genes that are actively being expressed, this technique provides a powerful method of analyzing gene expression. SEE ALSO BIOTECHNOLOGY; CLONING GENES; DNA POLYMERASES; GEL ELECTROPHORESIS; GENETIC TESTING; HOMOLOGY; HUMAN GENOME PROJECT; NUCLEOTIDE; REPLICATION; SEQUENCING DNA.

Paul J. Muhlrad

reverse transcriptase
enzyme that copies RNA into DNA

Bibliography

Alberts, Bruce, et al. *Molecular Biology of the Cell*, 4th ed. New York: Garland Publishing, 2002.

Lodish, Harvey, et al. *Molecular Cell Biology*, 4th ed. New York: W. H. Freeman, 2000.

Micklos, David A., and Greg A. Freyer. *DNA Science: A First Course in Recombinant DNA Technology*. Cold Spring Harbor, NY: Cold Spring Harbor Laboratory Press, 1990.

Watson, James D., et al. *Recombinant DNA*, 2nd ed. New York: Scientific American Books, 1992.

Polymorphisms

Genetic polymorphisms are different forms of a DNA sequence. "Poly" means many, and "morph" means form. Polymorphisms are a type of genetic diversity within a population's gene pool. They can be used to map (locate) genes such as those causing a disease, and they can help match two samples of DNA to determine if they come from the same source. Depending on its exact nature, a polymorphism may or may not affect biological function.

A short tandem repeat (STR) polymorphism at a particular locus on a chromosome.

Coding and Noncoding Sequences

The amino acid sequence of proteins is directed by the information found in genes, which in turn are made up of DNA. Genes that have different DNA sequences are said to be polymorphic. These different gene forms are called alleles, exemplified by the alleles that control eye color. When alleles result in differences in the amino acid sequence of a protein, the proteins encoded by alleles are called isoforms. The position of the gene on a chromosome is its locus (plural, loci). More generally, a locus refers to any position on a chromosome, whether or not a gene is located there.

Polymorphisms arise through mutation. The mutation may be due to a change from one type of **nucleotide** to another, an insertion or deletion (collectively known as indels), or a rearrangement of nucleotides. Once formed, a polymorphism can be inherited like any other DNA sequence, allowing its inheritance to be tracked from parent to child.

nucleotide a building block of RNA or DNA

Polymorphisms are also found outside of genes, in the vast quantity of DNA that does not code for protein. Indeed, regions of DNA that do not code for proteins tend to have more polymorphisms. This is because changes in DNA sequences that encode proteins may have a harmful effect on the individual that carries it. Polymorphisms that do not have any effect on the organism are said to be selectively neutral since they do not affect its ability to survive and reproduce.

Identifying Polymorphisms

The process of determining an individual's genetic polymorphisms is known as genotyping. One of the earliest methods used in genotyping looked not at genes but at polymorphic proteins known as isoenzymes, or isozymes. Isoenzymes are different forms of a protein, with slightly different amino acid compositions. Since a protein's amino acid composition is genetically programmed by the DNA sequence that encodes it, analysis of isoenzymes surveys genetic polymorphism. Because these differences in amino acid composition can cause proteins to have different electrical charges, isoenzyme polymorphisms are assessed by extracting an organism's proteins and separating them using gel electrophoresis—a technique also used to study DNA polymorphisms.

In gel electrophoresis, an electric field is applied across a gel matrix, and molecules move through the matrix in response to the electric field. The gel matrix is a porous material, similar to Jell-O, that acts as a sieve and slows down large molecules more than small molecules. Isoenzymes move through the gel matrix according to their electrical charge and size, and are separated from each other on this basis. In this way, different isoenzymes can be identified.

Many tools for assessing DNA polymorphisms are now available. Some of these methods assess length polymorphism (indels), sequence polymorphism (base changes or rearrangements), or combinations of the two. DNA

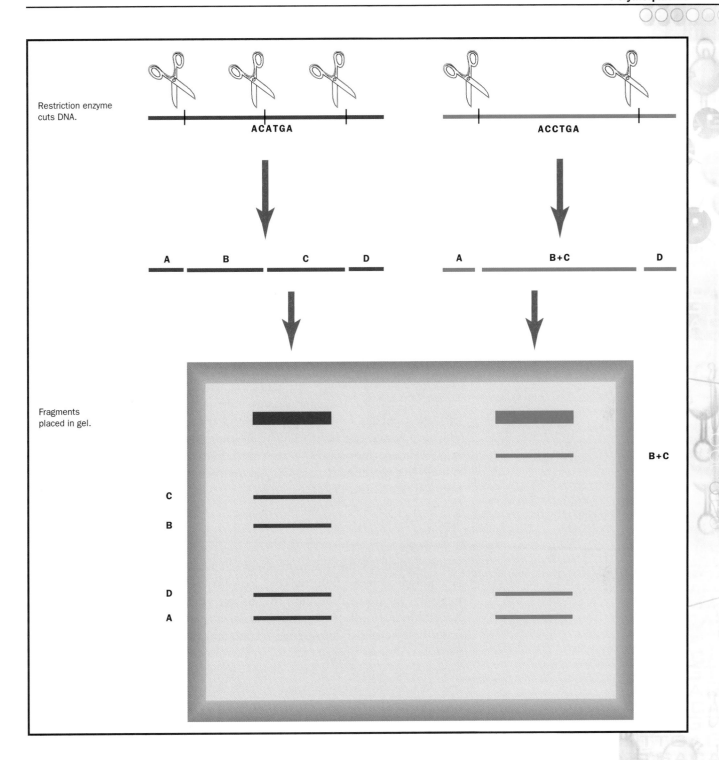

Restriction enzyme cuts DNA.

ACATGA

ACCTGA

A   B   C   D

A   B+C   D

Fragments placed in gel.

C

B

D

A

B+C

sequencing reveals all types of polymorphisms but is costly and labor-intensive. Gel electrophoresis is used in DNA analysis to both separate and sequence DNA. PCR (polymerase chain reaction) is frequently employed beforehand to produce large quantities of the DNA to be analyzed.

Restriction fragment length polymorphisms can be separated and identified by gel electrophoresis. A single nucleotide polymorphism (A→C) causes loss of a restriction site.

RFLPs

An early method of detecting DNA polymorphisms still in use employs restriction endonucleases. These bacterial enzymes cut DNA at specific recognition sequences. Restriction enzymes cleave DNA into a characteristic set

of fragments that can be separated by gel electrophoresis. Some polymorphisms alter recognition sequences, so that the enzyme no longer recognizes a site or recognizes a new site. This results in a new set of DNA fragments that can be compared to others to detect the differences. These differences are called restriction fragment length polymorphisms (RFLPs).

STRs, VNTRs, and SNPs

Repetitive genetic elements are an important class of polymorphic DNA. These sequences consist of several repeats of a simple DNA sequence pattern, and they typically do not encode a protein or have strict requirements of size and sequence. For example, the two **base pairs** cytosine (C) and adenine (A) may be found together multiple times, resulting in a "CACACACA" sequence. If another copy of this sequence were found as "CACA" (two CA pairs shorter), then this sequence would be polymorphic. Repetitive genetic elements include microsatellites or STRs (short tandem repeats) and the minisatellites or VNTRs (variable number of tandem repeats), which are distinguished primarily on the basis of size and repeat pattern: The repeated sequence in microsatellites range from two to six bases, while in a VNTR it ranges from eleven to sixty base pairs.

Differences in single base pairs, known as single nucleotide polymorphisms (SNPs), are a valuable class of polymorphism that can be detected by DNA sequencing, RFLP analysis, and other methods such as allele-specific PCR and allele-specific DNA hybridization. Many RFLPs are due to single nucleotide polymorphisms. There are hundreds of thousands of SNP loci throughout the human genome, making them especially valuable for mapping human disease genes.

Uses of Polymorphisms

The study of polymorphism has many uses in medicine, biological research, and law enforcement. Genetic diseases may be caused by a specific polymorphism. Scientists can look for these polymorphisms to determine if a person will develop the disease, or risks passing it on to his or her children. Besides being useful in identifying people at risk for a genetically based disease, knowledge of polymorphisms that cause disease can provide valuable insight into how the disease develops. Polymorphisms located near a disease gene can be used to find the gene itself, through mapping. In this process, researchers look for polymorphisms that are co-inherited with the disease. By finding linked polymorphisms on smaller and smaller regions of the chromosome, the chromosome region implicated in the disease can be progressively narrowed, and the responsible gene ultimately can be located.

A related use of polymorphism is widely employed in agriculture. If a polymorphism can be identified that is associated with a desirable characteristic in an agriculturally important plant or animal, then this polymorphism can be used as a genetic flag to identify individuals that have the desirable characteristic. Using this technique, known as marker-assisted selection, breeding programs aimed at improving agriculturally important plants and animals can be made more efficient, since individuals that have the desired trait can be identified before the trait becomes apparent.

base pairs two nucleotides (either DNA or RNA) linked by weak bonds

Polymorphisms can be used to illuminate fundamental biological patterns and processes. By studying polymorphisms in a group of wild animals, the familial relationships (brother, sister, mother, father, etc.) between them can be determined. Also, the amount of interbreeding between different groups of the same species (gene flow) can be estimated by studying the polymorphisms they contain. This information can be used to identify unique populations that may be important for survival of the species. Sometimes it is not immediately obvious if two different groups of organisms should be classified as different species. Comparing the genetic polymorphisms in the two groups aids in making a judgment as to whether they warrant classification as different species.

If enough polymorphisms are analyzed, it is possible to distinguish between individual humans with a high degree of confidence. This method is known as DNA profiling (or DNA fingerprinting) and provides an important tool in law enforcement. A person's genotype, or DNA profile, can be determined from very small samples, such as those that may be left at a crime scene (hair, blood, skin cells, etc.). The genotype of samples found at the crime scene can then be compared to a suspect's genotype. If they match, it is very likely that the suspect was present at the crime scene. Currently, the FBI uses thirteen different polymorphic loci for DNA fingerprinting. In a similar manner, analysis of polymorphisms can help prove or disprove fatherhood (paternity) in cases where responsibility for a child is disputed.

SEE ALSO GEL ELECTROPHORESIS; LINKAGE AND RECOMBINATION; MAPPING; MUTATION; REPETITIVE DNA ELEMENTS.

R. John Nelson

Bibliography

Avise, John C. *Molecular Markers, Natural History and Evolution.* New York: Chapman & Hall, 1994.

Weaver, Robert F., and Philip W. Hedrick. *Genetics,* 2nd ed. Dubuque, IA: William C. Brown, 1992.

Internet Resource

SNPs: Variations on a Theme. National Center for Biotechnology Information. <http://www.ncbi.nlm.nih.gov/About/primer/snps.html>.

Polyploidy

In eukaryotic organisms, chromosomes come in sets. The **somatic** cells, called soma, usually have a **diploid** chromosome number, which in scientific notation is abbreviated as 2N. The diploid state contains two sets of chromosomes, one set of which has been contributed by each parent. A single set of chromosomes composes the **haploid** chromosome number, which is abbreviated as N. The haploid set is found in reproductive cells or gametes (also called the germplasm). In humans the diploid number is 46, and is represented as 2N = 46. Human sperm or eggs, however, have a haploid number of 23, which is represented as N = 23. In some circumstances, however, an organism can have more than two chromosomal sets. This occurrence is called polyploidy.

One cause of polyploidy is polyspermy. If two sperm fertilize an egg, the resulting zygote or fertilized egg will have three sets of chromosomes,

somatic nonreproductive; not an egg or sperm

diploid possessing pairs of chromosomes, one member of each pair derived from each parent

haploid possessing only one copy of each chromosome

Ploidy refers to the number of chromosomes in an organism's genome. A haploid organism has just one copy of each chromosome (A, B, C, and D). Diploids have two of each, and may have sex chromosomes (X and Y). Triploids have three, which prevents proper pairing and separation of chromosomes in meiosis.

chromosomes	A B C D	A A B B C C D D	A A A B B B C C C D D D
	haploid	diploid	triploid

A A B B C C X X
diploid female

A A B B C C X Y
diploid male

A A A B B B C C C X X X ——————→ **meiosis** ——————→ A A B C C X + A B B C X X

triploid female

Incorrect chromosome numbers leads to few offspring

triploid possessing three sets of chromosomes

and thus have a **triploid** number (3N). When this occurs in humans, 3N = 96. Triploidy in humans and most other animals is incompatible with life. Triploid individuals abort or fail to survive the first days of life after birth. Polyploidy is more common in plants, and polyploid forms often survive to produce much larger cells and plant organs. Ferns, which may have up to 1,500 chromosomes, are frequently polyploid, as are varieties of domesticated cereal plants. Most often, polyploids run in sets of three to eight (triploid to octoploid).

Polyploidy in Animals

Geneticist Hermann Muller argued that polyploidy is more rare in animals than plants because animals have a more complex development, with more organ systems that are fine-tuned to dosages of genes. Any given gene is represented three times in a triploid. If the amount (dosage) of gene product causes a heart, brain, or other vital organ not to form, the embryo will abort. When these developmental genes produce too much or too little of the products that induce organ formation, as they might if there are too many or too few copies of the genes, events occur too soon or too late to be coordinated. Muller raised the possibility that the sex chromosomes serve as a barrier to polyploidy in most animals. Plants, by contrast, do not usually have sex chromosomes, and thus this sexual reproductive barrier is not a problem for them.

autosomes chromosomes that are not sex determining (not X or Y)

Muller noted that most animals use a sex-chromosome mechanism for sex determination. In fruit flies and humans, diploid males have the sex chromosomes XY, whereas diploid females have XX. A triploid fly or human would have three chromosomes along with three sets of **autosomes**. In such a triploid, XXX will result in a female. However, a zygote having XXY XYY may not produce a male. Rather, it may result in an intersex organism, with abnormal mixed male and female reproductive organs.

While human triploids do not survive, this is not the case for fruit flies. The XXY or XYY is an intersex, sterile form, but the triploid female is fertile. If the 3N female is mated to a 2N XY male, however, only a relatively few offspring will emerge, because many of the eggs will have an incorrect number of chromosomes. This state of excesses or deficits of chromosomes in an otherwise diploid or triploid cell is called aneuploidy. Aneuploid embryos rarely survive in humans or other animals, although there are exceptions (such as infants born with Down syndrome).

This seedless watermelon is the product of fourteen years of research at the University of Florida's Institute of Food and Agricultural Sciences. Seedless watermelon cells are triploids, the result of crossing tetraploid with diploid hybrids.

Human triploid embryos are a major reason for first-trimester spontaneous abortions (popularly called miscarriages). Polyploid amphibians, on the other hand, have evolved an alternate means of sex determination that allows them to have fertile triploid or tetraploid (4N) forms. As with polyploid plants, these forms are generally larger in size than their diploid relatives. It is not yet known why stillborn or short-lived human triploids do not display this enlarged size.

Polyploidy in Plants

Polyploidy can be induced with chemicals such as colchicine, as O. J. Eigsti first demonstrated in 1935. His work extended that done by F. A. Blakeslee, and the technique he used has been adopted commercially to produce products such as seedless watermelon. The seeds are missing because the embryos abort from aneuploidy before they can form seeds.

In nature there are different kinds of polyploids. An autopolyploid plant has all its chromosomes derived from one haploid set. An allopolyploid plant has its sets derived from two different plant species. In general, allopolyploids

gamete reproductive cell, such as sperm or egg

apoptosis programmed cell death

are fertile and survive, whereas autopolyploids are sterile and must be propagated as clones (identical twins), by cuttings.

The difference between autopolyploidy and allopolyploidy can be appreciated by an example. No one knows the reasons for mitotic failure leading to spontaneous tetraploids, but artificial ones are induced by mitotic poisons, like colchicine, that prevent spindle fiber formation. If one species has chromosomes ABCD in a (haploid) **gamete**, and a related species has chromosomes FGHI, the resulting (diploid) zygote will have a chromosome set consisting of ABCDFGHI. If that collection of chromosomes undergoes a spontaneous doubling, the resulting plant is AABBCCDDFFGGHHII. Such a plant will produce ABCDFGHI gametes and by self-pollination, which is common in many flowering plants, the new allopolyploid will be fertile.

In the case of autopolyploids, by contrast, the chromosomes ABCD become triplicated (3N: AAABBBCCCDDD) or quadruplicated (4N: AAAABBBBCCCCDDDD). This may lead to nondisjunctional separations during meiosis, wherein the chromosomes will divide improperly or incompletely. In the 3N plant many of the gametes may be AABCDD or ABCCD or other variations of aneuploidy that will disturb embryonic development.

Among familiar plant polyploids are strains of wheat with chromosome numbers of 14 (2N), 28 (4N), and 42(6N), all of which are based on an ancestral form whose haploid number was 7. Chrysanthemums have a series of varieties with a range of chromosome numbers: 18, 36, 54, 72, and 90. The ancestral haploid is assumed to be 9. About half of all flowering plant species are believed to have polyploid varieties. If an accidental doubling of the zygote chromosome number is the major mechanism involved, most of these forms are tetraploid.

Genetic Analysis

The transmission of genetic traits in polyploids is more difficult to calculate than in diploids because a gene for a recessive trait in a triploid, for example, would have to appear in the same location on all three of its homologous chromosomes in order for it to be phenotypically apparent. Such calculations, when done for diploids, rely upon binomial equations and generate a familiar ratio of 9AB:3aB:3Ab:1ab, whereas the calculations for polyploid plants require the use of trinomial equations for triploids and quadrinomial equations for tetraploids, instead of the traditional binomial $(A + B)^2$ that generates the familiar 9AB:3aB:3Ab:1ab ratio for diploids. Thus for a trinomial (three gene) the equation will be the expansion of $(A + B + C)^3$.

The use of polyploids in laboratory research has allowed research into the function of specific genes. For instance, triploid female fruit flies crossed to diploid males were used to create a diploid offspring with a chromosome of a sibling species. In this experiment, the tiny fourth chromosome of *Drosophila simulans* was inserted into an otherwise diploid *D. melanogaster* offspring. This permitted analysis of the genes shared in common (most of them) as well as gene differences that led to visible malformations in the hybrid fly. Triploid flies have also been crossed to irradiated diploid males to prove that X rays induce breaks in chromosomes, causing **apoptosis** and

embryonic abortion. SEE ALSO CHROMOSOMAL ABERRATIONS; MEIOSIS; MULLER, HERMANN; SEX DETERMINATION; X CHROMOSOME, Y CHROMOSOME.

Elof Carlson

Bibliography

Muller, H. J. "Why Polyploidy is Rarer in Animals than in Plants." *The American Naturalist* LIX (1925): 346–353.

Dobzhansky, Theodosius. "Patterns of Evolution." In *Genetics and the Origin of Species*, Theodosius Dobzhansky, ed. New York: Columbia University Press, 1951.

Population Bottleneck

A population bottleneck is a significant reduction in the size of a population that causes the extinction of many genetic lineages within that population, thus decreasing genetic diversity. Population bottlenecks have occurred in the evolutionary history of many species, including humans. Present-day bottlenecks are seen in endangered species such as the Yangtze River dolphin, whose numbers have dwindled to less than 100. Endangered species that do not become extinct may expand their numbers later on, but with a limited amount of genetic diversity with which to adapt to changing conditions. The genomes of future populations will reflect the narrowing of genetic possibility for thousands of years.

Reconstructing Genealogies

The genomes of living organisms record both genealogical and population histories. Our own genome tells a remarkable story of events in recent human evolution. Relatedness of individuals within and between populations and species can be determined by measuring the number of genetic differences between two individuals. When applied to segments of the genome that accumulate mutations at relatively constant rates over time, they can provide information about the time that has elapsed since the existence of their last common ancestor. Research shows that human and chimpanzee lineages diverged about six million years ago, that neanderthals and anatomically modern humans diverged 500 thousand years ago, and that all living humans can trace their ancestry to a maternal lineage that lived in Africa about 130 thousand years ago. Figure 2 illustrates differentiation of lineages and the effects of bottlenecks on diversity.

Reconstructing Ancient Population Sizes

Knowledge of mutation rates also permits reconstruction of past population sizes. A small number of genetic differences between individuals in a population or species may indicate either a recent origin, or a population bottleneck. Which of these two possible causes is responsible can be determined by measuring the number of so-called pairwise differences (mismatch distributions) in the DNA sequences that occur between individuals. Population expansion times are earlier for populations with higher average pairwise differences. Irregular mismatch distributions indicate long-term populations that have been stable for long times.

As shown in Figure 3, humans have remarkably little genetic diversity, especially in comparison to our closest living relative, the chimpanzee.

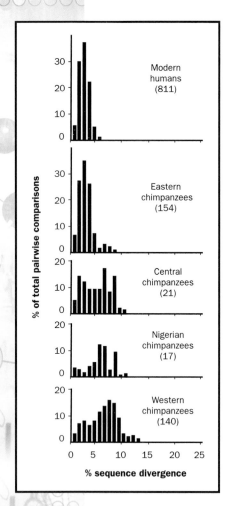

Figure 1. Pairwise sequence divergence distributions in humans and four geographic populations (subspecies) of chimpanzees. The broad and uneven mismatch distributions of central, Nigerian and Western chimpanzees indicate they have maintained relatively large effective population sizes for a very long time. The low genetic diversities of eastern chimpanzees and humans reflect recent population expansions following the loss of many lineages during population bottlenecks. The small peak on the right side of the Eastern chimpanzee distribution corresponds to a population expansion 69 thousand years ago, and the larger peak reflects expansion around 20,000 years ago, near the end of the last ice age.

Indeed, there is substantially more genetic difference among individuals within chimpanzee troops in West Africa than among all living humans on earth. As shown in Figure 1, this is due to a series of bottlenecks in human evolutionary history. Geneticists studying many different parts of the human genome have concluded that the past effective population size (that is, the number of reproducing females) averaged only 10,000 individuals over the last one million years, and was as low as 5,000 around 70,000 years ago. Compare this to the approximately one billion reproducing females alive today, and it becomes clear just how narrow these bottlenecks were.

Population Bottlenecks and Expansions in Human Evolution

The genetic structure of human populations suggests four bottlenecks in our lineage. Stanley Ambrose has proposed that two bottlenecks may be related to past environmental changes. Marta Lahr has attributed bottlenecks to migrations of small populations across geographic barriers, a phenomenon variously referred to as the founder effect or colonization bottlenecks.

Bottleneck 1. When traced backward in time, all human lineages coalesce to an ancestral lineage that lived in Africa about 130 thousand years ago. This date coincides with the end of the penultimate glacial period (190 to 130 thousand years ago). Populations were probably very small during this ice age. Expansion (bottleneck release) occurred during the last interglacial (130 to 71 thousand years ago), when warm climates and higher rainfall returned. Other lineages probably existed at that time, but they left no modern descendants.

Bottleneck 2. A severe bottleneck around 70,000 years ago may have reduced the effective population size in Africa to only 5,000 females. This date coincides with the super-eruption of Toba, a volcano located in northern Sumatra. Toba blasted over 800 cubic kilometers of volcanic ash and millions of tons of sulfur gas into the atmosphere. The volcanic ash settled relatively quickly, but the sulfur formed a long-lasting stratospheric haze that reflected sunlight and may have caused rapid global cooling. Annual layers of ice in the Greenland ice sheet suggest that this haze lasted six years, causing a "volcanic winter." This was followed by 1,000 years of the coldest temperatures of the last ice age. Analysis of air trapped in these ice layers suggests that temperatures dropped 16 °C over Greenland during this "instant ice age." Drought and famine during this cataclysmic event undoubtedly decimated populations in most parts of Africa.

Bottleneck 3. Analysis of Y chromosomes shows that all modern populations in southern Australasia can trace their ancestry to a small founding population from the Horn of northeast Africa (Ethiopia and Somalia) around 60,000 to 70,000 years ago. Increases in windblown dust in Greenland ice indicate a rapid drop in sea level to more than 100 meters lower than at present. This would have greatly facilitated dispersal from Africa to the Arabian Peninsula. Expansion around the perimeter of the Indian Ocean culminated in the colonization of Australia about 60,000 years ago.

Bottleneck 4. Analyses of gene sequences provide evidence of a possible second exodus from Africa by a small founding population that traveled over-

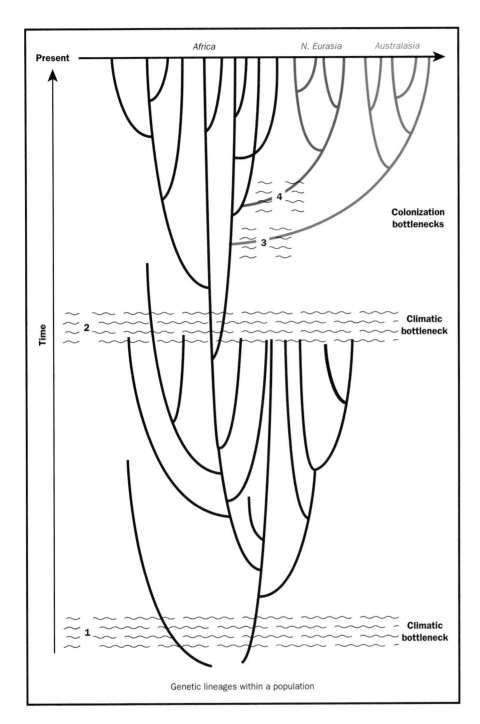

Figure 2. Differentiation of human genetic lineages within a population through time, and the effects of environmental and colonization bottlenecks (founder events) on survival of human lineages.

land via the shoreline of the Red Sea. This colonization bottleneck occurred during a period of milder climate about 50,000 years ago, and also coincides with the appearance of advanced stone tool technologies. Expansion continued into Europe and northern Asia. All living humans outside of Africa can thus trace their ancestry to these colonizing populations.

Technological and Social Influences on Past Population Size

Social and technological innovations in Africa during the later Middle Stone Age and early Later Stone Age (50,000 to 70,000 years ago) may

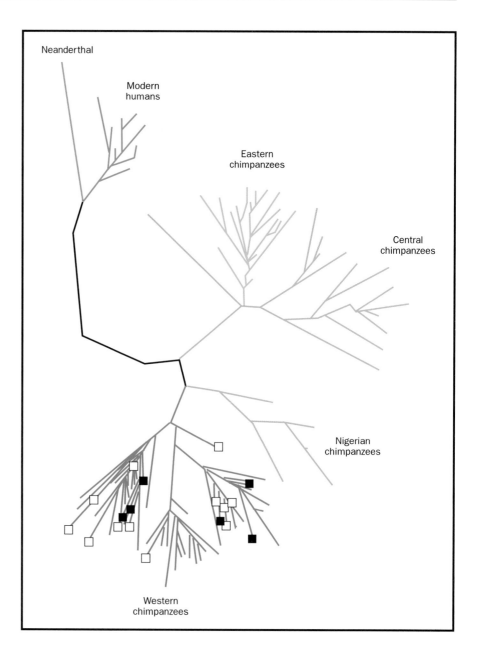

Figure 3. Unrooted family tree diagram of mitochondrial DNA control region sequences of a worldwide sample of 811 humans, and 332 chimpanzees from four regions of Africa. Branch length is proportional to the amount of genetic difference between individuals, and to the time since they shared a common ancestor. Symbols indicate individual chimpanzees belonging to the same social group. West African chimpanzees have three common ancestral lineages. Genetic differences between individuals in the same group substantially exceed differences between all living humans, and between humans and Neanderthals.

have facilitated population expansions and colonizations by enhancing survival in arid, unpredictable ice age environments. New stone tool technologies may have increased foraging efficiency and food supply. A system of mutual reliance and cooperation between distant foraging groups, mediated by reciprocal gift exchange, may have also increased humans' ability to survive in unpredictable environments. Further social and technological innovations may have facilitated population expansion within Africa, dispersals out of Africa, and the replacement of archaic populations, including Neanderthals, by anatomically modern humans outside of Africa.

Low levels of modern human diversity thus reflect our recent African ancestry and the effects of several population bottlenecks. In a similar fashion, colonization bottlenecks promoted rapid differentiation of northwestern Eurasians and southeastern Australasians. SEE ALSO CONSERVATION BIOLOGY:

GENETIC APPROACHES; FOUNDER EFFECT; HARDY-WEINBERG EQUILIBRIUM; MOLECULAR ANTHROPOLOGY; POPULATION GENETICS; Y CHROMOSOME.

Stanley Ambrose

Bibliography

Ambrose, Stanley H. "Late Pleistocene Human Population Bottlenecks, Volcanic Winter, and the Differentiation of Modern Humans." *Journal of Human Evolution* 34 (1998): 623–651.

Harpending, Henry, and Alan R. Rogers. "Genetic Perspectives on Human Origins and Differentiation." *Annual Review of Genomics and Human Genetics* 1 (2000): 361–385.

Harpending, Henry C., et al. "The Genetic Structure of Ancient Human Populations." *Current Anthropology* 34 (1993): 483–496.

Jorde, Lynn B., Michael Bamshad, and Alan R. Rogers. "Using Mitochondrial and Nuclear DNA Markers to Reconstruct Human Evolution." *BioEssays* 20 (1998): 126–136.

Ke, Yuehai, et al. "African Origin of Modern Humans in East Asia: A Tale of 12,000 Y Chromosomes." *Science* (2001): 1151–1153.

Lahr, Marta. *The Evolution of Modern Human Diversity.* Cambridge, U.K.: Cambridge University Press, 1996.

Underhill, Peter A., et al. "Y Chromosome Sequence Variation and the History of Human Populations." *Nature Genetics* 26 (2000): 358–361.

Population Genetics

Population genetics is the study of the genetic structure of populations, the frequencies of **alleles** and **genotypes**. A population is a local group of organisms of the same species that normally interbreed. Defining the limits of a population can be somewhat arbitrary if neighboring populations regularly interbreed. All the humans in a small town in the rural United States could be defined as a population, but what about the humans in a suburb of Los Angeles? They can interbreed directly with nearby populations, and, indirectly, with populations extending continuously north and south for a hundred or more miles. In addition, a large human population often consists of subpopulations that do not readily interbreed because of differences in education, income, and ethnicity. Despite these complexities, one can make some simple definitions.

alleles particular forms of genes

genotypes sets of genes present

Gene Pool and Genetic Structure

All of the alleles shared by all of the individuals in a population make up the population's gene pool. In diploid organisms such as humans, every gene is represented by two alleles. The pair of alleles may differ from one another, in which case it is said that the individual is "heterozygous" for that gene. If the two alleles are identical, it is said that the individual is "homozygous" for that gene. If every member of a population is homozygous for the same allele, the allele is said to be fixed. Most human genes are fixed and help define humans as a species.

The most interesting genes to geneticists are those represented by more than one allele. Population genetics looks at how common an allele is in the whole population and how it is distributed. Imagine, for example, an allele "*b*" that when homozygous, "*bb*," produces blue-eyed individuals. Allele *b*

A caribou herd is a population: a group of organisms of the same species that inhabit an area and interbreed. Individuals may migrate between populations, causing gene flow.

might have an overall frequency in the population of 20 percent; that is, 20 percent of all the eye-color alleles are *b*.

However, not everyone who has the *b* allele will be homozygous for *b*. Some people will have *b* combined with another allele, "*B*," which gives them brown eyes (because *B* is dominant and *b* is recessive). Others won't have the *b* allele at all and instead will be homozygous for *B*.

The frequency of each genotype—whether *bb*, *Bb*, or *BB*—in the population is also of interest to population geneticists. The frequency of alleles and genotypes is called a population's genetic structure. Populations vary in their genetic structure. For example, the same allele may have a frequency of 3 percent among Europeans, 10 percent among Asians, and 94 percent among Africans. Blood types vary across different ethnic groups in this way. The frequency of genotypes depends partly on the overall allele frequencies, but also on other factors.

Hardy-Weinberg Theorem

Large, isolated populations whose members mate randomly and do not experience any selection pressure will tend to maintain a frequency of genotypes predicted by a simple equation called the Hardy-Weinberg Theorem. For example, if *b* has a frequency of 20 percent and *B* has a frequency of 80 percent, we can predict the frequency of the three genotypes (*bb*, *Bb*, and *BB*). The total of all the genotype frequencies is 100 percent ($b + B$), and the frequencies of each are given by $(b + B)^2 = 100$ percent. This can be restated as the following equation:

$$100\% = b^2 + 2(bB + B^2).$$

And we can calculate the genotype frequencies as:

$$100\% = (20\%)^2 + 2(20\% \times 80\%) + (80\%)^2 = 4\% + 32\% + 64\%.$$

So even though 20 percent of all the genes in this imaginary population are *b* alleles, only 4 percent of the population is homozygous for *b* and actually

has blue eyes. Furthermore, this same distribution will be maintained over time, as long as the conditions of the Hardy-Weinberg Theorem are met.

However, few, if any, natural populations (including human ones) actually conform to the assumptions of Hardy-Weinberg, so both genotype frequencies and allele frequencies can and do change from generation to generation. For example, humans do not mate randomly. Instead, they tend to take partners of similar height and intelligence. And even in modern human populations, genetic diseases such as Tay-Sachs kill children long before they grow up and reproduce. A difference in survival and reproduction due to differences in genotype is called selection. Even subtle selection can change gene frequencies over long periods of time.

Another assumption of the Hardy-Weinberg theorem is that individuals from different populations do not mate, so that gene flow, the passage of new genetic information from one gene pool into another, is zero. Such isolation does characterize many animal and plant populations, but almost no modern human populations are isolated from all other populations. Instead, humans travel to different countries, inter-

Trains and buses ferry more than seven million commuters to and from Bombay every day. Demographers predict that the population of Bombay and its surrounding regions will exceed twenty-seven million by 2015.

marrying and producing children who reflect the novel intermingling of unusual alleles.

Genetic Drift

In very small populations, rare alleles can become common or disappear because of genetic drift—random changes in gene frequency that are not due to selection, gene mutation, or immigration. We can explain this as follows. When flipping a coin 1,000 times, it is likely to get 50 percent heads and 50 percent tails (if it's a fair coin). But flip it only five or ten times, and it is unlikely to get exactly half heads and half tails. Chances are good that the results will be something quite different. In the same way, if 10,000 people mate and produce children, the *bb* genotype will pretty much conform to the Hardy-Weinberg equation described above (provided the other assumptions are approximately true). For example, in a sample of just twenty people, instead of getting a group of children of whom 4 percent have blue eyes, the result could end up none with blue eyes, or maybe half having blue eyes. It all depends on how the alleles happen to combine when eggs meet sperm.

Because of genetic drift, small, isolated populations often have unusual frequencies of a few alleles. Although similar to other people in most important respects, such isolated populations may harbor high frequencies of one or more alleles that are rare in most other populations. For example, in 1814, fifteen people founded a British colony on a group of small islands in the mid-Atlantic, called Tristan de Cunha. They brought with them a rare recessive allele that causes progressive blindness, and the disease, extraordinarily rare in most places, is common on Tristan de Cunha. Such "inbreeding" produces more homozygotes than usual and increases the probability of children born with genetic diseases. The Old Order Amish have a high frequency of Ellis-van Creveld syndrome, and Ashkenazi Jews were, until a few years ago, susceptible to Tay-Sachs disease. Fortunately, genetic testing has greatly reduced the incidence of Tay-Sachs and many other such genetic diseases.

Population genetics also provides information about evolution. It is known, for example, that populations that have unusual allele frequencies must have been isolated from other populations. And we can surmise that populations that share similar frequencies of certain rare alleles may have interbred at some point in the past. Human populations in sub-Saharan Africa show the greatest diversity of all human populations. On the basis, in part, of this diversity, one theory of human evolution suggests that all humans originated in Africa, and then emigrated to Asia, Europe, and the rest of the world. SEE ALSO FOUNDER EFFECT; GENETIC DRIFT; HARDY-WEINBERG EQUILIBRIUM; MOLECULAR ANTHROPOLOGY; POPULATION BOTTLE-NECK; SELECTION; TAY-SACHS DISEASE.

Jennie Dusheck

Bibliography

Jones, J. S. "How Different Are Human Races?" *Nature* 293 (1981): 188–190.

Klug, W. S., and M. R. Cummings. *Concepts of Genetics*, 6th ed. Upper Saddle River, NJ: Prentice Hall, 2000.

Lewontin, R. *Human Diversity*. Redding, CT: W. H. Freeman, 1982.

Population Screening

As scientific research reveals more information about treating diseases and maintaining good health, it has become increasingly important to identify diseases in their early stages in order to treat them most effectively. Thus, researchers have developed tests for some diseases to identify people at high risk for the disease before the symptoms of the disease actually appear. These tests are routinely administered to individuals in a defined population who have no apparent symptoms of the disease being screened. This process is called population screening. A primary goal of population screening is to predict with high accuracy which individuals in this group are at significant risk of developing or transmitting a disease. Once individuals at high risk for a disease are identified, confirmatory (diagnostic) tests are then performed to detect the screened-for disease with greater certainty.

Screening versus Diagnostic Tests

Examples of routine population screening currently used in the health care field include Pap smears for women to predict their risk for cervical cancer, mammograms for women to predict their risk for breast cancer, the PPD skin test to predict exposure to tuberculosis mycobacterium (TB) in health care workers, and the prostatic antigen screening (PSA) test for men to predict their risk for prostate cancer.

However, screening tests have limits. A screening test only indicates who, in a given population, is most likely to be at higher risk for developing a disease. As a result, screening tests will have both false positives and false negatives. A false positive occurs when a test misidentifies individuals as being higher risk, when they are actually not at higher risk. A false negative occurs when individuals with a higher risk for the disorder are not identified by the screening test. For this reason, a diagnostic test is done following a positive screening test. A diagnostic test determines with relative certainty whether an individual has a disorder and thus rules out false positives. Diagnostic tests are typically more expensive and/or more invasive and therefore cannot be used as part of the screening process.

Criteria for a Screening Program

Despite limitations, population screening can still be very advantageous for improving public health. However, for population screening to be beneficial, a disease should meet several conditions. There should be a high rate of occurrence of the disorder or a significant number of individuals who carry the disease in the high-risk population; the disorder could potentially have a harmful impact on an individual's health, if not identified and treated; the disorder must be treated or prevented though not necessarily cured; testing should be minimally invasive, easily carried out, and relatively inexpensive; and the testing method must have been studied and scientifically demonstrated to be accurate, reliable, and confirmed by follow-up testing. In addition, testing should be performed only after informed consent is obtained from the individual or responsible party.

Screening for Inherited Disorders

The disorders that are screened for, above, may or may not be genetic. However, with the numerous advances in genetic research, public health con-

cerns have shifted to include the growing number of recognized inherited disorders. Screening for inherited disorders began in the early 1960s, with testing for phenylketonuria (PKU), a metabolic disease that causes severe mental deficiency. Since that time, developments such as the completion of the Human Genome Project (HGP) has resulted in an increase in the number of genetic screening tests available. Types of population screening for inherited disorders can include newborn screening and carrier screening of individuals within populations known to be at high risk for certain inherited disorders.

Newborn Screening. In the United States, every state requires that certain screening tests be done on all newborn infants. Interestingly, individual states vary significantly as to which screening tests they require. For instance, while it is mandatory to screen for PKU in every state, other relatively common inherited disorders, such as medium-chain acyl-CoA dehydrogenase deficiency and congenital adrenal hyperplasia, are only screened for in selected states. Furthermore, some private organizations offer an expanded selection of testing for genetic disorders in newborns beyond what individual states mandate.

Newborn screening is the only population-based type of screening for inherited disorders. One public health benefit of population-based screening means that everyone is tested. This is especially useful for studying inherited disorders, since it permits scientists to determine with great accuracy how frequently some inherited disorders occur in the general population.

In simple terms, there are two types of testing done for newborn screening: non-DNA-based testing and DNA-based testing. With DNA-based testing, an individual's DNA is tested directly. With non-DNA-based testing, two indirect methods are used: Enzymatic or electrophoretic testing methods are used to figure out whether an individual has an inherited disorder.

The most well-known newborn screening test is for PKU. PKU is an **autosomal**, recessively inherited metabolic disorder characterized by a lack or defect of phenylalanine hydroxylase, an **enzyme** involved in the metabolism of an amino acid called phenylalanine. Because this enzyme does not function properly, persons with PKU have a buildup of phenylalanine in their bodies, which results in severe mental retardation if untreated. Treatment currently involves dietary restriction of phenylalanine. Because phenylalanine is an essential amino acid (meaning that our bodies cannot make it, and it is essential for life), people with PKU cannot eliminate the substance entirely from their diet. Instead they must take care to modulate the amount they consume, because they cannot metabolize extra phenylalanine. Therefore, persons with PKU are usually under the care of a dietitian, who helps them to eat a balanced diet with the right amount of phenylalanine. Most importantly, by following the proper diet, individuals with PKU will not develop mental retardation.

Screening for PKU is done using the Guthrie test, a bacterial inhibition assay, a non-DNA-based (enzymatic) laboratory test. Another example of a non-DNA-based test is hemoglobin electrophoresis, which is done to determine the types of hemoglobin an individual carries. This information is a direct reflection of an individual's genetic makeup. Some other examples of disorders screened for, using non-DNA-based tests, include hypothy-

autosomal describes a chromosome other than the X and Y sex-determining chromosomes

enzyme a protein that controls a reaction in a cell

roidism, congenital adrenal hyperplasia and hemoglobinopathies (such as sickle cell disease and thalassemias). In contrast, DNA-based genetic testing is a relatively new clinical tool in newborn screening. In at least one state, DNA-based newborn screening is done for cystic fibrosis.

Carrier Testing. In contrast to newborn screening, which is done on all newborns regardless of family history or ethnicity, carrier testing is aimed at a specific population that is viewed to be at high risk for a given disorder. In addition, carrier testing is always DNA-based, whereas newborn screening is typically non-DNA-based. Carrier testing is offered to determine if individuals carry a single non-working copy of a gene for a genetic disorder.

In general, individuals who carry one copy of a non-working gene will not have symptoms or signs of the disorder. When two individuals who are carriers for the same inherited disorder have a child, that child is at a 25 percent risk of inheriting a copy of the non-working gene from each parent, and will usually show symptoms and signs of the inherited disorder. Diseases that may be inherited in this way are called autosomal recessive disorders, and include sickle cell disease, Tay-Sachs disease, and cystic fibrosis.

In the late 1990s, a panel convened at the National Institutes of Health and recommended that carrier screening for cystic fibrosis be offered to a number of populations, including adults with a family history of the disease, the partners of people who had the disease, couples who were planning a pregnancy, and couples seeking prenatal testing. Newborn screening and general population screening, however, were not recommended.

Ethical Considerations

Before carrier screening can be offered to a high-risk population, the population must be educated about the disorder being screened, the basic **tenets** of carrier screening, and the potential benefits and risks of carrier screening. Screening for inherited disorders raises many complex issues. For example, screening raises a number of psychosocial issues, such as how an individual's self-esteem might be affected if he or she was found to carry a non-working gene). There are also implications for nonscreened family members if an individual is identified as a carrier. For instance, how will other family members be notified that they are also at risk for being a carrier. In addition, there is the potential for discrimination; other people may inaccurately infer that an individual who is a carrier has the inherited disorder.

tenets generally accepted beliefs

Finally, and most importantly, no discussion of any type of genetic testing is complete without raising the topic of eugenics and the atrocities of the past that were associated with the abuse of genetic information. Today, each individual has the right to choose whether or not they want to know if they carry genes that predispose them to an inherited disorder. No one should be forced to learn about their carrier status, even by another family member. Moreover, the goal of any type of an inherited disorders screening program is not to eliminate genes that cause disease. Eliminating disease-causing genes would mean eliminating the human race. It is estimated that every human carries at least three to four genes that are associated with inherited diseases. Furthermore, there is a presumed constant rate of change (mutation) in human genetic material. These changes ensure variation amongst individuals. However, variation in human DNA occurs randomly, with some changes

leading to beneficial effects, and some to detrimental effects like an inherited disorder. Therefore, genetic disorders will continue to occur and be a part of human genetic makeup. Although general population screening and population-targeted carrier screening raise many complex issues, they will ultimately allow society to better prepare for working with these disorders. SEE ALSO CYSTIC FIBROSIS; EUGENICS; GENETIC COUNSELING; GENETIC COUNSELOR; GENETIC DISCRIMINATION; GENETIC TESTING; GENETIC TESTING: ETHICAL ISSUES; HEMOGLOBINOPATHIES; METABOLIC DISEASES; TAY-SACHS DISEASE.

Chantelle Wolpert

Bibliography

Levy H. L., and S. Alber. "Genetic Screening of Newborns." *Annual Review of Genomics and Human Genetics* 1 (2000): 139–177.

Internet Resources

American College of Medical Genetics. "Principles of Screening: Report of the Subcommittee of the American College of Medical Genetics Clinical Practice Committee." 1997. Policy Statement. <http://www.faseb.org/genetics/acmg/>.

Hall B., and S. Durham. "NIH Consensus Panel Makes Recommendations for Offering Genetic Testing for Cystic Fibrosis." NIH news release: April 16, 1997. <http://odp.od.nih.gov/consensus/cons/106/106_intro.htm>.

Post-translational Control

Post-translational control can be defined as the mechanisms by which protein structure can be altered after translation. Proteins are polymers of amino acids, and there are twenty different amino acids. Both the order and identity of these amino acids are important for the role that the protein plays in the cell. In some cases, the chemical identity of these amino acids is changed after translation. Alternatively, the sequence or number of the amino acids in a protein can be altered. These changes can alter the structure or function of the protein, or they can target it for destruction.

Alterations of Amino Acids

Post-translational control of protein function or structure can be accomplished by chemical alteration of an amino acid side chain or by modification of the ends of the protein backbone. While there are many diverse chemical modifications of amino acids, three common examples are phosphorylation, glycosylation, and ubiquitination.

Phosphorylation involves the addition of phosphate to an amino acid side chain, usually to the side chain hydroxyl (–OH) of serine, threonine, or tyrosine. This modification results from the action of a protein known as a **kinase** and uses **ATP** as the source of phosphate. The phosphate can be removed by another enzyme, a phosphatase. Phosphorylation can alter protein function and is relevant in cellular signaling pathways. Aberrant phosphorylation can lead to disruption of the cell cycle and the induction of cancer.

Glycosylation involves the addition of one or more sugar **monomers** to the side chains of amino acids, either during or after translation, to make a **glycoprotein**. Sugars are attached either to the side-chain nitrogen of the

kinase an enzyme that adds a phosphate group to another molecule, usually a protein

ATP adenosine triphosphate, a high-energy compound used to power cell processes

monomers "single parts"; monomers are joined to form a polymer

glycoprotein protein to which sugars are attached

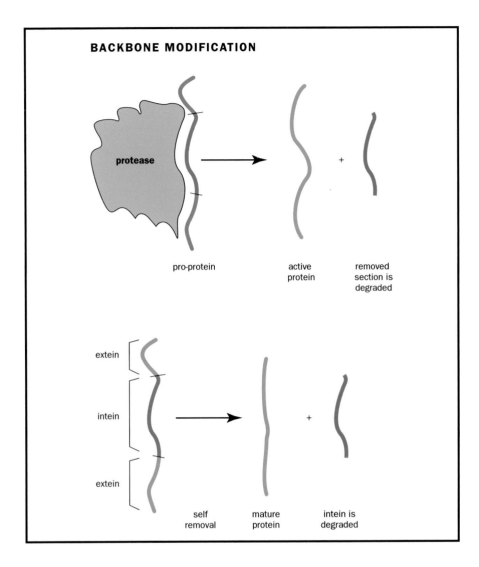

BACKBONE MODIFICATION

protease

pro-protein

active
protein

removed
section is
degraded

extein

intein

extein

self
removal

mature
protein

intein is
degraded

Removal of some
portions of the
translateral polypeptide
may be needed to form a
functional protein.

amino acid asparagine or to the hydroxyl of the amino acids serine or threonine. The structure of these carbohydrates can be complex and variable and often does not affect the function of proteins directly. However, glycosylation can affect the protein's solubility, its targeting to a particular part of the cell, its folding into a three-dimensional structure, its lifetime before it is degraded, and its interaction with other proteins.

The addition of ubiquitin (a protein composed of seventy-six amino acids) to another protein can render the target protein susceptible to degradation by the 26S proteosome, which is a large protease (a protein that cleaves other proteins). Ubiquitin is ubiquitous in the cell (hence its name) and varies little between organisms as diverse as yeast and humans. It is attached via the side chain of the amino acid lysine, and often additional ubiquitin proteins are added to the first to make a chain.

Alteration of the Polypeptide Backbone

Control of protein function by post-translational modification may also occur by altering the order of the amino acids in the protein backbone. These modifications may be promoted by other proteins or they may be self-directed.

Three mechanisms of post-translational control.

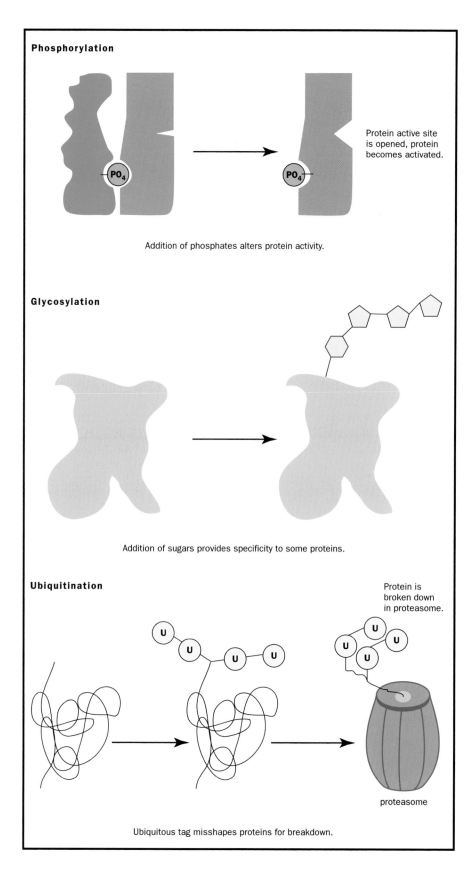

Phosphorylation

Protein active site is opened, protein becomes activated.

Addition of phosphates alters protein activity.

Glycosylation

Addition of sugars provides specificity to some proteins.

Ubiquitination

Protein is broken down in proteasome.

proteasome

Ubiquitous tag misshapes proteins for breakdown.

Certain proteins are synthesized as larger precursor proteins and are activated by cleavage of their peptide backbone by proteases. Many of these large protein precursors, called zymogens or proproteins, are synthesized with an N-terminal signal sequence that instructs the cell to export the protein. (Proteins have an amino group at one end and a carboxyl group at the other. The amino-group end is called the N-terminus, the carboxyl end is called the C-terminus.) The N-terminal signal sequences are then cleaved, but the exported protein may still be inactive until cleaved again by another protease. Proteins activated by this mechanism include digestive proteases such as trypsin, the activity of which must be controlled before export by the cell. Serum albumin is also processed in this manner, as are the peptide hormones insulin, vasopressin, and oxytocin. Post-translational cleavage is also responsible for controlling the process of blood clotting.

Inteins

Proteins can also direct the rearrangement of their own polypeptide backbones. For instance, proteins called inteins facilitate a process known as protein splicing. Inteins interrupt the amino acid sequence, and probably the function, of other proteins. Examples include an intein in yeast that interrupts an ATPase, one in a mycobacteria that interrupts the RecA protein (which is involved in DNA repair and recombination), and one in a pyrococcus species that interrupts a DNA polymerase.

Inteins promote their own excision from their target protein as well as the ligation of the flanking protein segments, which are called exteins. It is possible that some inteins play a role in regulating gene expression, but it is also possible that they are vestiges of ancient control mechanisms or simply molecular parasites.

Inteins are analogous to introns in an RNA transcript. Introns interrupt a gene in DNA. The introns are excised after transcription from the surrounding RNA sequences. These flanking sequences, called exons, are then spliced together. Unlike protein splicing, RNA splicing is usually aided by other proteins and RNA molecules.

While it is unclear if protein splicing has a regulatory role, a class of proteins evolutionarily related to inteins, the hedgehog proteins, are involved in embryonic patterning and segmentation. Hedgehog proteins promote their own internal cleavage, coupled to the addition of cholesterol to the C-terminal end of the N-terminal fragment, which is probably important for anchoring the hedgehog protein in the cell membrane. SEE ALSO RNA PROCESSING; SIGNAL TRANSDUCTION; TRANSCRIPTION.

Kenneth V. Mills

Bibliography

Creighton, Thomas E. *Proteins: Structure and Molecular Properties.* New York: W. H. Freeman, 1993.

Neurath, Hans, and Kenneth A. Walsh. "Role of Proteolytic Enzymes in Biological Regulation (A Review)." *Proceedings of the National Academy of Sciences (USA)* 73 (1976): 3825–3832.

Paulus, Henry. "Protein Splicing and Related Forms of Protein Autoprocessing." *Annual Review of Biochemistry* 69 (2000): 447–496.

Schwartz, Alan L., and Aaron Ciechanover. "The Ubiquitin-Proteasome Pathway and Pathogenesis of Human Diseases." *Annual Review of Medicine* 50 (1999): 57–74.

Stryer, Lubert. *Biochemistry*. New York: W. H. Freeman, 1995.

Uy, Rosa, and Finn Wold. "Posttranslational Covalent Modification of Proteins." *Science* 198 (1977): 890–896.

Progeria *See Accelerated Aging: Progeria*

Prenatal Diagnosis

The future health of a new individual can be predicted, to an extent, from clues that are apparent before birth. Prenatal diagnosis is the identification of a medical condition in a developing embryo or fetus. Prenatal testing can sample fetal cells to examine DNA sequences that correspond to specific disease-causing genes or chromosomes (the structures that carry the genes). Biochemicals obtained with the fetal cell samples can also hold clues to health. Other prenatal tests analyze a pregnant woman's blood serum for telltale biochemicals that indicate the fetus faces an elevated risk of a particular condition. Ultrasound scans provide views of many aspects of fetal anatomy. Preimplantation genetic diagnosis is a technique that is actually pre-prenatal. It provides a health check on very early embryos grown in a laboratory dish, enabling parents to select those that are most likely to develop into healthy infants.

Because prenatal tests that sample fetal cells are invasive, they carry a risk of the test causing miscarriage. Therefore, these procedures are typically offered only to those pregnant women whose risk of carrying a fetus with a detectable condition is greater than the risk of miscarriage. Reasons include already having had a child or family history with a detectable genetic or chromosomal condition, or "advanced maternal age." After age thirty-five, a woman's risk of carrying a fetus with an extra or missing chromosome exceeds the risk that the procedure will cause miscarriage.

Viewing Chromosomes

Biologists first tried to visualize the chromosomes in a human cell in the late nineteenth century, with estimates of the total number ranging from 30 to 80. As methods to untangle and stain chromosomes improved, the count narrowed to 46 or 48, and by 1956 was confirmed as 46, or 23 pairs. By 1959, the first chromosomal abnormalities were identified using size and crude staining patterns to distinguish the chromosomes. In the 1970s, vastly improved staining techniques enabled **cytogeneticists** to much more easily distinguish chromosomes, and they began amassing databases of specific chromosomal abnormalities and the clinical syndromes that they cause.

cytogeneticists scientists who study chromosome structure and behavior

Also in the 1970s, general staining began to be replaced with *in situ* hybridization, an approach that links a radioactive molecule to a short sequence of DNA called a DNA probe, chosen to match a known gene of interest. When the DNA probe binds to its complementary sequence among a sample of chromosomes spread against a piece of photographic film, the radioactivity exposes the film exactly where the probed DNA sequence resides. In the 1990s, fluorescent molecules replaced the radioactive tags, and a procedure called fluorescence *in situ* hybridization (FISH) was born. A flash of light matches probe to chromosome. Today, FISH can use combinations of fluorescent labels and computer analysis to individually label

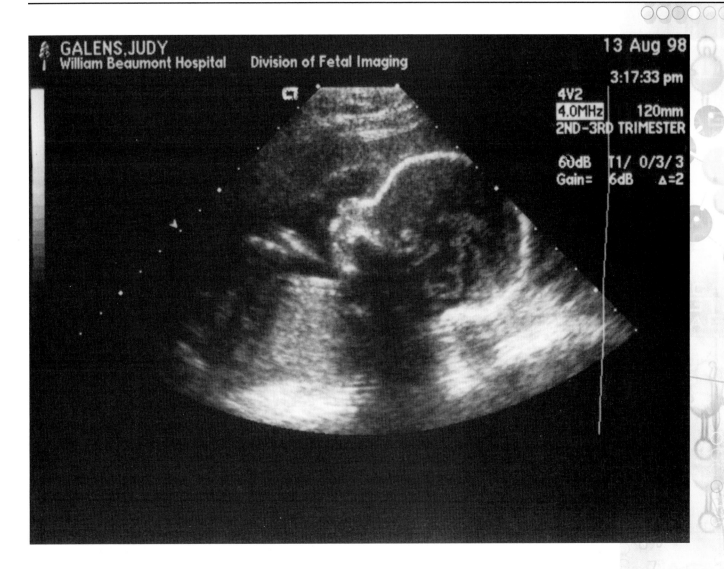

each chromosome. The technique is called chromosome painting or spectral karyotyping. A karyotype is a picture of a person's chromosomes displayed in size-ordered pairs. FISH can be used to highlight chromosomes obtained by amniocentesis, CVS, or fetal cell sorting, described next.

Viewing fetal chromosomes requires obtaining cells from the fetus. The most common procedure is amniocentesis, first successfully performed in 1966. In amniocentesis, a needle is used to remove a sample of the amniotic fluid that surrounds the fetus. This is usually done after the fifteenth week of pregnancy. The fluid sample contains skin cells that the fetus has shed, and these are analyzed for their chromosomal content. Results from amniocentesis typically are available within two weeks. FISH is not routinely offered, but in the labs that do offer it, some preliminary information may be available more quickly than is possible with other testing procedures.

Aberrations of chromosomes 13, 18, 21, X, and Y are seen most commonly. This is not necessarily because they are affected more often, but because problems in other **autosomes** are so severe that development ceases long before prenatal testing can be done.

Biochemicals in the amniotic fluid can also be analyzed for signs of metabolic disorders, though this procedure is not commonly performed unless

Ultrasound monitors fetal development late in the pregnancy. This image was taken late in the second trimester.

autosomes chromosomes that are not sex-determining (not X or Y)

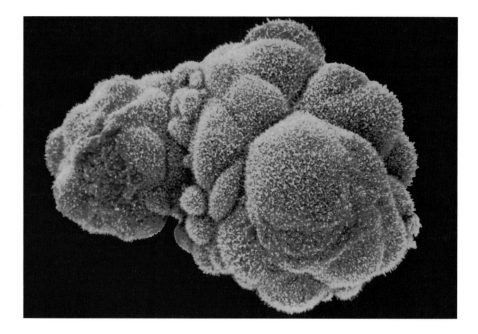

A human embryo at the blastula stage (day 6). This embryo could have been tested earlier in its formation (by in vitro fertilization) for various birth defects before being implanted into the female to develop.

there is already a suspicion that one may be present. Chemical markers may also be sought for neural tube defects (NTDs), which are abnormalities in brain or fetal spinal cord development. Risk of amniocentesis causing miscarriage is about 1 in 200.

Chorionic villus sampling (CVS) can be performed earlier than FISH, from the tenth to twelfth week of pregnancy. A physician removes a small sample of the chorionic villi, reached either through the vagina or the abdominal wall. The chorionic villi are fingerlike projections of cells that form part of the placenta, which provides nutrients to the developing fetus. Because the chorionic villi originate from the fertilized ovum, their chromosomes and genes should be the same as those in fetal cells. However, in practice, sometimes a mutation affects only the chorionic villi, leading to a false positive test result, or only the fetus, leading to a false negative result. Maternal cells may also contaminate the sample. Because of these uncertainties, follow-up testing such as amniocentesis is required for clarification.

CVS has been linked to a fatal limb defect, and carries a risk of miscarriage of about 1 percent. It is typically recommended for women over the age of thirty-five, for those who have already had a child with a detectable genetic or chromosomal defect, if there is a family history of a genetic disorder, or when abnormalities are detected by ultrasound. For example, CVS is often used if there is a family history of Duchenne muscular dystrophy or Tay-Sachs disease. Unlike amniocentesis, CVS cannot detect NTDs because it does not sample biochemicals. Its advantage is that it can be performed earlier in the pregnancy.

A third technique, called fetal cell sorting, is being studied and may eventually replace amniocentesis and CVS in obtaining fetal cells. This approach isolates the rare fetal cells that enter the mother's blood stream and analyzes them for gene and chromosome abnormalities. A device called a fluorescence-activated cell sorter detects and isolates fetal cells by their different surface features compared to cells from the pregnant woman.

Because fetal cell sorting requires only a blood sample from the pregnant woman, it cannot endanger the fetus.

Less Invasive Methods

An ultrasound scan bounces soundwaves off of the fetus to create an image. A scan is often performed after the sixteenth week of pregnancy, and the anatomy and size of the fetus is measured to see if it is growing and developing normally. The scan can often detect major structural problems, such as a malformed heart or spine. An unusual finding on an ultrasound scan can be a warning to investigate further. However, not all birth defects can be detected by ultrasound.

An ultrasound is sometimes done at weeks five or six to confirm that a pregnancy is present. This early, the embryo looks like a lima bean with a pulsating blip in the middle, which is the beating heart. Ultrasound performed late in pregnancy can provide clues to the approaching birth date. New three-dimensional ultrasound scans offer spectacular views of the fetus.

Another noninvasive method to detect fetuses at risk for some birth defects is maternal serum marker screening. A sample of blood from a pregnant woman taken at approximately weeks 15 to 18 is analyzed for the amount of several substances, including alpha fetoprotein (AFP); a form of estrogen called unconjugated estriol; and human chorionic gonadotropin (hCG), a hormone produced only during pregnancy.

Maternal serum screening began in the 1970s with the AFP test, invented by a man whose son was born with a neural tube defect. High levels of AFP in a woman's blood indicate an increased risk for a neural tube defect in the fetus. The neural tube forms by approximately day 28 of gestation, when a portion of the flat embryo (the neural plate) folds to form a tube that will develop into the brain and spinal cord. The tube normally closes up like a zipper starting at several points along its length. If a hole remains, the brain and spinal cord underneath are exposed, causing damage.

Several years after the AFP test was developed for neural tube defects, researchers noted that low AFP correlates to an increased risk that a fetus will have an extra chromosome, particularly at positions 18 or 21. This condition is called a trisomy. Trisomy 21, an extra chromosome 21, is the most common cause of Down syndrome. Over the years, analysis of other substances have been added to refine this test, which is now offered routinely to pregnant women. Abnormal results on maternal serum screening tests indicate that amniocentesis should be done to diagnose a neural tube or chromosome defect, and that genetic counseling should be offered.

Preimplantation Genetic Diagnosis

Amniocentesis, CVS, and maternal serum screening are performed after a pregnancy is confirmed or in progress. In contrast, preimplantation genetic diagnosis (PGD) occurs before the embryo implants in the womb. This technique is performed on an embryo that has been derived from in vitro fertilization (IVF) and is growing in a laboratory dish. At about the 8-cell (day 3) stage, a cell is removed and the DNA and chromosomes are checked using FISH or a probe for a specific gene. If the cell is free of the defects being probed, the remaining 7-celled embryo is implanted into the woman, where it continues development.

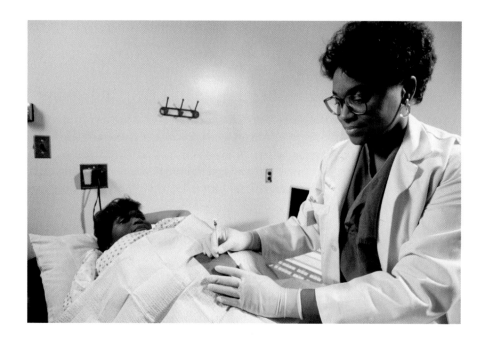

A doctor at Baltimore University Hospital has inserted the needle that will extract amniotic fluid from this patient's placenta. The fluid obtained from the amniocentesis will be analyzed to detect genetic disorders.

The first PGD was done in 1989, when it was used to enable families with X-linked disorders to select a girl, who would not be affected by the condition. Then it was used to conceive Chloe O'Brien, a youngster free of the cystic fibrosis that affected her brother. PGD attracted widespread public attention in 2001, when a Minnesota couple, Lisa and Jack Nash, conceived their son Adam so that his umbilical cord stem cells could be used to cure his sister Molly's Fanconi anemia. Adam not only had not inherited Fanconi anemia, but he was also a tissue match for Molly, saving her life.

PGD has been used to eliminate embryos with a variety of single-gene disorders, including metabolic disorders, dwarfism, cystic fibrosis, hemophilia, muscular dystrophies, and several other genetically inheritable diseases. The technique is being increasingly used in couples for whom IVF has repeatedly failed because they manufacture eggs or sperm that have abnormal numbers of one or more chromosomes. PGD enables physicians to sort through embryos to identify and transfer those few that have normal chromosomes. PGD has about a 66 percent success rate for identification of genetic disorders.

Genetic Counseling and the Ethics of Prenatal Diagnosis

A genetic counselor helps educate individuals, couples, and families about prenatal tests, and helps them to understand and cope with the results. The couselor also informs the prospective parents of the limitations of the tests, explaining that they can rule out certain conditions but cannot guarantee a healthy baby.

Ethical issues can arise in the decision to undergo prenatal testing. For example, the Nash family received criticism for their decision to intentionally conceive one child to save another. Some people also question the use of prenatal tests or PGD to reject embryos because of a gene that causes an adult-onset disease, such as Alzheimer's disease. In a more general sense, picking and choosing offspring based on genes can be considered eugenic, with the caveat that the intent is not to improve the gene pool, but to prevent suffering. This may mean terminating a pregnancy in which the fetus has a very bleak prognosis, which people opposed to abortion might find unethical. Opponents to this view point out that "letting nature take its course" can be painful for the fetus and may endanger the life of the woman.

The ethics of prenatal diagnosis becomes more complicated when the goal is not to prevent suffering, but to choose a child of a particular sex. Doctors have long reported patients using CVS or amniocentesis to learn the sex of the fetus, then terminating the pregnancy if the outcome is not what is desired. PGD is sometimes used for this purpose, too. Some people have compared this practice to a high-tech version of the ancient practice of leaving girl babies outside city walls to perish. The American Society for Reproductive Medicine endorses the use of PGD for sex selection to avoid passing on an X-linked disease, but discourages use for family planning as "inappropriate use and allocation of medical resources."

The ethical concerns that arose with the ability to foretell the sex of a child are certain to mushroom as data from the Human Genome Project continue to lengthen the list of disorders that can be detected before birth. Physicians and parents-to-be in the future will have to decide just how much they want to know about their offspring and how they will use that information. SEE ALSO CHROMOSOMAL ABERRATIONS; DOWN SYNDROME; EUGENICS; GENETIC COUNSELING; GENETIC COUNSELOR; GENETIC DISEASE; GENETIC TESTING; *IN SITU* HYBRIDIZATION; MOSAICISM; REPRODUCTIVE TECHNOLOGY: ETHICAL ISSUES; TAY-SACHS DISEASE.

Ricki Lewis

Bibliography

Ethics Committee of the American Society for Reproductive Medicine. "Preconception Gender Selection for Nonmedical Reasons." *Fertility and Sterility* 75, no. 5 (May 2001): 861–864.

Gottlieb, Scott. "Scientists Screen Embryo for Genetic Predisposition to Cancer." *British Medical Journal* 322 (June 23, 2001): 1505.

Josefson, D. "Couple Selects Healthy Embryo to Provide Stem Cells for Sister." *British Medical Journal* 321 (October 14, 2000): 917.

Lewis, Ricki. "Preimplantation Genetic Diagnosis: The Next Big Thing?" *Scientist* 14, no. 22 (November 13, 2000): 16.

Prion

In 1997 Stanley Prusiner was awarded the Nobel Prize in physiology or medicine for a revolutionary theory about the mechanisms of infection. His theory, the "prion hypothesis," concerns an unusual protein, the prion, which occurs in the complete absence of DNA and RNA. According to Prusiner's theory, the prion differs from other well-known infections agents including bacteria and viruses. While the latter rely on **nucleic acid** for survival and replication, the prion is made of a protein and lacks nucleic acid. Both the existence of the prion and the underlying mode of infection are unprecedented in medical sciences. While several critical issues remain to be addressed, the prion hypothesis may furnish a plausible framework to understand the **pathogenesis** of several deadly brain diseases of the central nervous system.

nucleic acid DNA or RNA

pathogenesis pathway leading to disease

Stanley Prusiner won the 1997 Nobel prize in physiology or medicine for his research on prions.

scrapie prion disease of sheep and goats

glycoprotein protein to which sugars are attached

polypeptide chain of amino acids

enzymes proteins that control a reaction in a cell

aggregate stick together

A New Infectious Agent

Prion is an acronym for "proteinaceous infectious particle," a term coined by Prusiner in the early 1980s to describe the nature of the agent causing the fatal brain disorders known as transmissible spongiform encephalopathies (TSE), also called prion diseases. Well-known examples of prion diseases include scrapie in sheep and goats, bovine spongiform encephalopathy (BSE, or "mad cow" disease) in cattle, and Creutzfeldt-Jakob disease (CJD) in humans. Prion diseases are infectious and can also be transmitted to healthy animals by inoculating them with extracts of diseased brain.

In the mid-1960s, Tikvah Alper and colleagues reported that nucleic acid was unlikely to be a component of the infectious agent that causes **scrapie**. In 1967 J. S. Griffith speculated that the scrapie agent might be a protein capable of "self replication" without nucleic acid. However, Prusiner was the first, in the early 1980s, to successfully purify the infectious agent and to show that it consisted mostly of protein (technically speaking it is a **glycoprotein**, because it has a sugar group attached). He chose to name the new agent "prion" to distinguish it from viruses or viroids.

The essential protein component of prion was later identified in 1984 as prion protein (PrP), which is encoded by a chromosome gene in the host genome. Researchers concluded that the prion is a new infectious agent that consists mostly of PrP. This view is often referred to as the "protein only" or prion hypothesis. Some scientists find this notion hard to accept and have argued that nucleic acid is needed to carry information necessary for infection. However, no one has been able to demonstrate that either DNA or RNA play a direct role in prion replication.

In 1992 Charles Weissmann and colleagues obtained conclusive evidence for the central role of PrP in the transmission of prion diseases, when they created transgenic mice devoid of the PrP gene. These so-called PrP knockout mice were found to be completely resistant to infection when inoculated with scrapie brain preparations. When the PrP gene was reintroduced into the knockout mice, they once again became susceptible to prion infection.

Role of Protein Conformation

How can a protein such as PrP made by a cellular gene become an infectious agent? Prusiner and associates had found that PrP could exist in two forms, a normal or cellular form (PrP^C) normally expressed at low levels in neurons and other cell types, and an abnormal or scrapie form (PrP^{Sc}) built up in diseased brain. PrP^C is a cell-surface glycoprotein, the function of which has yet to be established. PrP^C consists of a single **polypeptide** chain folded into predominantly spiral conformations known as α-helices. These structures give rise to a globular shape that is soluble and can be cleared from the cell by degrading **enzymes** called proteases.

In contrast, PrP^{Sc} that has been isolated from diseased brain is rich in an alternative conformation that resembles extended strands. These structures are known as β-sheets. The β-sheet rich PrP^{Sc} tends to **aggregate** and is resistant to heat and degradation by proteases. It is assumed that PrP^{Sc} can initiate the infection process by binding to predominantly α-helical PrP^C and converting it into more stable PrP^{Sc} with β-sheet conformation. This will set off a chain reaction leading to accumulation of large amounts of

The illustration on the left is a computer-generated image of a healthy human prion protein. That on the right is a model of the disease-causing prion protein. The blue sections are β-strands, the green are α-helices, and the yellow are the chains connecting these regions.

PrP^Sc to levels that result in brain tissue damage. The conformational conversion from α-helices to β-sheets transforms the benign PrP^C into disease-causing PrP^{Sc}. This model of conformational conversion provides useful insights into the pathogenesis of prion diseases.

Prion Diseases

Historically, prion diseases have been given distinct names. Scrapie is a naturally occurring prion disease of sheep and goats that was first documented in Iceland during the eighteenth century. BSE or mad cow disease is a prion disease of cattle and is believed to be acquired through scrapie-contaminated foodstuffs. Kuru, a prion disease found among the Fore tribe of New Guinea, was shown by D. Carleton Gajdusek to be transmitted by the consumption of human tissue, particularly brain tissue, during funerary rituals. Gajdusek was awarded the 1976 Nobel Prize in physiology or medicine for this contribution. The early symptom of Kuru is a loss of coordination, followed by mental confusion and, ultimately, death. It has virtually disappeared since 1958, when the practice of eating human tissue was more or less eradicated in New Guinea.

CJD is the most common human prion disease, affecting about one in a million people. The main symptom is **dementia**, along with other neurological signs. There are three forms of CJD. Sporadic CJD, the cause of which has yet to be found, is a spontaneous disease that accounts for a majority of CJD cases. Familial CJD affects people who carry a mutation in the PrP gene on chromosome 20. The third form, called iatrogenic CJD, is the result of accidental transmission during medical treatments. A newly emerged CJD phenotype, commonly called variant CJD, has occurred in the United Kingdom since 1985. Variant CJD has a unique disease profile, and may result from the consumption of BSE-contaminated meat products. It has been diagnosed mostly in young people who initially seek treatment for psychiatric symptoms. Gertsmann-Sträussler-Scheinker (GSS) syndrome

dementia neurological illness characterized by impaired thought or awareness

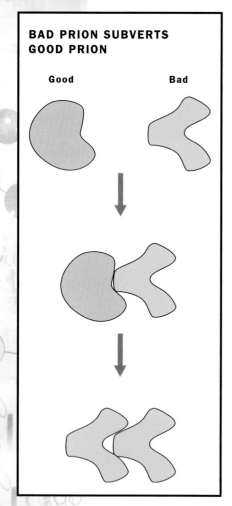

**BAD PRION SUBVERTS
GOOD PRION**

Good   Bad

A schematic model for the action of prions, showing how the "bad" prion alters the shape of the "good" prion.

phenotypes observable characteristics of organisms

is a familial prion disease resulting from a mutation in the PrP gene. The main symptom of GSS is the loss of coordination and dementia. Fatal familial insomnia (FFI) is another a familial prion disease in which fatal dementia follows the loss of physiological sleep.

Although human prion diseases manifest as three etiologically different forms—spontaneously (sporadic CJD), through inheritance (familial CJD, GSS, and FFI), and by infection (iatrogenic CJD, kuru, and possibly the new variant CJD), they nonetheless share a common pathogenetic event. Within the framework of "protein only" hypothesis, they all involve the protein conformational change that converts PrP^C to pathogenic PrP^{Sc}. Such a structural change in PrP may be triggered by a rare spontaneous event leading to a sporadic disease, a mutation that causes a familial disease, or exposure to foreign PrP^{Sc}, leading to an acquired disease. The "protein only" hypothesis provides a plausible mechanism underlying the pathogenesis of all forms of prion diseases. Moreover, it also helps explain the tremendous variability in prion-associated disease **phenotypes**. Structurally distinct variants of PrP^{Sc} may accumulate in different regions of the brain and initiate pathogenic changes that may eventually lead to distinct pathology in different areas of the brain, and subsequently the particular disease symptoms.

The concept of the prion and the role of protein conformation in disease pathogenesis have renewed inquiry into the causes of other and more common neurodegenerative disorders, such as Alzheimer's disease, Huntington's disease, and Parkinson's disease. A common hallmark of all these diseases, as in prion diseases, is the conversion of an otherwise soluble and functional neuronal protein into a β-sheet rich and protease-resistant protein that has a higher tendency to aggregate and is harmful to the brain. These common pathogenetic features raise the hope that therapeutic interventions based on the same principles may be effective in all these diseases. SEE ALSO PROTEIN.

Pierluigi Gambetti and Shu G. Chen

Bibliography

Cohen, F. E., and S. B. Prusiner. "Pathologic Conformations of Prion Proteins." *Annual Review in Biochemistry* 67 (1998): 793–819.

Prusiner, S. B. "Molecular Biology of Prion Diseases." *Science* 252 (1991): 1515–1522.

———. "The Prion Diseases." *Scientific American* (1995): 48–57.

———. *Prion Biology and Diseases.* New York: Cold Spring Harbor Laboratory Press, 1999.

Privacy

As more diagnostic, screening, and monitoring tests based on genetic data become available, privacy issues are becoming increasingly important. There are concerns that the results of genetic tests showing a person to be predisposed to a particular disease will fall into the hands of commercial medical suppliers or financial, legal, insurance, or government agencies, all of which control important products or services.

The confidentiality of medical information is of paramount importance to most consumers and patients. However, maintaining confidentiality is

made difficult by the use of large medical record databases and other electronically stored records, to which any number of individuals may have access.

The Potential for Misuse of Medical Records

Medical records can be misused unless they are coded to hide patients' identifying information. If a patient has been treated for a particular disease and his or her medical records are not held in confidence, a company selling products related to the disease could directly contact the patient. Although this may not pose a problem in most cases, in some situations, such as if a patient was treated for a sexually transmitted disease, the patient might not want family members or others with access to his or her mail to know about the treatment. There are also concerns about the potential for discrimination arising from the use of these data in determining a patient's eligibility for employment, housing, or other services.

In the United States, legislation has been passed to deal with issues surrounding genetic and health information. The Health Insurance Portability

Important personal information, such as a person's DNA, can be gleaned from blood samples. Proper labeling and secure storage of blood samples helps to protect patient privacy.

and Accountability Act of 1996 was enacted to address privacy issues related to personal health information. This act requires that health care providers, health plans, and health care clearinghouses implement certain privacy standards regarding health information.

Although the act protects "all medical records and other individually identifiable health information," there is some concern that it does not provide sufficient protection for the privacy of genetic information. In 2001 additional protection was proposed in at least three bills in the U.S. Congress. These bills were intended to prohibit discrimination on the basis of genetic information with respect to health insurance. The area of privacy and genetic information continues to develop, with additional legislation on the federal and state levels certain to arise.

European countries have addressed issues of privacy and personal information in a Directive on Data Protection. This directive, which became effective in October 1998, established a comprehensive legal regime in the European Union that governs the collection and use of personal information.

Privacy questions abound when it comes to genetic testing to determine if a person carries particular genes. One concern is that patients affected by genetic diseases, as well as those potentially at risk of disease, could be discriminated against. Another is that genetic information could also lead to discrimination against the children of those directly affected by a genetic disease.

Genetic Information in the Justice System

Many similar concerns arise in the context of criminal law, including the potential uses of DNA databases. There are issues relating to the collection and maintenance of DNA samples or information from everyone who is arrested, whether or not they are convicted. There are issues relating to the collection and maintenance of DNA samples and/or information collected from individuals upon arrest. For example, the DNA and/or information obtained from certain individuals may be saved, even if the person is not convicted. Indeed, prosecutors have issued many arrest warrants in old cases based solely on stored DNA data. These warrants have resulted in successful prosecutions, but the question being asked in the courts is whether it is legal to base arrests solely on "cold hit identification" using DNA evidence.

In contrast to medically oriented genetic tests, the DNA tests used in criminal law generally do not test for the presence or absence of a particular gene, since the noncoding regions of a person's DNA can be distinguished much more easily from the DNA of other individuals. Different individuals have different DNA sequences in these noncoding regions because there is no evolutionary penalty for mutations in such regions, as they are not used to produce proteins.

This helps provide the high level of discrimination required in criminal cases, enabling a jury to say that, based in part on the DNA evidence, an accused person is guilty beyond a reasonable doubt. An important caveat however, and one not always understood by prosecutors or juries, concerns what a DNA match actually proves. While nonmatching DNA proves inno-

cence, matching DNA does not prove guilt. In any large city, there will be at least a handful of people with similar DNA profiles. Even if DNA is found to be matching, a conviction must rely on other evidence, such as other physical evidence or eyewitness testimony.

Although the use of DNA data can assist investigations, there is an element of "big brother is watching" in its use. There are also concerns that by instituting wide programs of DNA collection based on arrests, not necessarily convictions, the practice will expand to other areas. For example, providing a DNA sample could be required, at some point, for obtaining a driver's license, marriage license, or social security number. There is also a question of what entities, including police departments, governmental agencies, employers, financial institutions, credit reporting businesses, and insurance carriers, would have access to the data. There is concern that by having genetic information recorded in a criminal record database, citizens would be subject to a wide variety of discrimination. SEE ALSO DISEASE, GENETICS OF; DNA PROFILING; GENETIC DISCRIMINATION; GENETIC TESTING; GENETIC TESTING: ETHICAL ISSUES; HUMAN DISEASE GENES, IDENTIFICATION OF; LEGAL ISSUES.

Kamrin T. MacKnight

Probability

Probability measures the likelihood that something specific will occur. For example, a tossed coin has an equal chance, or probability, of landing with one side up ("heads") or the other ("tails"). If you drive without a seat belt, your probability of being injured in an accident is much higher than if you buckle up. Probability uses numbers to explain chance.

If something is absolutely going to happen, its probability of occurring is 1, or 100 percent. If something absolutely will not happen, its probability of occurring is 0, or 0 percent.

Probability is used as a tool in many areas of genetics. A clinical geneticist uses probability to determine the likelihood that a couple will have a baby with a specific genetic disease. A statistical geneticist uses probability to learn whether a disease is more common in one population than in another. A computational biologist uses probability to learn how a gene causes a disease.

The Clinical Geneticist and the Punnett Square

A Punnett square uses probability to explain what sorts of children two parents might have. Suppose a couple knows that cystic fibrosis, a debilitating respiratory disease, tends to run in the man's family. The couple would like to know how likely it is that they would pass on the disease to their children.

A clinical geneticist can use a Punnett square to help answer the couple's question. The clinical geneticist might start by explaining how the disease is inherited: Because cystic fibrosis is a **recessive** disease caused by a single gene, only children who inherit the disease-causing form of the cystic fibrosis gene from both parents display symptoms. On the other hand, because the cystic fibrosis gene is a recessive gene, a child who inherits only

recessive requiring the presence of two alleles to control the phenotype

Both parents are carriers of the recessive CF gene (f) in this Punnett square. The probability of having a child with CF, if both parents are carriers, is 25 percent.

PUNNETT SQUARE FOR THE INHERITANCE OF CYSTIC FIBROSIS

father

	F	f
F	FF	Ff
f	Ff	ff

mother

Summary: 1 FF child   = 1 of the 4 children = 1/4 = 25%
2 Ff children = 2 of the 4 children = 2/4 = 50%
1 ff child   = 1 of the 4 children = 1/4 = 25%

The Bottom Line: normal children =   FF and Ff   = 25% + 50% = 75%
children with CF =   ff   = 25%

one copy of a defective gene, along with one normal version, will not have the disease.

Suppose the recessive, disease-causing form of the gene is referred to as "f" and the normal form of the gene is referred to as "F." Only individuals with two disease-causing genes, ff, would have the disease. Individuals with either two normal copies of the gene (FF) or one normal copy and one mutated copy (Ff) would be healthy.

If the clinical geneticist tests the parents and finds that each carries one copy of the cystic fibrosis gene, f, and one copy of the normal gene, F, what would be the probability that a baby of theirs would be born with the cystic fibrosis disease? To answer this question, we can use the Punnett square shown in the figure above. A Punnett square assumes that there is an equal probability that the parent will pass on either of its two gene forms ("alleles") to each child.

The parents' genes are represented along the edges of the square. A child inherits one gene from its mother and one from its father. The combinations of genes that the child of two Ff parents could inherit are represented by the boxes inside the square.

Of the four combinations possible, three involve the child's inheriting at least one copy of the **dominant**, healthy gene. In three of the four combinations, therefore, the child would not have cystic fibrosis. In only one of the four combinations would the child inherit the **recessive** allele from both parents. In that case, the child would have the disease. Based on the Punnett square, the counselor can tell the parents that there is a 25 percent probability, or a one-in-four chance, that their baby will have cystic fibrosis.

dominant controlling the phenotype when one allele is present

recessive requiring the presence of two alleles to control the phenotype

The Statistical Geneticist and the Chi-Square Test

Researchers often want to know whether one particular gene occurs in a population more or less frequently than another. This may help them determine, for example, whether the gene in question causes a particular disease. For a dominant gene, such as the one that causes Huntington's disease, the frequency of the disease can be used to determine the frequency of the gene, since everyone who has the gene will eventually develop the disease. How-

ever, it would be practically impossible to find every case of Huntington's disease, because it would require knowing the medical condition of every person in a population. Instead, genetic researchers sample a small subset of the population that they believe is representative of the whole. (The same technique is used in political polling.)

Whenever a sample is used, the possibility exists that it is unrepresentative, generating misleading data. Statisticians have a variety of methods to minimize sampling error, including sampling at random and using large samples. But sampling errors cannot be eliminated entirely, so data from the sample must be reported not just as a single number but with a range that conveys the precision and possible error of the data. Instead of saying the prevalence of Huntington's disease in a population is 10 per 100,000 people, a researcher would say the prevalence is 7.8–12.1 per 100,000 people.

The potential for errors in sampling also means that statistical tests must be conducted to determine if two numbers are close enough to be considered the same. When we take two samples, even if they are both from exactly the same population, there will always be slight differences in the samples that will make the results differ.

A researcher might want to determine if the prevalence of Huntington's disease is the same in the United States as it is in Japan, for example. The population samples might indicate that the prevalences, ignoring ranges, are 10 per 100,000 in the United States and 11 per 100,000 in Japan. Are these numbers close enough to be considered the same? This is where the Chi-square test is useful.

First we state the "null hypothesis," which is that the two prevalences are the same and that the difference in the numbers is due to sampling error alone. Then we use the Chi-square test, which is a mathematical formula, to test the hypothesis.

The test generates a measure of probability, called a p value, that can range from 0 percent to 100 percent. If the p value is close to 100 percent, the difference in the two numbers is almost certainly due to sampling error alone. The lower the p value, the less likely the difference is due solely to chance.

Scientists have agreed to use a cutoff value of 5 percent for most purposes. If the p value is less than 5 percent, the two numbers are said to be significantly different, the null hypothesis is rejected, and some other cause for the difference must be sought besides sampling error. There are many statistical tests and measures of significance in addition to the Chi-square test. Each is adapted for special circumstances.

Another application of the Chi-square test in genetics is to test whether a particular genotype is more or less common in a population than would be expected. The expected frequencies can be calculated from population data and the Hardy-Weinberg Equilibrium formula. These expected frequencies can then be compared to observed frequencies, and a p value can be calculated. A significant difference between observed and expected frequencies would indicate that some factor, such as natural selection or migration, is at work in the population, acting on allele frequencies. Population geneticists use this information to plan further studies to find these factors.

The Computational Biologist and BLAST

Genetic counseling lets potential parents make an informed decision before they decide to have a child. Geneticists, however, would like to be able to take this one step further: They would like to be able to cure genetic diseases. To be able to do so, scientists must first understand how a disease-causing gene results in illness. Computational biologists created a computer program called BLAST to help with this task.

To use BLAST, a researcher must know the DNA sequence of the disease-causing gene or the protein sequence that the gene encodes. BLAST compares DNA or protein sequences. The program can be used to search many previously studied sequences to see if there are any that are similar to a newly found sequence. BLAST measures the strength of a match between two sequences with a p value. The smaller the p value, the lower the probability that the similarity is due to chance alone.

If two sequences are alike, their functions may also be alike. For BLAST to be most useful to a researcher, there would be a gene that has already been entered in the library that resembles the disease-causing gene, and some information would be known about the function of the previously entered gene. This would help the researcher begin to hypothesize how the disease-causing gene results in illness. SEE ALSO BIOINFORMATICS; CLINICAL GENETICIST; COMPUTATIONAL BIOLOGIST; CYSTIC FIBROSIS; HARDY-WEINBERG EQUILIBRIUM; HOMOLOGY; INTERNET; MENDELIAN GENETICS; METABOLIC DISEASE; STATISTICAL GENETICIST.

Rebecca S. Pearlman

Bibliography

Nussbaum, Robert L., Roderick R. McInnes, and Huntington F. Willard. *Thompson & Thompson Genetics in Medicine*, 6th ed. St. Louis, MO: W. B. Saunders, 2001.

Purves, William K., et al. *Life: The Science of Biology*, 6th ed. Sunderland, MA: Sinauer Associates, 2001.

Seidman, Lisa, and Cynthia Moore. *Basic Laboratory Methods for Biotechnology: Textbook and Laboratory Reference*. Upper Saddle River, NJ: Prentice-Hall, 2000.

Tamarin, Robert H. *Principles of Genetics*, 7th ed. Dubuque, IA: William C. Brown, 2001.

Internet Resources

The Dolan DNA Learning Center. Cold Spring Harbor Laboratory. <http://vector.cshl.org>.

The National Center for Biotechnology Information. <http://www.ncbi.nlm.nih.gov>.

Protein Sequencing

The molecules that give cells and entire organisms their shape as well as their ability to move, grow, and reproduce are the proteins. Although they come in an almost infinite variety of shapes and sizes, they have all been designed by the process of evolution to serve a defined and useful function in the processes of life. Some proteins, like actin and collagen, help to give a cell its physical shape. Other proteins, like lactase and pepsin, help in the digestion of food. Others transport signals between cells, help us fight off

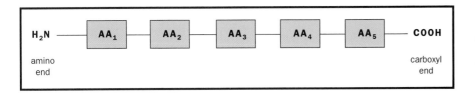

Schematic diagram of protein primary structure. Amino acids are linked head to tail, so that at one end there is a free amino group, and at the other a free carboxyl group. Proteins are typically 50–500 amino acids in length.

disease, or repair damaged DNA. For almost every job in a cell, there is a protein designed to do it.

The Building Blocks of Proteins

The building blocks of proteins are amino acids. There are twenty different amino acids used by living cells to build proteins. They are linked together in a long, linear chain during the process of translation, which is carried out by the ribosomes inside cells. Proteins begin to take on their characteristic three-dimensional shape even while they are being made, folding and twisting as each new amino acid added to the chain tugs or pushes at the others added before it. Each amino acid has an amino group ($-NH_3^+$) and a carboxyl group ($-COOH$). Peptide bonds link the carboxyl group of one amino acid to the amino group of the next amino acid. On one end of a protein, therefore, there is a free amino group called the N-terminus, and on the other end is a free carboxyl group, called the C-terminus.

The process of determining a protein's order of amino acids is called protein sequencing. A protein's sequence can easily be deduced from its gene sequence, since the order of bases on a DNA strand specifies the order in which the amino acids are linked together during translation. The chemistry involved in DNA sequencing is less complex than that which is involved in determining the order of each amino acid in an amino acid chain. There are two primary reasons why effort would be put into sequencing a protein. The first is to provide the information needed to design a synthetic DNA probe that can be used to locate the gene that codes for the protein. The second is to prove that a protein that has been isolated or manufactured in the laboratory is what it is believed to be.

Sequencing Techniques

The most widely used technique for sequencing proteins is the Edman degradation, a procedure developed by Pehr Edman in the 1950s. The reaction steps used for this method have since been completely automated by machine. The procedure uses special reagents under alternating basic and acidic conditions to remove one amino acid at a time from the protein's N-terminus. As each amino acid is released during each cycle of degradation, it is identified by chromatography, a separation technique that relies on an amino acid's unique size and electrical charge to distinguish it from the other nineteen amino acids.

In many automated approaches, high-performance liquid chromatography (HPLC) is used to tell which amino acid has been released; the amount of time it takes to travel through an HPLC column is unique to each amino acid. Up to fifty amino acids from the N-terminus can be identified using Edman degradation. If a scientist is trying to identify a previously sequenced protein, usually only the first fifteen to twenty amino acids of the purified

enzymes proteins that control a reaction in a cell

genome the total genetic material in a cell or organism

polymers molecules composed of many similar parts

protein need to be sequenced. That information can then be entered into a database and matched with known proteins having identical or related sequences.

Sequencing a protein from its C-terminus is particularly challenging, and there are no techniques that are as robust as Edman degradation. However, some limited amino acid sequence information can be obtained using **enzymes** called carboxypeptidases, which remove individual C-terminal amino acids. These enzymes, however, tend to cleave only specific amino acids from the C-terminus.

Carboxypeptidase B, isolated from cow pancreas, for example, can release the amino acids arginine and lysine from the C-terminus of a protein. Carboxypeptidase A, also isolated from cow pancreas, fails to release arginine, lysine, or proline, but can cleave off the other seventeen amino acids. Carboxypeptidases isolated from citrus leaves and yeast can cleave off any amino acid from the C-terminus of a protein, although the rate at which they do this depends on the particular amino acid. If one amino acid is released slowly and the next within the chain is released very quickly, they might appear to be cleaved at the same time, making it difficult to establish their order. C-terminal amino acid identification using enzymes, therefore, is not practical beyond the first several positions.

Another method of protein sequencing, called mass spectrometry, uses electric current to break individual amino acids from a protein. In a mass spectrometer, the released amino acids are collected in a detector and are each identified by their unique mass.

Sequencing of the human **genome** has allowed a giant leap in the understanding of how the human species evolved and how genetic diseases arise. Advances made in DNA sequencing technology lead to this grand accomplishment. The next frontier is to decipher how all the proteins encoded by the genome interact to carry out the processes of life. This is the study of proteomics. Advances in mass spectrometry and protein sequencing instrumentation are bringing this challenging problem closer to its resolution. SEE ALSO HPLC: HIGH-PERFORMANCE LIQUID CHROMATOGRAPHY; MASS SPECTROMETRY; PROTEINS; SEQUENCING DNA.

Frank H. Stephenson and Maria Cristina Abilock

Bibliography

Creighton, Thomas E. *Proteins: Structures and Molecular Properties.* New York: W. H. Freeman, 1993.

Proteins

Proteins are **polymers** of amino acids that provide structure and control reactions in all cells. When humans think of expressing the meaning of life, they often resort to words. From poems to sonnets to short stories to novels, words tell the stories of life. But in biological terms, the words of life are proteins. While DNA holds the code of life, proteins are the language in which that code is expressed.

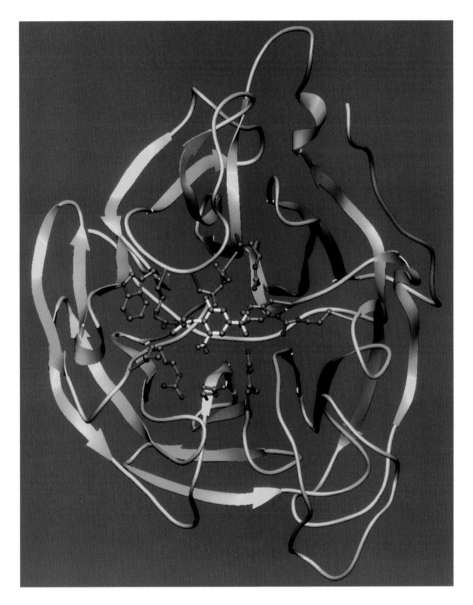

Computer generated representation of the enzyme neuraminidase (ribbons and strings) with an inhibitor (ball and stick) in the active site, a cleft on the enzyme's surface. This tertiary structure is built up from the primary structure (amino acid structure) and secondary structure. The arrows represent the beta-pleated sheet, a type of secondary structure.

To observe the mosaic of proteins in life is to observe nature in its finest array. The feathers of a bird and the silk of a spider's web are both almost pure protein. The most numerous proteins in an animal are the collagen proteins joining animal body parts. Other proteins include the positively charged histone proteins that condense the cell's negatively charged DNA and the transcription factor proteins that control which genes are expressed (made into proteins) and which remain silent. A plant traps CO_2 to make sugar with Earth's most abundant protein, the **enzyme** ribulose 1,5-biphosphate carboxylase. The protein hemoglobin transports gases through the bloodstream necessary for the metabolism of life. Other proteins store minerals (ferritin) or fats (ovalbumin), contract muscles (myosin), protect against infection (antibodies), or act as toxins (botulinum) or hormones (insulin).

enzyme a protein that controls a reaction in a cell

Properties of Amino Acids

The English language consists of thousands of words, created from any of twenty-six letters arranged in a precise order. In an analogous fashion,

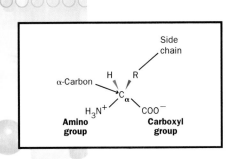

Amino acid structure. Adapted from Robinson, 2001.

hydrophilic "water-loving"

polar partially charged, and usually soluble in water

hydrophobic "water hating," such as oils

nonpolar without charge separation; not soluble in water

basic having the properties of a base; opposite of acidic

acidic having the properties of an acid; the opposite of basic

codons sequences of three mRNA nucleotides coding for one amino acid

hydrogen bonding weak bonding between the hydrogen of one molecule or group and a nitrogen or oxygen of another

proteins are made up of twenty common amino acids in a precise order dictating the protein's structure and function. Every amino acid has a common structure, in which a central carbon is covalently bonded to a carboxyl group (COOH), an amino group (NH₂), a hydrogen, and a variable "R" group.

The chemical properties of the R group are what give an amino acid its character. The R group can be **hydrophilic** (attracted to water and other **polar** molecules) or **hydrophobic** (attracted to **nonpolar** molecules and repelled by water or other polar molecules). Hydrophilic R groups can have **basic** charges, as in the amino acid valine, or **acidic**, as in glutamic acid, or they may even be an uncharged polar group such as –OH (alcohol) or –NH₂ (amino), as in serine. A nonpolar or hydrophobic R group can be a hydrocarbon chain, as in leucine. There are also three special amino acids: cysteine, glycine, and proline. Cysteine has a reactive sulfhydryl R group that forms disulfide bridges (S–S) between regions of the protein chain. These bridges increase toughness and resistance to unfolding of the protein structure. Glycine is the smallest amino acid, with hydrogen as its R group, and it fits into tight places within a protein's structure. Proline has a cyclic ring involving the central carbon, and it causes kinks to occur in a protein chain. Both proline and glycine are common at the corner of turns in the protein foldings.

Primary Structure

The unique sequence of amino acids in a protein is termed the primary structure. When amino acids form a protein chain, a unique bond, termed the peptide bond, exists between two amino acids. The sequence of a protein begins with the amino of the first amino acid and continues to the carboxyl end of the last amino acid.

The unique sequence of amino acids results from the translation of **codons** present in messenger RNA (mRNA). The mRNA, in turn, is a complementary copy of the gene that codes for that protein. Protein structure and function can change when "misspellings" occur in the order of amino acids during their transcription and translation. Sickle-cell hemoglobin, for example, is "misspelled" in only one amino acid; the sixth amino acid in the beta chain, where a valine is substituted for a glutamic acid. This occurs because the codon for valine, GUG, has replaced the codon for glutamic acid, GAG. This change from acidic to basic amino acid causes the hemoglobin molecules to stick to one another, forming long chains and blocking oxygen binding. These chains of hemoglobin precipitate in the cell, causing the red blood cells to assume a sickle shape. All of these structural and functional changes occur because of the mutation in the hemoglobin gene and a "misspelling" in the hemoglobin's amino acid sequence.

Secondary Structure and Motifs

The secondary structure of proteins is due to foldings that occur within their structure. These foldings are either in a helical shape, called the "alpha-helix" (which was first proposed by Linus Pauling), or a beta-pleated sheet shaped similar to the zig-zag foldings of an accordion. The turns of the alpha-helix are stabilized by **hydrogen bonding** between every fourth amino acid in the chain. The alpha-helix can cover specific regions of the protein

or it may involve the entire protein, as in the alpha-keratin found in claws and horns. The two sides of the alpha-helix may differ in polarity, with hydrophilic R groups projecting to the lining of the channel, while hydrophobic R groups project to the outside of the channel, where they embed in the hydrophobic membrane. This structure is exemplified in membrane channel proteins, proteins that channel ions across from one surface to another. The beta-pleated sheet is formed by folding successive planes. Each plane is five to eight amino acids long. The folds are stabilized by hydrogen bonding. The strength observed in silk fibers is due to their stacks of beta-pleated sheets.

Combinations of secondary structure form "motifs." A coiled-coil motif is common among proteins that associate with the DNA helix. The helix-loop-helix motif is a knobby structure, and the zinc finger projects outward like its name. These last two motifs allow associations between RNA and proteins that form the basis of their interactions.

Tertiary Structure and Protein Domains

Domains are large functional regions of the protein, such as an enzyme's active site, which binds the substrate to the enzyme. Myoglobin, the muscle protein that stores and releases oxygen, contains several alpha-helices wound around a central crevice. It is in this central crevice that the O_2 molecule binds. Just as words take on their meanings when completed, the functional domains unite to form the overall purpose of a protein. For example, a membrane protein stabilizes itself by anchoring itself with a hydrophilic **cytoplasmic** domain, then weaves its alpha-helices throughout the membrane domain and projects its carbohydrate hydrophilic side chains into the extracellular surface domain. Such membrane proteins often act as receptors, important for receiving signals such as hormones, or work in the immune system to recognize infected cells.

cytoplasmic describes the material in a cell, excluding the nucleus

The local foldings, evident in secondary structure, then combine into a single polypeptide chain. This chain is called the tertiary structure, or **conformation**. For example, the pancreatic enzyme ribonuclease, which aids in digestion of RNA in the diet, consists mainly of beta sheet folds, with three small alpha-helical regions. Tertiary structure is often stabilized by disulfide bonds between adjacent cysteine in different regions of the protein. For example, the tertiary structure of ribonuclease contains four disulfide bonds, located at specific sites. The stability of the tertiary structure of proteins is destroyed by toxic heavy metals such as mercury. Concentrations of mercury in the environment, for example, result in the displacement of hydrogen on the sulfur atom (SH), thereby blocking functional disulfide bonds.

conformation three-dimensional shape

Several other weak, noncovalent interactions also help stabilize tertiary structure. These noncovalent interactions can be disrupted by heating a protein or exposing it to extremes in pH (acidity or alkalinity), which alters the charge of polar groups on the amino acids. Such disruptions cause the protein to unfold, often exposing hydrophobic groups and leading to precipitation (clumping together) of the protein. If these disruptive factors are removed, some proteins can refold to their original conformation. This ability to refold confirms that protein folding is a self-assembly process that is dependent upon the sequence of amino acids.

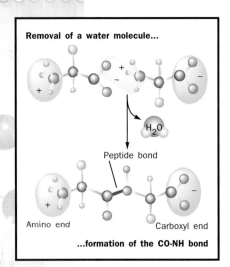

Removal of a water molecule...

H_2O

Peptide bond

Amino end   Carboxyl end

...formation of the CO-NH bond

Peptide bond formation. Adapted from Robinson, 2001.

multimeric describes a structure composed of many parts

antigen a foreign substance that provokes an immune response

heme iron-containing nitrogenous compound found in hemoglobin

plasma membrane outer membrane of the cell

endoplasmic reticulum network of membranes within the cell

lumen the space within the tubes of the endoplasmic reticulum

Quaternary Structure

Some proteins need to functionally associate with others as subunits in a **multimeric** structure. This is called the quaternary structure of the protein. This can also be stabilized by disulfide bonds and by noncovalent interactions with reacting substrates or cofactors. For example an antibody consists of two "light" polypeptide chains covalently linked to two longer "heavy" chains, forming a Y-shaped molecule with each branch able to bond to an identical **antigen**. The protein subunits of the single-stranded binding protein of *Escherichia coli* bind to DNA only as a tetramer (a multimeric form), acting to stabilize the separated DNA strands during replication.

Another excellent example of quaternary structure is that of hemoglobin. Adult hemoglobin consists of two alpha subunits and two beta subunits, held together by noncovalent interactions. Each of the four subunits contains a **heme** group that binds an oxygen molecule, O_2. This binding of oxygen is a cooperative process whereby the binding of one oxygen molecule occurs slowly, but once achieved then speeds the binding of the remaining three oxygen molecules. The fourth oxygen molecule binds 300 times faster than the first oxygen molecule. This cooperativity assures that maximum oxygen is captured and retained as it enters into the capillaries within the lungs.

The unloading of oxygen is also facilitated by cooperativity, such that after one oxygen molecule is released, the other three soon follow. This assures that the tissues will receive maximum oxygen once it is delivered. Alpha-hemoglobin by itself, or tetramers of all beta subunits, also bind oxygen, but not with the same cooperativity. Such evidence indicates that there is some form of molecular interaction between the subunits of the tetramer of adult hemoglobin.

Signal Sequences in Protein Synthesis

Protein must be delivered to the proper destination in the cell to function properly. Signal sequences within the protein itself act like "zip codes" to ensure correct delivery. The synthesis of secreted proteins like insulin and of proteins that will be integral to the **plasma membrane** occurs at a ribosome tethered to the **endoplasmic reticulum**, which is a system of membranes that transport materials within cells. The peptides formed there are then translocated into the **lumen**, or channel, of the endoplasmic reticulum, where they will be formed into a polypeptide chain. This translocation occurs because of a specific signal sequence that is formed by the first twenty or so amino acids in the protein. The core of this sequence consists of ten to fifteen amino acids that have hydrophobic side chains such as alanine, leucine, valine, isoleucine, and phenylalanine, which are usually cleaved from the protein later on. The nascent polypeptide chain is guided along this path by a signal receptor protein.

Proteins targeted for internal cellular functions are synthesized on ribosomal assemblages that float free in the cytoplasm. Such proteins also have their signal sequences. Proteins destined for the cell's nucleus have a specific nuclear signal sequence consisting of a small series of basic amino acids such as arginine and lysine bounded by proline. This nuclear signaling sequence can be located anywhere in the protein's sequence as long as it projects outward from the three-dimensional tertiary structure. Signal sequences for proteins targeted to be part of organelles such as the mito-

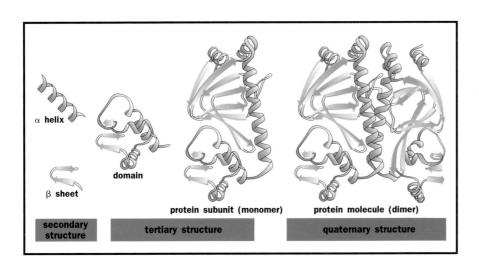

Protein structure components. Alpha helices and beta sheets are linked by less-structured loop regions to form domains. The domains combine to eventually form fully functional proteins. Adapted from Robinson, 2001.

chondria and chloroplasts are anywhere from twenty to seventy amino acids long and are mostly hydrophilic. This charged nature allows easy travel through the hydrophilic cytoplasm to the organelle.

Molecular Chaperones

Although the folding of the protein into its tertiary structure is determined by the primary order of amino acids, the process of folding occurs with the assistance of molecular chaperone proteins. These molecular chaperones often have pockets or tunnels that envelop the nascent polypeptide. This enveloping allows the folding of the protein to occur unhindered by unwanted interaction with other cellular components.

Chemical Modification and Processing of Proteins

Most proteins are structurally altered after synthesis through chemical modification or processing. These alterations help the cell determine a protein's fate, such as whether that protein is active or inactive, how long the protein will function, and to some degree the location where that protein will function. Chemical modifications, which are additions of chemical groups to the R groups in the amino acids, are made after translation. Such modifications may include the attachment of a phosphate group (phosphorylation) to the alcohol group on the amino acids of serine, threonine, or tyrosine. The amino acid proline in proteins such as collagen is often hydroxylated, which means that an alcohol group is attached. Other amino acids with amino groups in their R region, such as lysine or arginine, may be chemically modified through methylation, which is the addition of a methyl group ($-CH_3$), or through acetylation, in which an acetyl group ($-CH_3CO$) is added. Larger modifications, such as the addition of a carbohydrate group, occur to create **glycoproteins** in specialized organelles termed Golgi apparati.

Modifications change the charge of the protein, and often cause a change in the protein's activity level. For many DNA-associated proteins their regional acetylations cause them to "loosen" their grip on the DNA helix, thereby enabling **transcription factors** to enter, signaling gene activation. A cascade of internal protein phosphorylation (successive additions of a phosphate group) is a common mechanism for carrying a hormone's message

glycoproteins proteins to which sugars are attached

transcription factors proteins that increase the rate of gene transcription

from the membrane, where it docks into the cell and induces a metabolic change inside the target cell.

Processing results in cutting off specific parts of the protein (cleavage). Many digestive proteins such as pepsin and hormones such as insulin are processed. Pepsin, which is a digestive protein secreted into the lumen of the stomach, remains in an inactive form until stomach acid is also secreted. The timing of the acid secretion, pepsin activation, and entry of food coincide so that pepsin's activity will be directed toward the food and not the wall of the stomach.

Conformational Changes in Protein Structure

As noted above, a protein's activity can be regulated when it undergoes a change in its conformation. A dramatic and extensively studied model of protein conformational change is that of the Na^+/K^+ ATPase pump. This is an integral membrane protein with one side facing the exterior of the cell and the other facing the cytosol. It is used for the specific transport of sodium or potassium across the membrane, and one of its most important functions is the repolarization of a nerve fiber after it "fires."

ATP adenosine triphosphate, a high-energy compound used to power cell processes

The first step in the transport process is the binding of three Na^+ (sodium) ions to the inside face of the protein. This is followed by protein phosphorylation using **ATP**, which causes the protein to change its conformation. This moves the sodium ions from the cytosol to the exterior. This conformational change also opens up exterior binding sites, which tightly bind two potassium ions outside the cell. Following the potassium binding, the protein is dephosphorylated, losing its recently added phosphate group. This dephosphorylation then changes the protein back to the original conformation, causing the protein to loosen its binding of potassium and deliver those two ions to the cytosol. This process demonstrates that protein structure can be reversibly changed. The net result is that the inside of the cell develops a slight negative charge compared to the outside. The disruption of this "polarized" state constitutes nerve cell firings, which allow the cells of the nervous system to communicate with one another.

Proteomics

Proteomics is a new field of study that seeks to describe which proteins are expressed in a cell, when they are expressed, what consequences result from their expression, and how they fit into biochemical pathways. The first step in the study of proteomics is to define the language of protein structure. The field of proteomics promises to bring a complex understanding to the role of proteins in living cells. SEE ALSO CELL, EUKARYOTIC; CHAPERONES; GENETIC CODE; HEMOGLOBINOPATHIES; IMMUNE SYSTEM GENETICS; MUTATION; NUCLEASES; PROTEOMICS.

Paul K. Small

Bibliography

Fairbanks, Daniel, J., and W. Ralph Anderson. *Genetics: The Continuity of Life.* Pacific Grove, CA: Brooks/Cole, 1999.

Lodish, Harvey, et al. *Molecular Cell Biology,* 4th ed. New York: W. H. Freeman, 2000.

Sadava, David E. *Cell Biology: Organelle Structure and Function.* Boston: Jones and Bartlett, 1993.

Stryer, Lubert. *Biochemistry,* 3rd ed. New York: W. H. Freeman, 1988.

Proteomics

Proteomics is the science of studying the multitude of proteomes found in living organisms. A proteome is the entire collection of proteins expressed by a genome or in a tissue. The contents of a proteome can differ in various tissue types, and it can change as a result of aging, disease, drug treatment, or environmental effects.

This is contrary to the concept of a genome, which is an organism's complete collection of DNA. A genome's composition remains more or less constant from tissue to tissue, except for mutations and **polymorphisms** that can occur.

polymorphisms DNA sequence variants

The word "proteome" was first coined in late 1994. By 1997 there were a number of research conferences focusing on proteomics.

According to the first draft of the human genome, based on the work by the Human Genome Project and by Celera Inc., there are only between thirty thousand and seventy thousand genes in the human genome, many fewer than had been estimated previously. However, as of 2002 there were still groups that believed that there are at least 120,000 genes. Regardless of which of these estimates proves more accurate, the number of potential proteins in the human proteome is quite large. Although the first draft of the human genome reduced the estimates for the total number of human genes, it also predicted a greater amount of alternative splicing of genes, and therefore more distinct protein products per gene, than had been anticipated.

At its simplest level, proteomics is the study of protein expression in a proteome, or trying to understand the relative levels (amounts) of each protein within the mixture. Proteomics attempts to characterize proteins, compare variations in their expression levels in normal and disease states, study their interactions with other proteins, and identify their functional roles.

Unlike the traditional approach of studying individual proteins one at a time, proteomics uses an automated, high-throughput approach. High-throughput refers to the number of items (in this case, proteins) that can be analyzed or studied per unit of time. New technologies and substantial bioinformatics tools are required to compare entire proteomes. Expansion of the field of proteomics into the realm of "big science" (meaning many dollars invested by a large number of companies and universities) is several years behind the expansion of genomics. This is primarily because proteins are more difficult to work with in a laboratory setting than are nucleic acids such as DNA.

The development of protein analysis technologies is more difficult than the development of DNA analysis technologies for three reasons. First, the basic alphabet for encoding proteins consists of twenty amino acids, whereas there are only four different nucleotides, the alphabet of DNA. Second, the messenger RNA (mRNA) for some genes can be differentially spliced, meaning that multiple messages can be made from a single gene, resulting in multiple, distinct protein products. Finally, many proteins are modified once they have been synthesized. This is known as post-translational modification. There are a number of types of post-translational modifications, such as the addition of sugar, phosphate, sulfate, lipid, acetyl, or methyl groups.

Mass spectrometry systems are used to help scientists analyze the various proteomes within an organism.

Each of these modifications has the ability to change the functional activity of a protein.

The above issues have made the elucidation of reliable, high-throughput techniques for characterizing proteins, including their expression levels, on a proteome-wide level a major challenge. Hence, techniques for doing, for example, high-throughput DNA sequencing and gene expression studies have been developed and commercialized on a large scale sooner than similar protein analysis techniques. This is not to imply that all of the techniques involved in proteomics are new. Some, such as two-dimensional **gel electrophoresis**, have been around since the 1970s. However, the need to adapt these techniques to a large "proteome" scale brings with it a unique set of challenges.

gel electrophoresis technique for separation of molecules based on size and charge

For researchers involved in areas such as drug discovery, proteomics approaches will need to be used to obtain a greater understanding of disease mechanisms and drugs' mechanisms of action. Large-scale studies looking at gene expression via quantification of mRNA abundance are already possible and well commercialized. These technologies are very powerful, and the highest throughput approaches are capable of analyzing tens of thousands of genes per experiment. Sophisticated bioinformatics systems have been, and continue to be, developed to analyze these vast amounts of data. However, studies have shown that mRNA levels do not necessarily correlate well with protein levels.

Researchers must understand proteins and their roles, since proteins are the functional units within cells. As of 2002, the vast majority of drug targets were proteins. There are a handful of drugs, including some chemotherapeutic agents, that bind to DNA, but most drugs bind to specific protein targets. In the cases where the target is a protein, the drugs themselves are primarily small inorganic molecules or, in some cases, small proteins, such as **hormones**, that bind to a larger protein target in the body.

hormones molecules released by one cell to influence another

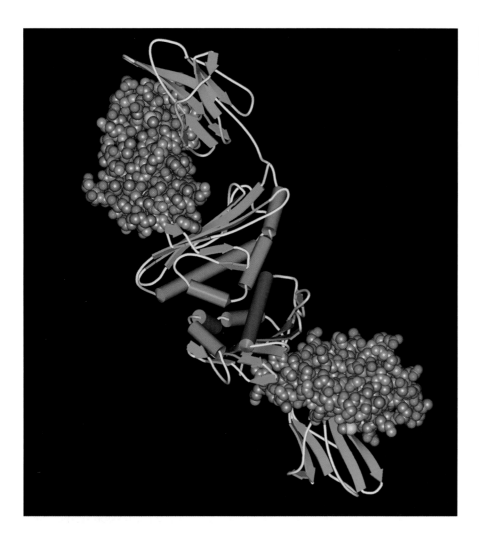

Proeomics can help researchers understand how proteins interact in cells.

Some drugs are actually therapeutic proteins that are delivered to the site of the disease.

Laboratory Techniques

The primary attributes used to identify proteins include the protein's mass and apparent mass, its isoelectric point, and its N- and C-terminal sequence tags. A protein's mass and its apparent mass are probably the most common characteristics used. Protein mass is determined by adding the total mass of all the amino acids in the protein to the mass of any molecules added through post-translational modification. A protein's isoelectric point is the pH at which it is neutrally charged. A protein's N- and C-terminal sequence tags are short sequences of amino acids on either end of the protein. Since there are twenty different possible amino acids at each position in a protein, a **peptide** of only four or five amino acids in length is likely to be unique to a specific protein. There are 160,000 (20^4) combinations of sequences that are four amino acids long.

peptide amino acid chain

The most commonly used laboratory techniques in proteomics are two-dimensional polyacrylamide gel electrophoresis (2-D PAGE) and mass spectrometry. These techniques have been modified for use in proteomics. Both can be used in combination with more traditional protein separation techniques, including column chromatography.

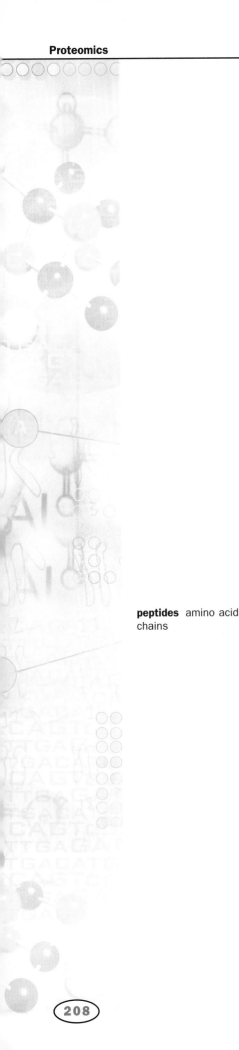

Starting in the late 1990s, several companies also started developing "protein chips," another strategy for studying proteomes and other complex protein mixtures. These chips allow a researcher to collect minute quantities of proteins that bind to specific molecules on their surface. By 2001, some companies announced they were developing "antibody chips" onto which antibodies will be attached. The antibodies can then be used as probes to capture and quantify specific proteins found in complex mixtures.

The use of 2-D PAGE allows the simultaneous separation of thousands of proteins, and the technique is still a key tool in proteomics technologies. The first dimension of protein separation on the gel is by isoelectric focusing, in which proteins are separated along a pH gradient until they reach a stationary position, where their net charge is zero.

The second dimension of separation on the gel is by molecular mass. Sodium dodecyl sulphate (SDS) is applied, and it binds to all the proteins. This provides the proteins with a uniform charge along their length, so that they will migrate across the gel according to their molecular mass when a current is applied. After the 2-D PAGE is run, the gel is stained. The result is a two-dimensional map consisting of hundreds or thousands of protein spots.

Since the early use of 2-D PAGE in the early 1970s, a number of modifications have been made to make gels more reproducible and more amenable to the higher-throughput use necessary for proteomics applications. However, 2-D PAGE is still something of an art form, and high-quality, reproducible results are difficult to obtain except in the hands of very experienced users. The technology needs to be further simplified to allow casual and novice users to obtain reproducible, quality results.

peptides amino acid chains

Mass spectrometry is an analytical technique that very accurately measures the mass of proteins and **peptides**. There are two common types of mass spectrometry. The first type, matrix-assisted laser desorption/ionization time-of-flight mass spectrometry, can be used to analyze proteins that are embedded in solid samples and measures their mass in a flight tube. The second type, electrospray ionization mass spectrometry, can be used to analyze proteins that are in a liquid solution and measures their mass in either a flight tube or in a device known as a quadrupole. There are also other variations on these techniques.

Mass spectrometry is commonly used for peptide mass fingerprinting. In this process, a protein sample is isolated by 2-D PAGE and cut with an enzyme that specifically targets particular amino acids. Mass spectrometry is used to measure the masses of the resulting cut pieces, or peptides. These masses can be thought of as a fingerprint that can be compared to the fingerprints of proteins whose amino acid sequences have already been analyzed and stored in a database.

To determine the fingerprints of proteins that have already been sequenced, a computer program determines the amino acid composition, and thus the masses, of the pieces that would result if those proteins were also cut by the same enzyme. A list of proteins is generated from the database, sorted by how many peptides they share with the unknown experimental protein.

There are also technologies, including the yeast two-hybrid system, that can be used to study interactions between proteins. These approaches complement 2-D PAGE and mass spectrometry data by helping to elucidate functional cellular pathways.

Databases and Computational Approaches

There is an ever-increasing number of protein and proteome databases being developed. The most comprehensive information about specific proteins is found in databases that store protein sequences. One of the first and probably the best known such database is SWISS-PROT, which was created in 1986.

SWISS-PROT is a curated database that provides not only protein sequences but also such information as descriptions of a protein's function, its domain structure, and post-translational modifications, as well as links to other related databases. Other sequence-based protein databases include the Yeast Proteome Database and Human PSD.

There are also a number of widely used pattern and profile databases that are used to reveal relationships among proteins based on the presence of particular groups of amino acids in the proteins' sequences. Such groups, known as patterns, motifs, domains, signatures, or fingerprints, are found in specific regions of proteins that are important to some function of the protein. They could be in an area that performs some type of enzymatic activity or that is the site of a certain post-translational modification. Both their sequence and structure are typically well conserved. Some of the best known pattern and profile databases are: PROSITE, Pfam, PRINTS, and BLOCKS. SEE ALSO ALTERNATIVE SPLICING; BIOINFORMATICS; GEL ELECTROPHORESIS; GENOME; HUMAN GENOME PROJECT; MASS SPECTROMETRY; POST-TRANSLATIONAL CONTROL; PROTEINS.

Anthony J. Recupero

Bibliography

Wilkins, Marc R., et al. eds. *Proteome Research: New Frontiers in Functional Genomics.* New York: Springer-Verlag, 1997.

Pseudogenes

Pseudogenes are defective copies of functional genes. These may be partial or complete duplicates derived from polypeptide-encoding genes or RNA genes. The DNA sequence of a pseudogene is characteristically very similar to its functional counterpart, but contains variant mutations that render the gene inactive. The functional polypeptide-encoding gene contains an open reading frame, a long stretch of nucleotides that are transcribed and subsequently translated into a series of amino acids uninterrupted by stop **codons**. In contrast, pseudogenes derived from polypeptide sequences generally are punctuated with stop codons, effectively rendering them incapable of producing a functional protein.

Pseudogenes may also contain frameshift **mutations**, yielding a change in the reading frame. Additionally, there may be mutations that inactivate regulatory elements or intron-splicing sites. In either case, the duplicated

codons sequences of three mRNA nucleotides coding for one amino acid

mutations changes in DNA sequences

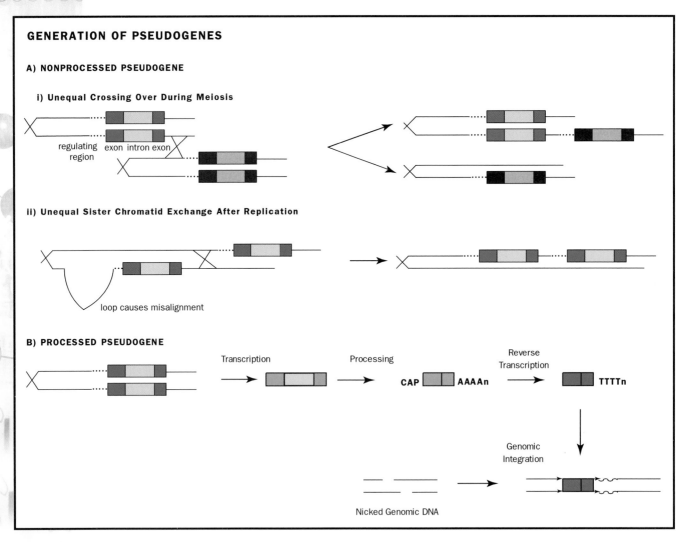

Figure 1. Generation of pseudogenes. (A) Direct duplication of a DNA sequence including regulatory elements and introns yields a nonprocessed pseudogene. (B) Duplication via an RNA intermediate, processed with a 3′ polyadenylated tail, 5′ cap (7-methylguanosine), and intron removal, yields a retropseudogene. The arrows indicate direct repeat sequences resulting from the integration. The A-T refers to a stretch of adenines and complement thymines in the newly derived pseudogene.

gene may be rendered nonfunctional. Genes and pseudogenes derived from the duplication of an ancestral gene are said to be paralogous.

Nonprocessed Pseudogenes

Gene duplication may occur by a direct increase of DNA content (nonprocessed) or via an RNA intermediate (processed). Nonprocessed pseudogenes can arise by unequal crossing over in homologous chromosomes (paired meiotic chromosomes containing the same genetic loci) or unequal sister **chromatid** exchange (crossover at improperly aligned sequences) after replication in a single chromosome (Figure 1A). Replication slippage also increases DNA content by looping of the synthesized strand during DNA replication, but typically involves short sequence stretches such as microsatellites (short repeated sequences).

A nonprocessed duplicated gene contains the introns and regulatory sequences of the original gene. This yields genetic redundancy, which allows

chromatid a replicated chromosome before separation from its copy

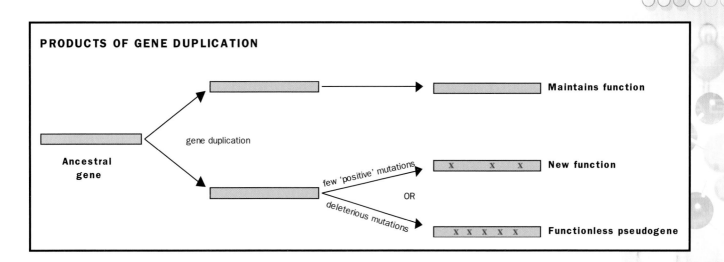

PRODUCTS OF GENE DUPLICATION

Ancestral gene

gene duplication

few 'positive' mutations

OR

deleterious mutations

Maintains function

New function

Functionless pseudogene

Figure 2. Gene duplication yields either the acquisition of a gene with novel function or a pseudogene. X refers to mutations.

one of the genes to acquire mutations, becoming a nonfunctional pseudogene (Figure 2). Occasionally, the duplicated gene acquires mutations yielding a gain-of-function that differs from the original gene (Figure 2). This may allow the evolution of new capabilities in the organism possessing it.

New genes generated by nonprocessed duplication are generally located in the vicinity of the ancestral (original) gene within the genome. However, it is possible that these genes may become separated from each other as a result of major chromosomal rearrangements (such as translocations). Examples of functional and nonfunctional duplicated genes in adjacent locations and on different chromosomes are exhibited by the globin gene superfamily.

Processed Pseudogenes

Pseudogenes generated via a messenger RNA (mRNA) intermediate demonstrate the features of processed RNA. These genes lack the flanking transcriptional regulatory sequences, do not contain introns, and typically have a polyadenylated 3′ (3-prime) region (adenine-containing nucleotides are added to the mRNA in eukaryotes). These sequences are converted to complementary DNA (cDNA) by the enzyme reverse transcriptase, and then integrated back in the genome at a new location (Figure 1B). These elements, therefore, are not necessarily in the chromosomal vicinity of the original sequence, and are essentially dead on arrival. Processed pseudogenes are also referred to as retropseudogenes.

Processed pseudogenes may also be derived from other RNA genes, such as tRNA (transfer RNA), rRNA (ribosomal RNA), snRNA (small nuclear RNA), and 7SL RNA. Evidence for this phenomenon includes the identification of nonfunctional tRNA genes containing a CCA sequence at the 3′ terminal. CCA is not part of the original DNA sequence, but is enzymatically added to the tRNA molecule following transcription. The gene for 7SL RNA is an integral component of the signal recognition particle complex involved in transmembrane protein transport. The 7SL RNA gene is thought to be the ancestral gene for the primate Alu and rodent B1 retroposons, based on sequence similarities. (Retroposons are a type of transposable genetic element that is found littered throughout the genome.) The *Alu* element differs from 7SL by having two internal sequence deletions,

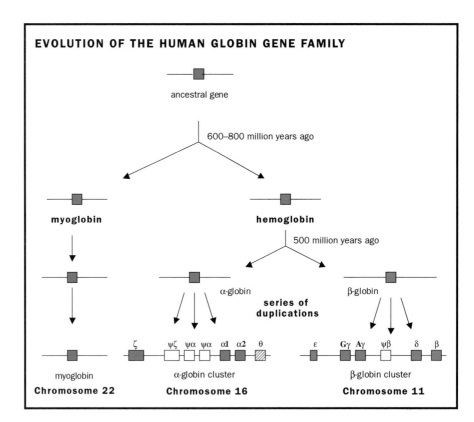

EVOLUTION OF THE HUMAN GLOBIN GENE FAMILY

ancestral gene

600–800 million years ago

myoglobin   hemoglobin

500 million years ago

α-globin   series of duplications   β-globin

ζ   ψζ ψα ψα α1 α2   θ   ε   Gγ Aγ ψβ   δ   β

myoglobin   α-globin cluster   β-globin cluster

Chromosome 22   **Chromosome 16**   **Chromosome 11**

duplication of the entire sequence, and numerous nucleotide substitutions. At some point in its evolutionary past a 7SL retropseudogene apparently integrated into a highly fortuitous location, as there are about 1.5 million *Alu* elements in the human genome, accounting for approximately 10 percent of our DNA. Most *Alu* elements are retropositionally incompetent pseudogenes, hence incapable of generating additional copies. Other retroposons are thought to be derived from processed duplicated tRNA genes (for example, rodent B2 and ID elements).

Pseudogene Examples

The globin gene superfamily provides an interesting example of the generation of both functional and nonfunctional duplicated genes (Figure 3). Based on nucleotide sequence data, it appears that a gene duplication occurred about 600 to 800 million years ago, yielding myoglobin and hemoglobin genes. Another duplication of the hemoglobin gene occurred about 500 million years ago, yielding **α**-globin and **β**-globin genes. (Adult human hemoglobin contains two α and two β strands.) These are all functional genes, found on three different human chromosomes. The α-globin and β-globin genes further duplicated, yielding both pseudogenes and functional genes. Possession of more than one globin gene provides a selective advantage because it compensates for the variation of oxygen in the prenatal versus postnatal environment.

The α-globin gene cluster consists of three functional genes and three pseudogenes. There is also an additional gene that is expressed but not incorporated into a hemoglobin molecule. In other words, this would be an example of an expressed pseudogene. The β-globin gene complex con-

α the Greek letter alpha

β the Greek letter beta

sists of five functional genes and one pseudogene. Examples of other processed polypeptide-encoding pseudogenes include those derived from actin, ferritin, and glyceraldehyde 3-phosphate dehydrogenase genes. SEE ALSO EVOLUTION OF GENES; GENE; GENE FAMILIES; HEMOGLOBINOPATHIES; READING FRAME; REPLICATION; RNA; RNA PROCESSING.

David H. Kass and Mark A. Batzer

Bibliography

Brown, Terence A. *Genomes*. New York: John Wiley & Sons, 1999.

Li, Wen-Hsiung. *Molecular Evolution*. Sunderland, MA: Sinauer Associates, 1997.

Strachan Tom, and Andrew P. Read. *Human Molecular Genetics*, 2nd ed. New York: John Wiley & Sons, 1999.

Psychiatric Disorders

Genetic studies of psychiatric disorders have become an important specialty area within medical genetics. Much of the progress in the area is the result of advances in molecular genetics techniques, The Human Genome Project, developments in the neurosciences, and recent genetic findings in complex brain disorders such as Alzheimer's and Huntington's disease.

Psychiatric Disorders with Genetic Involvement

It is well accepted that many of the psychiatric disorders listed in the *Diagnostic and Statistical Manual*, fourth edition, have hereditary predispositions. (The *DSM-IV* is the official diagnostic manual for mental disorders in the United States.) In the last two decades, clinical genetic (family, twin, and adoption) studies have provided some valuable information about both the genetic and environmental aspects that lead to the development of these psychiatric disorders. More recently, a growing number of molecular genetics studies have provided more data about the genetics of the disorders. However, there has been little success in the identification of specific genes involved in the **etiology** of these complex disorders.

etiology causation of disease, or the study of causation

The *DSM-IV* includes more than 350 diagnoses and at least one third of the disorders have had one or more clinical genetic study completed. Additionally, some of the disorders have been subjected to molecular genetic studies. Several of the most frequently examined *DSM-IV* psychiatric disorders from a "genetic study" perspective are listed below:

- Alcoholism
- Alzheimers
- Attention Deficit Hyperactivity Disorder (ADHD)
- Autism
- Bipolar Disorder
- Mental Retardation
- Obsessive-Compulsive Disorder (OCD)
- Panic Disorder
- Schizophrenia

- Social Phobia

- Substance Use Disorder

- Tourette's Syndrome

Major psychiatric disorders, such as schizophrenia and bipolar disorder, have been shown in numerous studies to have significant genetic factors involved in their etiology. The effects of environment and gene-environment interactions on the expression of the disorders are thought to be important, but need more study. Several childhood onset disorders, such as mental retardation, autism, and ADHD, also have strong genetic components in their etiology. All of these disorders are considered complex and are thought to have multiple genes acting together with nongenetic factors, with each gene typically contributing only small effects.

Inheritance Patterns and Linkage Studies

Complex disorders such as schizophrenia and bipolar disorder usually do not follow classic Mendelian inheritance patterns, but they can frequently mimic a pattern of autosomal dominance with reduced penetrance. Explanatory models for these complex disorders include multifactorial inheritance (multiple genes with nongenetic components) and epistasis (few genes acting jointly). But, since the mode of inheritance is unknown, a range of analytic methods must be used to study the genetic aspects of these disorders. Linkage and allelic association studies are frequently methods used to investigate possible causal genes for complex psychiatric disorders. However, in large populations, there are likely to be several causal or susceptibility genes and nongenetic causes, as well.

Historically, the first positive linkage study to have a major impact on psychiatry was the linkage of bipolar disorder to chromosome eleven in a large Amish family. However, a later assessment of the family failed to confirm the original linkage results. A similar situation occurred with the failure to replicate the findings of an early report of linkage of schizophrenia to a region on chromosome 5 in Icelandic and British families.

As a result of these early problems, it became clear that more rigorous methodology, such as more careful clinical phenotyping, the use of more genetic markers, and new analytic techniques would be necessary if molecular approaches were to be used in finding genes involved in psychiatric disorders.

Clinical genetic (family, twin, and adoptive) studies have been attempted in a large number of psychiatric disorders. However, these studies generally provide no information about what genes are involved in the disorder. Molecular studies are necessary to begin to elucidate this data. Schizophrenia and bipolar disorder are examples of psychiatric disorders that have been studied for nearly two decades.

Schizophrenia and Bipolar Disorder

Schizophrenia is a disorder characterized by psychotic symptoms such as hallucinations, delusions, and disordered thinking, as well as deficits in emotional and social behavior. Numerous family, twin, and adoption studies have provided substantial evidence for genetic factors in the **etiology** of this dis-

etiology causation of disease, or the study of causation

214

order. Nongenetic factors also appear to play an important role. The risk for relatives of an individual with schizophrenia is 6.5 percent, nearly eight times the population rate of the disorder. The concordance rate for monozygotic (identical) twins is 45 percent, and for dizygotic (fraternal) twins the rate is 12 percent. The higher rate for monozygotic twins, who share all their genes, is strong evidence that genes play a role in schizophrenia.

Molecular genetic (linkage and association) studies suggest that there are multiple genes with small effects that predispose one to the disorder. In an early study, chromosome 5q was implicated, but was not replicated. Areas on chromosome 6p and 8p have been replicated in at least one follow-up study. More recently, chromosome 1 has been implicated. While no causative genes have been identified, there have been molecular studies that implicate chromosomes 1q, 3p, 5, 6p, 6q, 8p, 10p, 13q, 18p and 22q in schizophrenia. Like other multifactoral diseases, the etiology of schizophrenia involves multiple genes and gene-environment interactions.

Bipolar disorder is characterized by episodes of depression and mania, elevated or irritable mood, and symptoms such as rapid thoughts, grandiose ideas, and reckless behavior. Population, twin, and adoption studies provide evidence for the role of genetics in the etiology of bipolar disorder. First-degree relatives have about a 7 to 10 percent risk of having the disorder once one family member is diagnosed. The concordance rate for monozygotic twins is 60 to 65 percent, and for dizygotic twins the rate is 10 to 15 percent, the same as for non-twin siblings.

One of the first bipolar disorder molecular genetic studies implicated chromosone 11, but this finding was not replicated in several other studies. A similar failure to replicate occurred with the initial reports of linkage on the chromosome X. Regions on chromosome four are reported to show strong evidence of linkage to some bipolar families. Major efforts in the last few years have been focused on chromosome seven and eighteen. Some of the studies have suggested a parent-of-origin effect, with maternal transmission more common than paternal. Both linkage and association studies have implicated chromosomes 4p, 6p, 12q, 13q, 16p, 18p, 18q, 21q, and 22q in bipolar disorder. Major depressive disorder has also been studied, but the underlying genetic factors have not been identified.

Tourette's Syndrome

Some psychiatric disorders, like Tourette's syndrome, have significant overlap (co-morbidity) with other psychiatric disorders (OCD and ADHD in the case of Tourette's). This may make the elucidation of genetic causes more difficult.

Tourette's syndrome is characterized by multiple motor tics and one or more vocal tics. It has onset before eighteen years and is one and one-half to three times more common in males. Twin, adoption, and segregation analysis studies support a genetic etiology for Tourette's syndrome. An **autosomal** dominant inheritance was initially suggested, but other inheritance patterns have been recently reported. There appears to be a relation with ADHD (up to 50 percent of individuals with Tourette's also have ADHD) and OCD (up to 40 percent of individuals with Tourette's have OCD). First-degree relatives are at high risk for developing tics and obsessive compulsive disorders.

autosomal describes a chromosome other than the X and Y sex-determining chromosomes

Studying psychiatric disorders for genetic factors in their etiology is difficult because of the phenotype definition, co-morbidity, and multiple causal factors. The summary finding from the genetic studies to date suggest that there are multiple genes of small to moderate effect underlying the predisposition to these complex psychiatric disorders. SEE ALSO ALZHEIMER'S DISEASE; ATTENTION DEFICIT HYPERACTIVITY DISORDER; COMPLEX TRAITS; GENE DISCOVERY; INHERITANCE PATTERNS.

Harry H. Wright and Ruth Abramson

Bibliography

Malhotra, A. K. "The Genetics of Schizophrenia." *Current Opinions in Psychiatry* 14, no. 1 (2001): 3–7.

Potash, J. B., and J. R. DePaulo. "Searching High and Low: A Review of the Genetics of Bipolar Disorder." *Bipolar Disorders* 2, no. 1 (2000): 8–26.

Roy, M. A., C. Merette, and M. Maziado. "Introduction to Psychiatric Genetics: Progress in the Search for Psychiatric Disorder Dusceptibility Genes." *Canadian Journal of Psychiatry* 46, no. 1 (2001): 52–60.

Public Health, Genetic Techniques in

As of 2002 more than ten thousand genes have been discovered, and it is estimated that 30,000 to 70,000 human genes will be identified as a result of the Human Genome Project over the following few years. Tests for more than 600 gene variants are already available in medical practice.

Genetic variants, or polymorphisms, are a normal part of genetic viability that may or may not be associated with an increase or decrease in disease risk. With advances in biotechnology, newly characterized genetic variants are being identified at a rapid rate. The challenge will be to ensure the appropriate use of genetic information to improve health and prevent disease in individuals, families, and communities.

The broad mission of public health is to act in society's interest to assure conditions in which people can be healthy. Public health genetics is the application of advances in genetics and molecular **biotechnology** to improve the public's health and prevent disease. Rapid progress in biotechnology, in sequencing the human genome, and in characterizing gene expression have generated high hopes of finding new ways to improve the public's health through the prevention and treatment of diseases.

At the same time, concerns have been raised about the premature application of such technologies and knowledge. Genetic testing and the understanding of how factors such as **hormones**, diet, and the environment influence risk in susceptible individuals and human populations present important challenges in the field of public health genetics. Ethical issues, health care policy priorities, risks of discrimination in employment and insurance, and complex psychological aspects within families are also crucial issues in this field.

Public Health Approaches in Genetics

Disease-related genetic variants that have been characterized include those associated with rare diseases as well as those that increase susceptibility to

biotechnology production of useful products

hormones molecules released by one cell to influence another

common chronic diseases such as cancer and heart disease. Risk for almost all human diseases results from the interactions between inherited gene variants and environmental factors, including chemical, physical, and infectious agents, as well as behavioral or nutritional factors. Thus it appears reasonable to direct disease-prevention and health-promotion efforts toward individuals at high risk because of their genetic makeup.

To function effectively, public health genetics needs to meet several challenges, including (1) the implementation of research, (2) the evaluation of genetic information and tests, (3) the development, implementation, and evaluation of population interventions, and (4) effective communication and information dissemination. With the realization that all human disease is the result of interactions between genetic variation and the environment (dietary, infectious, chemical, physical, and social factors), it becomes evident that it is important to identify the modifiable risk factors for disease that interact with the genetic variation. This approach can be used to help develop preventive strategies. To be able to deliver appropriate genetic tests and services for disease prevention and health promotion, it is important to integrate genetic services into disease-prevention and health-promotion activities.

Applied Research

Public-health assessment relies on scientific approaches such as surveillance and **epidemiology** to assess the impact of discovered gene variants on the health of communities. Surveillance, which involves the systematic gathering, analysis, and dissemination of population data, is needed to determine population frequencies of genetic variants that predispose people to specific diseases. This information can be used to assess the population morbidity and mortality associated with such diseases. Surveillance can also be used to determine the prevalence and effect of environmental factors known to interact with given genotypes. Additionally, surveillance can aid in the evaluation of a genetic test in terms of its safety, effectiveness, and cost-effectiveness.

epidemiology study of incidence and spread of diseases in a population

Another example where surveillance can come into play in public health genetics is with infectious diseases and DNA fingerprinting. The ability to characterize **pathogenic** organisms phenotypically and genotypically can be a powerful approach that provides information important for an antimicrobial surveillance program. It also is a means of providing information that may be useful for understanding pathogenic microorganisms worldwide. With such a program in place, the clonal spread of multiresistant pathogens among patients can be identified.

pathogenic disease-causing

Epidemiology is the study of the distribution and determinants of health-related states or events in populations and is used to investigate risk factors for various diseases and identify high-risk populations. Epidemiological information can be used to target populations that could benefit most from prevention and intervention actions. Epidemiology is also a tool for evaluating the effect of health programs and services on a population's health. Population-based epidemiological studies are increasingly needed to quantify the impact of gene variants on the risk of disease, death, and disability, and to identify and quantify the impact of modifiable risk factors that interact with gene variants.

metabolites molecules involved in metabolic pathways

Evaluation of Genetic Information and Tests

Genetic tests include the analysis of human DNA, RNA, chromosomes, proteins, and certain **metabolites** to detect a person's genotype for clinical purposes, including predicting the risk of disease, identifying gene mutation carriers, and establishing prenatal and clinical diagnoses or prognoses. Successful implementation of genetic tests to improve public health requires careful assessment of how and when genetic tests can and should be used to promote health and diagnose and prevent disease. This assessment must include the development of standards and guidelines for assuring quality genetic testing, and the consideration of ethical and legal issues.

Genetic tests need to be evaluated on the basis of several parameters before they can be taken from research laboratory to clinic. It is necessary to assess (1) how good the test is in predicting the underlying genotype, (2) how good the test is in diagnosing or predicting the phenotype or disease, and (3) the benefits and risks of the genetic test and ensuing interventions. Genetic test validity is quantified in terms of sensitivity (the probability of testing positive for the genetic test if there is a gene mutation or if disease occurs), specificity (the probability of testing negative if the genetic mutation is not present or if the disease does not occur), and predictive value (the test's ability to accurately predict disease).

All clinical laboratories in the United States that provide information to referring physicians are certified under the Clinical Laboratory Improvement Act (CLIA) amendments of 1988. The CLIA standards for quality control, proficiency testing, personnel, and other quality assurance practices apply to all genetic tests.

Development, Implementation, and Evaluation of Population Interventions

Recent advances in human genetics have brought high expectations for implementing prevention strategies among genetically susceptible individuals. Yet the clinical use of this information poses risks as well. It is the role of public health to develop intervention strategies for diseases with a genetic component, implement pilot demonstration programs, and evaluate the impact of the intervention on reducing morbidity and mortality in the population. This evaluation includes conducting a needs assessment of genetic services, studying the impact of genetic counseling on public health, and applying prevention-effectiveness principles to genetics programs. Policy analysis of **informed consent** to genetic testing, stigmatization of individuals and groups, discrimination in employment, and access to insurance need to be considered.

informed consent knowledge of risks involved

The two recently identified susceptibility genes *BRCA1* and *BRCA2*, which are associated with a high risk of developing breast and ovarian cancers, illustrate some of the complexities individuals from high-risk families face. Studies to determine the efficacy of prophylactic surgeries, chemoprophylactic strategies, and other preventive measures have not been conclusive, leaving individuals from high-risk families—and those who are carriers of mutations—with complex decisions concerning genetic testing and medical intervention.

Newborn screening illustrates the evolution of effective genetic-screening programs. Recognizing the potential importance of phenylketonuria

screening over forty years ago, the Children's Bureau (now called the Maternal and Child Health Bureau) sponsored a multistate urine-screening program. The initial outcome was that 30 percent of infants remained untested due to inadequate specimen-collection and delivery. There was sometimes a false-positive reading. Improvements were made, and today a single drop of blood, rather than a urine sample, is sufficient for eight to ten assays for metabolic-disease indicators, along with genetic and infectious information about the mother.

Another example of an evolving genetic-screening program involves the common abnormal hemoglobin "S," or sickle hemoglobin, detected in newborn-screening programs in the United States. This hemoglobin is the defining characteristic for sickle cell disease. The first statewide screening program was established in 1975. Widespread acceptance and implementation was lacking until after a 1986 study showed the efficacy of daily oral penicillin prophylaxis in preventing infection among young children with sickle cell anemia. The efficacy of sickle cell screening was demonstrated through epidemiological efforts to evaluate pediatric outcomes after newborn screening, by demonstrating that mortality rates declined from 1968 to 1992, particularly in cohorts of sickle cell patients.

Communication and Information Dissemination

The fourth public-health function in genetics is developing and applying communication principles and strategies related to advances in human genetics, interventions, and genetic tests and services, as well as in interventions, the ethical, legal, and social issues related to these topics. Public health agencies can play a role in translating the very complex information related to genetics and disease prevention to health care workers and the public. An appropriate mix of mechanisms should be used to disseminate information, including distance-based interactive meetings, information centers, and electronic communication.

Summary

In summary, four public health functions for genetics have been outlined that underscore the complexities involved in public health genetics. All are carried out by efforts among various groups, including partnerships and coordinated efforts among federal, state, and local agencies, the public and private sectors, and the public-health, medical, and academic sectors, with various levels of community and consumer involvement. Annual national meetings on genetics and public health facilitate these efforts. SEE ALSO ANTIBIOTIC RESISTANCE; CANCER; DIABETES; DISEASE, GENETICS OF; GENETIC DISCRIMINATION; GENETIC TESTING; HEMOGLOBINOPATHIES; METABOLIC DISEASE; PRENATAL DIAGNOSIS.

Joellen M. Schildkraut

Bibliography

Centers for Disease Control and Prevention. "Regulations for Implementing Clinical Laboratory Improvements Amendments of 1988: A Summary." *Morbidity and Mortality Weekly Report* 41, no. RR-2 (1992): 1–17.

Condit, C. M., R. L. Parrott, and B. O'Grady. "Principles and Practices of Communication Processes for Genetics in Public Health." In *Genetics and Public Health in the Twenty-First Century: Using Genetic Information to Improve Health and Prevent*

Disease. Muin J. Khoury, Wylie Burke, and Elizabeth J. Thomson, eds. New York: Oxford University Press, 2000.

Coughlin, S. S. "The Intersection of Genetics, Public Health, and Preventive Medicine." *American Journal of Preventive Medicine* 16, no. 2 (1999): 89–90.

Davis, H., et al. "National Trends in the Mortality of Children with Sickle Cell Disease, 1968 through 1992." *American Journal of Public Health* 87 (1997): 1317–1322.

Khoury, Muin J. "Genetic Epidemiology and the Future of Disease Prevention and Public Health." *Epidemiologic Reviews* 19, no. 1 (1997): 175–189.

Khoury, Muin J., Wylie Burke, and Elizabeth J. Thomson. "A Framework for the Integration of Human Genetics into Public Health Practice." In *Genetics and Public Health in the Twenty-First Century: Using Genetic Information to Improve Health and Prevent Disease.* Muin J. Khoury, Wylie Burke, and Elizabeth J. Thomson., eds. New York: Oxford University Press, 2000.

Lee, A., et al. "Improved Survival in Homozygous Sickle Cell Disease: Lessons from a Cohort Study." *British Medical Journal,* 311 (1995): 1600–1602.

Leikin, S. L., et al. "Mortality in Children and Adolescents with Sickle Cell Disease. Cooperative Study of Sickle Cell Disease." *Pediatrics* 84 (1989): 500–508.

Pass, K. A. "Lessons Learned from Newborn Screening for Phenylketonuria." In *Genetics and Public Health in the Twenty-First Century: Using Genetic Information to Improve Health and Prevent Disease.* Muin J. Khoury, Wylie Burke, and Elizabeth J. Thomson., eds. New York: Oxford University Press, 2000.

Pfaller, M. A., et al. "Integration of Molecular Characterization of Microorganisms in a Global Antimicrobial Resistance Surveillance Program." *Clinical Infectious Diseases* 32, supplement 2 (2001): S156–S167.

Schildkraut, J. M. "Examining Genetic Interactions." *Approaches to Gene Mapping in Complex Diseases,* Jonathan L. Haines, Margaret A. Pericak-Vance, eds. New York: John Wiley & Sons, 1998.

Terris, M. "The Society of Epidemiologic Research and the Future of Epidemiology." *American Journal of Epidemiology* 136 (1992): 909–915.

Internet Resources

GeneTests. Roberta A. Pagon and Peter Tarczy-Hornoch, eds. University of Washington. <http://www.genetests.org/>.

Public Health Genetics in the Context of Law, Ethics and Policy. Institute for Public Health Genetics, University of Washington School of Public Health and Community Medicine. <http://depts.washington.edu/phgen/>.

Purification of DNA

Many procedures in molecular biology require an initial pure sample of DNA. These procedures include the polymerase chain reaction, sequencing, gene cloning, blotting, and DNA profiling. Purification of DNA involves removing it and other constituents from the cell, separating it from the various other cell constituents, and protecting it from degradation by cellular **enzymes**. Isolation procedures must also be gentle enough that the long DNA strands are not sheared by mechanical stress.

enzymes proteins that control a reaction in a cell

DNA can be isolated from almost any cellular source. White blood cells and cheek cells taken directly from humans are most commonly used for diagnostic purposes, but skin, hair follicles, semen, and other tissues can be used for **forensic** analysis. Cells grown in petri dishes or in suspension can also be used. The cells are isolated from any surrounding fluid (such as blood serum) by centrifuging them—spinning them at high speeds—and then are resuspended in a buffer solution. The buffer prevents rapid or dramatic changes in pH, which can interfere with subsequent reactions. To break open the cell membranes, a detergent is added to the buffer. Sodium dodecyl sulfate (SDS) is often used for this purpose. The detergent also helps remove proteins and lipids in the cell.

forensic related to legal proceedings

The buffer also contains ethylenediaminetetraacetic acid (EDTA), which is a chelator. Chelators are molecules that act as scavengers for metal ions in solution. This is important because DNase, an enzyme that digests DNA, is present in the cell and would destroy the long DNA strands if it was active. DNase activity requires magnesium ions, and EDTA removes them from solution, preventing DNase from cutting up the DNA. RNase is also present in the buffer at this step, to break up the RNA present in the cells.

The solution is then treated with proteinase K, a highly effective enzyme that inactivates all types of proteins. This enzyme can also be used earlier in the procedure to break apart clumped cells. Unlike many proteins, proteinase K remains active at elevated temperatures, so the solution can be heated to about 55 °C to aid protein inactivation and removal by the detergent. This step may last between two and sixteen hours.

Once the cells are broken open and the RNA, proteins, and lipids have been dissolved in the buffer, the DNA must be separated from these materials. One standard technique uses phenol to remove the proteins, leaving DNA and other water-soluble materials behind. The DNA is then extracted from the water phase using chloroform and precipitated from the chloroform using ethyl alcohol mixed with sodium acetate salt. The DNA is then removed either by spooling the long threads onto a glass rod, or by spinning it out of solution using a centrifuge. The DNA is then resuspended in **buffer**.

Another technique for separating the DNA from the mixture avoids the use of phenol and chloroform, which are toxic. Instead, the proteins are "salted out" by adding a concentrated salt solution and then removed by centrifuging. In another method, DNA is adsorbed onto very small glass beads, in the presence of "chaotropic" salts, which disrupt protein structure. The beads are removed (or washed in place), and the DNA is released by changing the salt concentration.

Once a pure sample of DNA has been obtained, the fragments it contains may be separated on the basis of size by **gel electrophoresis**. Specific sequences can be identified by **Southern blotting** using probes with complementary sequences, and they can then be cut out of the gel for further use, such as cloning. If the sample is small, the DNA can be amplified by the **polymerase** chain reaction.

A crude preparation of DNA can be made in the kitchen with simple ingredients, including baking soda (which acts as a buffer), laundry detergent, table salt, and rubbing alcohol. Meat tenderizer (an enzyme preparation) can be used to destroy proteins. SEE ALSO BLOTTING; CLONING GENES; DNA; DNA PROFILING; GEL ELECTROPHORESIS; POLYMERASE CHAIN REACTION; SEQUENCING DNA.

Richard Robinson

buffers substances that counteract rapid or wide pH changes in a solution

gel electrophoresis technique for separation of molecules based on size and charge

Southern blotting separating DNA fragments by electrophoresis and then identifying a target fragment with a DNA probe

polymerase enzyme complex that synthesizes DNA or RNA from individual nucleotides

Bibliography

Bloom, Mark V., Greg A. Freyer, and David A. Micklos. *Laboratory DNA Science : An Introduction to Recombinant DNA Techniques and Methods of Genome Analysis.* Menlo Park, CA: Benjamin Cummings, 1996.

Carlson, Shawn. "Spooling the Stuff of Life." *Scientific American* September (1998): 74–75.

Rapley, Ralph, and John M. Walker. *Biomolecular Methods.* Totowa, NJ: Humana Press, 1998.

Photo Credits

Unless noted below or within its caption, the illustrations and tables featured in *Genetics* were developed by Richard Robinson, and rendered by GGS Information Services. The photographs appearing in the text were reproduced by permission of the following sources:

Volume 1

Accelerated Aging: Progeria (p. 2), Photo courtesy of The Progeria Research Foundation, Inc. and the Barnett Family; *Aging and Life Span* (p. 8), Fisher, Leonard Everett, Mr.; *Agricultural Biotechnology* (p. 10), © Keren Su/Corbis; *Alzheimer's Disease* (p. 15), AP/Wide World Photos; *Antibiotic Resistance* (p. 27), © Hank Morgan/Science Photo Library, Photo Researchers, Inc.; *Apoptosis* (p. 32), © Microworks/Phototake; *Arabidopsis thaliana* (p. 34), © Steinmark/ Custom Medical Stock Photo; *Archaea* (p. 38), © Eurelios/Phototake; *Behavior* (p. 47), © Norbert Schafer/Corbis; *Bioinformatics* (p. 53), © T. Bannor/Custom Medical Stock Photo; *Bioremediation* (p. 60), Merjenburgh/ Greenpeace; *Bioremediation* (p. 61), AP/Wide World Photos; *Biotechnology and Genetic Engineering, History of* (p. 71), © Gianni Dagl Orti/Corbis; *Biotechnology: Ethical Issues* (p. 67), © AFP/Corbis; *Birth Defects* (p. 78), AP/Wide World Photos; *Birth Defects* (p. 80), © Siebert/ Custom Medical Stock Photo; *Blotting* (p. 88), © Custom Medical Stock Photo; *Breast Cancer* (p. 90), © Custom Medical Stock Photo; *Carcinogens* (p. 98), © Custom Medical Stock Photo; *Cardiovascular Disease* (p. 102), © B&B Photos/Custom Medical Stock Photo; *Cell, Eukaryotic* (p. 111), © Dennis Kunkel/ Phototake; *Chromosomal Aberrations* (p. 122), © Pergement, Ph.D./Custom Medical Stock

Photo; *Chromosomal Banding* (p. 126), Courtesy of the Cytogenetics Laboratory, Indiana University School of Medicine; *Chromosomal Banding* (p. 127), Courtesy of the Cytogenetics Laboratory, Indiana University School of Medicine; *Chromosomal Banding* (p.128), Courtesy of the Cytogenetics Laboratory, Indiana University School of Medicine; *Chromosome, Eukaryotic* (p. 137), Photo Researchers, Inc.; *Chromosome, Eukaryotic* (p. 136), © Becker/Custom Medical Stock Photo; *Chromosome, Eukaryotic* (p. 133), Courtesy of Dr. Jeffrey Nickerson/University of Massachusetts Medical School; *Chromosome, Prokaryotic* (p. 141), © Mike Fisher/Custom Medical Stock Photo; *Chromosomes, Artificial* (p. 145), Courtesy of Dr. Huntington F. Williard/University Hospitals of Cleveland; *Cloning Organisms* (p. 163), © Dr.Yorgos Nikas/Phototake; *Cloning: Ethical Issues* (p. 159), AP/Wide World Photos; *College Professor* (p. 166), © Bob Krist/Corbis; *Colon Cancer* (p. 169), © Albert Tousson/Phototake; *Colon Cancer* (p. 167), © G-I Associates/Custom Medical Stock Photo; *Conjugation* (p. 183), © Dennis Kunkel/Phototake; *Conservation Geneticist* (p. 191), © Annie Griffiths Belt/ Corbis; *Delbrück, Max* (p. 204), Library of Congress; *Development, Genetic Control of* (p. 208), © JL Carson/Custom Medical Stock Photo; *DNA Microarrays* (p. 226), Courtesy of James Lund and Stuart Kim, Standford University; *DNA Profiling* (p. 234), AP/Wide World Photos; *DNA Vaccines* (p. 254), Penny Tweedie/Corbis-Bettmann; *Down Syndrome* (p. 257), © Custom Medical Stock Photo.

Volume 2

Embryonic Stem Cells (p. 4), Courtesy of Dr. Douglas Strathdee/University of Edinburgh, Department of Neuroscience; *Embryonic Stem Cells* (p. 5), Courtesy of Dr. Douglas Strathdee/University of Edinburgh, Department of Neuroscience; *Escherichia coli* (*E. coli* bacterium) (p. 10), © Custom Medical Stock Photo; *Eubacteria* (p. 14), © Scimat/ Photo Researchers; *Eubacteria* (p. 12), © Dennis Kunkel/Phototake; *Eugenics* (p. 19), American Philosophical Society; *Evolution, Molecular* (p. 22), OAR/National Undersea Research Program (NURP)/National Oceanic and Atmospheric Administration; *Fertilization* (p. 34), © David M. Phillips/Photo Researchers, Inc.; *Founder Effect* (p. 37), © Michael S. Yamashita/Corbis; *Fragile X Syndrome* (p. 41), © Siebert/Custom Medical Stock Photo; *Fruit Fly:* Drosophila (p. 44), © David M. Phillips, Science Source/Photo Researchers, Inc.; *Gel Electrophoresis* (p. 46), © Custom Medical Stock Photo; *Gene Therapy* (p. 75), AP/Wide World; *Genetic Counseling* (p. 88), © Amethyst/ Custom Medical Stock Photo; *Genetic Testing* (p. 98), © Department of Clinical Cytogenetics, Addenbrookes Hospital/Science Photo Library/ Photo Researchers, Inc.; *Genetically Modified Foods* (p. 109), AP/Wide World; *Genome* (p. 113), Raphael Gaillarde/Getty Images; *Genomic Medicine* (p. 119), © AFP/Corbis; *Growth Disorders* (p. 131), Courtesy Dr. Richard Pauli/U. of Wisconsin, Madison, Clinical Genetics Center; *Hemoglobinopathies* (p. 137), © Roseman/Custom Medical Stock Photo; *Heterozygote Advantage* (p. 147), © Tania Midgley/Corbis; *HPLC: High-Performance Liquid Chromatography* (p. 167), © T. Bannor/Custom Medical Stock Photo; *Human Genome Project* (p. 175), © AFP/ Corbis; *Human Genome Project* (p. 176), AP/ Wide World; *Individual Genetic Variation* (p. 192), © A. Wilson/Custom Medical Stock Photo; *Individual Genetic Variation* (p. 191), © A. Lowrey/Custom Medical Stock Photo; *Inheritance, Extranuclear* (p. 196), © ISM/ Phototake; *Inheritance Patterns* (p. 206), photograph by Norman Lightfoot/National Audubon Society Collection/Photo Researchers, Inc.; *Intelligence* (p. 208), AP/Wide World Photos.

Volume 3

Laboratory Technician (p. 2), Mark Tade/Getty Images; *Maize* (p. 9), Courtesy of Agricultural Research Service/USDA; *Marker Systems* (p. 16), Custom Medical Stock Photo; *Mass Spectrometry* (p. 19), Ian Hodgson/© Rueters New Media; *McClintock, Barbara* (p. 21), AP/Wide World Photos; *McKusick, Victor* (p. 23), The Alan Mason Chesney Medical Archives of The Johns Hopkins Medical Institutions; *Mendel, Gregor* (p. 30), Archive Photos, Inc.; *Metabolic Disease* (p. 38), AP/Wide World Photos; *Mitochondrial Diseases* (p. 53), Courtesy of Dr. Richard Haas/University of California, San Diego, Department of Neurosciences; *Mitosis* (p. 58), J. L. Carson/Custom Medical Stock Photo; *Model Organisms* (p. 61), © Frank Lane Picture Agency/Corbis; *Molecular Anthropology* (p. 66), © John Reader, Science Photo Library/ PhotoResearchers, Inc.; *Molecular Biologist* (p. 71), AP/Wide World Photos; *Morgan, Thomas Hunt* (p. 73), © Bettmann/Corbis; *Mosaicism* (p. 78), Courtesy of Carolyn Brown/ Department of Medical Genetics of University of British Columbia; *Muller, Hermann* (p. 80), Library of Congress; *Muscular Dystrophy* (p. 85), © Siebert/Custom Medical Stock Photo; *Muscular Dystrophy* (p. 84), © Custom Medical Stock Photo; *Nature of the Gene, History* (p. 103), Library of Congress; *Nature of the Gene, History* (p. 102), Archive Photos, Inc.; *Nomenclature* (p. 108), Courtesy of Center for Human Genetics/ Duke University Medical Center; *Nondisjunction* (p. 110), © Gale Group; *Nucleotide* (p. 115), © Lagowski/Custom Medical Stock Photo; *Nucleus* (p. 120), © John T. Hansen, Ph.D./ Phototake; *Oncogenes* (p. 129), Courtesy of National Cancer Institute; *Pharmacogenetics and Pharmacogenomics* (p. 145), AP/Wide World Photos; *Plant Genetic Engineer* (p. 150), © Lowell Georgia/Corbis; *Pleiotropy* (p. 154), © Custom Medical Stock Photo; *Polyploidy* (p. 165), AP/Wide World Photos; *Population Genetics* (p. 173), AP/Wide World Photos; *Population Genetics* (p. 172), © JLM Visuals; *Prenatal Diagnosis* (p. 184), © Richard T. Nowitz/Corbis; *Prenatal Diagnosis* (p. 186), © Brigham Narins; *Prenatal Diagnosis* (p. 183), © Dr. Yorgos Nikas/Phototake; *Prion* (p. 188), AP/Wide World Photos; *Prion* (p. 189), AP/

Wide World Photos; *Privacy* (p. 191), © K. Beebe/Custom Medical Stock Photo; *Proteins* (p. 199), NASA/Marshall Space Flight Center; *Proteomics* (p. 207), © Geoff Tompkinson/ Science Photo Library, National Aubodon Society Collection/Photo Researchers, Inc.; *Proteomics* (p. 206), Lagowski/Custom Medical Stock Photo.

Volume 4

Quantitative Traits (p. 2), Photo by Peter Morenus/Courtesy of University of Connecticut; *Reproductive Technology* (p. 20), © R. Rawlins/Custom Medical Stock Photo; *Reproductive Technology: Ethical Issues* (p. 27), AP/Wide World Photos; *Restriction Enzymes* (p. 32), © Gabridge/Custom Medical Stock Photo; *Ribosome* (p. 43), © Dennis Kunkel/ Phototake; *Rodent Models* (p. 61), © Reuters NewMedia Inc./Corbis; *Roundworm:*

Caenorhabditis elegans (p. 63), © J. L. Carson/Custom Medical Stock Photo; *Sanger, Fred* (p. 65), © Bettmann/Corbis; *Sequencing DNA* (p. 74), © T. Bannor/Custom Medical Stock Photo; *Severe Combined Immune Deficiency* (p. 76), © Bettmann/Corbis; *Tay-Sachs Disease* (p. 100), © Dr. Charles J. Ball/ Corbis; *Transgenic Animals* (p. 125), Courtesy Cindy McKinney, Ph.D./Penn State University's Transgenic Mouse Facility; *Transgenic Organisms: Ethical Issues* (p. 131), © Daymon Hartley/Greenpeace; *Transgenic Plants* (p. 133), © Eurelios/Phototake; *Transplantation* (p. 140), © Reuters New Media/Corbis; *Twins* (p. 156), © Dennis Degnan/Corbis; *X Chromosome* (p. 175), © Gale Group; *Yeast* (p. 180), © Dennis Kunkel; *Zebrafish* (p. 182), Courtesy of Dr. Jordan Shin, Cardiovascular Research Center, Massachusetts General Hospital.

Glossary

α the Greek letter alpha

β the Greek letter beta

γ the Greek letter gamma

λ the Greek letter lambda

σ the Greek letter sigma

E. coli the bacterium *Escherichia coli*

"-ase" suffix indicating an enzyme

acidic having the properties of an acid; the opposite of basic

acrosomal cap tip of sperm cell that contains digestive enzymes for penetrating the egg

adenoma a tumor (cell mass) of gland cells

aerobic with oxygen, or requiring it

agar gel derived from algae

agglutinate clump together

aggregate stick together

algorithm procedure or set of steps

allele a particular form of a gene

allelic variation presence of different gene forms (alleles) in a population

allergen substance that triggers an allergic reaction

allolactose "other lactose"; a modified form of lactose

amino acid a building block of protein

amino termini the ends of a protein chain with a free NH_2 group

amniocentesis removal of fluid from the amniotic sac surrounding a fetus, for diagnosis

amplify produce many copies of, multiply

anabolic steroids hormones used to build muscle mass

anaerobic without oxygen or not requiring oxygen

androgen testosterone or other masculinizing hormone

anemia lack of oxygen-carrying capacity in the blood

aneuploidy abnormal chromosome numbers

angiogenesis growth of new blood vessels

anion negatively charged ion

anneal join together

anode positive pole

anterior front

antibody immune-system protein that binds to foreign molecules

antidiuretic a substance that prevents water loss

antigen a foreign substance that provokes an immune response

antigenicity ability to provoke an immune response

apoptosis programmed cell death

Archaea one of three domains of life, a type of cell without a nucleus

archaeans members of one of three domains of life, have types of cells without a nucleus

aspirated removed with a needle and syringe

aspiration inhalation of fluid or solids into the lungs

association analysis estimation of the relationship between alleles or genotypes and disease

asymptomatic without symptoms

ATP adenosine triphosphate, a high-energy compound used to power cell processes

ATPase an enzyme that breaks down ATP, releasing energy

attenuation weaken or dilute

atypical irregular

autoimmune reaction of the immune system to the body's own tissues

autoimmunity immune reaction to the body's own tissues

autosomal describes a chromosome other than the X and Y sex-determining chromosomes

autosome a chromosome that is not sex-determining (not X or Y)

axon the long extension of a nerve cell down which information flows

bacteriophage virus that infects bacteria

basal lowest level

base pair two nucleotides (either DNA or RNA) linked by weak bonds

basic having the properties of a base; opposite of acidic

benign type of tumor that does not invade surrounding tissue

binding protein protein that binds to another molecule, usually either DNA or protein

biodiversity degree of variety of life

bioinformatics use of information technology to analyze biological data

biolistic firing a microscopic pellet into a biological sample (from biological/ballistic)

biopolymers biological molecules formed from similar smaller molecules, such as DNA or protein

biopsy removal of tissue sample for diagnosis

biotechnology production of useful products

bipolar disorder psychiatric disease characterized by alternating mania and depression

blastocyst early stage of embryonic development

brackish a mix of salt water and fresh water

breeding analysis analysis of the offspring ratios in breeding experiments

buffers substances that counteract rapid or wide pH changes in a solution

Cajal Ramon y Cajal, Spanish neuroanatomist

carcinogens substances that cause cancer

carrier a person with one copy of a gene for a recessive trait, who therefore does not express the trait

catalyst substance that speeds a reaction without being consumed (e.g., enzyme)

catalytic describes a substance that speeds a reaction without being consumed

catalyze aid in the reaction of

cathode negative pole

cDNA complementary DNA

cell cycle sequence of growth, replication and division that produces new cells

centenarian person who lives to age 100

centromere the region of the chromosome linking chromatids

cerebrovascular related to the blood vessels in the brain

cerebrovascular disease stroke, aneurysm, or other circulatory disorder affecting the brain

charge density ratio of net charge on the protein to its molecular mass

chemotaxis movement of a cell stimulated by a chemical attractant or repellent

chemotherapeutic use of chemicals to kill cancer cells

chloroplast the photosynthetic organelle of plants and algae

chondrocyte a cell that forms cartilage

chromatid a replicated chromosome before separation from its copy

chromatin complex of DNA, histones, and other proteins, making up chromosomes

ciliated protozoa single-celled organism possessing cilia, short hair-like extensions of the cell membrane

circadian relating to day or day length

cleavage hydrolysis

cleave split

clinical trials tests performed on human subjects

codon a sequence of three mRNA nucleotides coding for one amino acid

Cold War prolonged U.S.-Soviet rivalry following World War II

colectomy colon removal

colon crypts part of the large intestine

complementary matching opposite, like hand and glove

conformation three-dimensional shape

congenital from birth

conjugation a type of DNA exchange between bacteria

cryo-electron microscope electron microscope that integrates multiple images to form a three-dimensional model of the sample

cryopreservation use of very cold temperatures to preserve a sample

cultivars plant varieties resulting from selective breeding

cytochemist chemist specializing in cellular chemistry

cytochemistry cellular chemistry

cytogenetics study of chromosome structure and behavior

cytologist a scientist who studies cells

cytokine immune system signaling molecule

cytokinesis division of the cell's cytoplasm

cytology the study of cells

cytoplasm the material in a cell, excluding the nucleus

cytosol fluid portion of a cell, not including the organelles

de novo entirely new

deleterious harmful

dementia neurological illness characterized by impaired thought or awareness

demography aspects of population structure, including size, age distribution, growth, and other factors

denature destroy the structure of

deoxynucleotide building block of DNA

dimerize linkage of two subunits

dimorphism two forms

diploid possessing pairs of chromosomes, one member of each pair derived from each parent

disaccharide two sugar molecules linked together

dizygotic fraternal or nonidentical

DNA deoxyribonucleic acid

domains regions

dominant controlling the phenotype when one allele is present

dopamine brain signaling chemical

dosage compensation equalizing of expression level of X-chromosome genes between males and females, by silencing one X chromosome in females or amplifying expression in males

ecosystem an ecological community and its environment

ectopic expression expression of a gene in the wrong cells or tissues

electrical gradient chemiosmotic gradient

electrophoresis technique for separation of molecules based on size and charge

eluting exiting

embryogenesis development of the embryo from a fertilized egg

endangered in danger of extinction throughout all or a significant portion of a species' range

endogenous derived from inside the organism

endometriosis disorder of the endometrium, the lining of the uterus

endometrium uterine lining

endonuclease enzyme that cuts DNA or RNA within the chain

endoplasmic reticulum network of membranes within the cell

231

endoscope tool used to see within the body

endoscopic describes procedure wherein a tool is used to see within the body

endosymbiosis symbiosis in which one partner lives within the other

enzyme a protein that controls a reaction in a cell

epidemiologic the spread of diseases in a population

epidemiologists people who study the incidence and spread of diseases in a population

epidemiology study of incidence and spread of diseases in a population

epididymis tube above the testes for storage and maturation of sperm

epigenetic not involving DNA sequence change

epistasis suppression of a characteristic of one gene by the action of another gene

epithelial cells one of four tissue types found in the body, characterized by thin sheets and usually serving a protective or secretory function

Escherichia coli common bacterium of the human gut, used in research as a model organism

estrogen female horomone

et al. "and others"

ethicists a person who writes and speaks about ethical issues

etiology causation of disease, or the study of causation

eubacteria one of three domains of life, comprising most groups previously classified as bacteria

eugenics movement to "improve" the gene pool by selective breeding

eukaryote organism with cells possessing a nucleus

eukaryotic describing an organism that has cells containing nuclei

ex vivo outside a living organism

excise remove; cut out

excision removal

exogenous from outside

exon coding region of genes

exonuclease enzyme that cuts DNA or RNA at the end of a strand

expression analysis whole-cell analysis of gene expression (use of a gene to create its RNA or protein product)

fallopian tubes tubes through which eggs pass to the uterus

fermentation biochemical process of sugar breakdown without oxygen

fibroblast undifferentiated cell normally giving rise to connective tissue cells

fluorophore fluorescent molecule

forensic related to legal proceedings

founder population

fractionated purified by separation based on chemical or physical properties

fraternal twins dizygotic twins who share 50 percent of their genetic material

frontal lobe one part of the forward section of the brain, responsible for planning, abstraction, and aspects of personality

gamete reproductive cell, such as sperm or egg

gastrulation embryonic stage at which primitive gut is formed

gel electrophoresis technique for separation of molecules based on size and charge

gene expression use of a gene to create the corresponding protein

genetic code the relationship between RNA nucleotide triplets and the amino acids they cause to be added to a growing protein chain

genetic drift evolutionary mechanism, involving random change in gene frequencies

genetic predisposition increased risk of developing diseases

genome the total genetic material in a cell or organism

genomics the study of gene sequences

genotype set of genes present

geothermal related to heat sources within Earth

germ cell cell creating eggs or sperm

germ-line cells giving rise to eggs or sperm

gigabase one billion bases (of DNA)

glucose sugar

glycolipid molecule composed of sugar and fatty acid

glycolysis the breakdown of the six-carbon carbohydrates glucose and fructose

glycoprotein protein to which sugars are attached

Golgi network system in the cell for modifying, sorting, and delivering proteins

gonads testes or ovaries

gradient a difference in concentration between two regions

Gram negative bacteria bacteria that do not take up Gram stain, due to membrane structure

Gram positive able to take up Gram stain, used to classify bacteria

gynecomastia excessive breast development in males

haploid possessing only one copy of each chromosome

haplotype set of alleles or markers on a short chromosome segment

hematopoiesis formation of the blood

hematopoietic blood-forming

heme iron-containing nitrogenous compound found in hemoglobin

hemolysis breakdown of the blood cells

hemolytic anemia blood disorder characterized by destruction of red blood cells

hemophiliacs a person with hemophilia, a disorder of blood clotting

herbivore plant eater

heritability proportion of variability due to genes; ability to be inherited

heritability estimates how much of what is observed is due to genetic factors

heritable genetic

heterochromatin condensed portion of chromosomes

heterozygote an individual whose genetic information contains two different forms (alleles) of a particular gene

heterozygous characterized by possession of two different forms (alleles) of a particular gene

high-throughput rapid, with the capacity to analyze many samples in a short time

histological related to tissues

histology study of tissues

histone protein around which DNA winds in the chromosome

homeostasis maintenance of steady state within a living organism

homologous carrying similar genes

homologues chromosomes with corresponding genes that pair and exchange segments in meiosis

homozygote an individual whose genetic information contains two identical copies of a particular gene

homozygous containing two identical copies of a particular gene

hormones molecules released by one cell to influence another

hybrid combination of two different types

hybridization (molecular) base-pairing among DNAs or RNAs of different origins

hybridize to combine two different species

hydrogen bond weak bond between the H of one molecule or group and a nitrogen or oxygen of another

hydrolysis splitting with water

hydrophilic "water-loving"

hydrophobic "water hating," such as oils

hydrophobic interaction attraction between portions of a molecule (especially a protein) based on mutual repulsion of water

hydroxyl group chemical group consisting of -OH

hyperplastic cell cell that is growing at an increased rate compared to normal cells, but is not yet cancerous

hypogonadism underdeveloped testes or ovaries

hypothalamus brain region that coordinates hormone and nervous systems

hypothesis testable statement

identical twins monozygotic twins who share 100 percent of their genetic material

immunogenicity likelihood of triggering an immune system defense

immunosuppression suppression of immune system function

immunosuppressive describes an agent able to suppress immune system function

in vitro "in glass"; in lab apparatus, rather than within a living organism

in vivo "in life"; in a living organism, rather than in a laboratory apparatus

incubating heating to optimal temperature for growth

informed consent knowledge of risks involved

insecticide substance that kills insects

interphase the time period between cell divisions

intra-strand within a strand

intravenous into a vein

intron untranslated portion of a gene that interrupts coding regions

karyotype the set of chromosomes in a cell, or a standard picture of the chromosomes

kilobases units of measure of the length of a nucleicacid chain; one kilobase is equal to 1,000 base pairs

kilodalton a unit of molecular weight, equal to the weight of 1000 hydrogen atoms

kinase an enzyme that adds a phosphate group to another molecule, usually a protein

knocking out deleting of a gene or obstructing gene expression

laparoscope surgical instrument that is inserted through a very small incision, usually guided by some type of imaging technique

latent present or potential, but not apparent

lesion damage

ligand a molecule that binds to a receptor or other molecule

ligase enzyme that repairs breaks in DNA

ligate join together

linkage analysis examination of co-inheritance of disease and DNA markers, used to locate disease genes

lipid fat or wax-like molecule, insoluble in water

loci/locus site(s) on a chromosome

longitudinally lengthwise

lumen the space within the tubes of the endoplasmic reticulum

lymphocytes white blood cells

lyse break apart

lysis breakage

macromolecular describes a large molecule, one composed of many similar parts

macromolecule large molecule such as a protein, a carbohydrate, or a nucleic acid

macrophage immune system cell that consumes foreign material and cellular debris

malignancy cancerous tissue

malignant cancerous; invasive tumor

media (bacteria) nutrient source

meiosis cell division that forms eggs or sperm

melanocytes pigmented cells

meta-analysis analysis of combined results from multiple clinical trials

metabolism chemical reactions within a cell

metabolite molecule involved in a metabolic pathway

metaphase stage in mitosis at which chromosomes are aligned along the cell equator

metastasis breaking away of cancerous cells from the initial tumor

metastatic cancerous cells broken away from the initial tumor

methylate add a methyl group to

methylated a methyl group, CH₃, added

methylation addition of a methyl group, CH₃

microcephaly reduced head size

microliters one thousandth of a milliliter

micrometer 1/1000 meter

microsatellites small repetitive DNA elements dispersed throughout the genome

microtubule protein strands within the cell, part of the cytoskeleton

miscegenation racial mixing

mitochondria energy-producing cell organelle

mitogen a substance that stimulates mitosis

mitosis separation of replicated chromosomes

molecular hybridization base-pairing among DNAs or RNAs of different origins

molecular systematics the analysis of DNA and other molecules to determine evolutionary relationships

monoclonal antibodies immune system proteins derived from a single B cell

monomer "single part"; monomers are joined to form a polymer

monosomy gamete that is missing a chromosome

monozygotic genetically identical

morphologically related to shape and form

morphology related to shape and form

mRNA messenger RNA

mucoid having the properties of mucous

mucosa outer covering designed to secrete mucus, often found lining cavities and internal surfaces

mucous membranes nasal passages, gut lining, and other moist surfaces lining the body

multimer composed of many similar parts

multinucleate having many nuclei within a single cell membrane

mutagen any substance or agent capable of causing a change in the structure of DNA

mutagenesis creation of mutations

mutation change in DNA sequence

nanometer 10^{-9}(exp) meters; one billionth of a meter

nascent early-stage

necrosis cell death from injury or disease

nematode worm of the Nematoda phylum, many of which are parasitic

neonatal newborn

neoplasms new growths

neuroimaging techniques for making images of the brain

neurological related to brain function or disease

neuron nerve cell

neurotransmitter molecule released by one neuron to stimulate or inhibit a neuron or other cell

non-polar without charge separation; not soluble in water

normal distribution distribution of data that graphs as a bell-shaped curve

Northern blot a technique for separating RNA molecules by electrophoresis and then identifying a target fragment with a DNA probe

Northern blotting separating RNA molecules by electrophoresis and then identifying a target fragment with a DNA probe

nuclear DNA DNA contained in the cell nucleus on one of the 46 human chromosomes; distinct from DNA in the mitochondria

nuclear membrane membrane surrounding the nucleus

nuclease enzyme that cuts DNA or RNA

nucleic acid DNA or RNA

nucleoid region of the bacterial cell in which DNA is located

nucleolus portion of the nucleus in which ribosomes are made

nucleoplasm material in the nucleus

nucleoside building block of DNA or RNA, composed of a base and a sugar

nucleoside triphosphate building block of DNA or RNA, composed of a base and a sugar linked to three phosphates

nucleosome chromosome structural unit, consisting of DNA wrapped around histone proteins

nucleotide a building block of RNA or DNA

ocular related to the eye

oncogene gene that causes cancer

oncogenesis the formation of cancerous tumors

oocyte egg cell

open reading frame DNA sequence that can be translated into mRNA; from start sequence to stop sequence

opiate opium, morphine, and related compounds

organelle membrane-bound cell compartment

organic composed of carbon, or derived from living organisms; also, a type of agriculture stressing soil fertility and avoidance of synthetic pesticides and fertilizers

osmotic related to differences in concentrations of dissolved substances across a permeable membrane

ossification bone formation

osteoarthritis a degenerative disease causing inflammation of the joints

osteoporosis thinning of the bone structure

outcrossing fertilizing between two different plants

oviduct a tube that carries the eggs

ovulation release of eggs from the ovaries

ovules eggs

ovum egg

oxidation chemical process involving reaction with oxygen, or loss of electrons

oxidized reacted with oxygen

pandemic disease spread throughout an entire population

parasites organisms that live in, with, or on another organism

pathogen disease-causing organism

pathogenesis pathway leading to disease

pathogenic disease-causing

pathogenicity ability to cause disease

pathological altered or changed by disease

pathology disease process

pathophysiology disease process

patient advocate a person who safeguards patient rights or advances patient interests

PCR polymerase chain reaction, used to amplify DNA

pedigrees sets of related individuals, or the graphic representation of their relationships

peptide amino acid chain

peptide bond bond between two amino acids

percutaneous through the skin

phagocytic cell-eating

phenotype observable characteristics of an organism

phenotypic related to the observable characteristics of an organism

pheromone molecule released by one organism to influence another organism's behavior

phosphate group PO_4 group, whose presence or absence often regulates protein action

phosphodiester bond the link between two nucleotides in DNA or RNA

phosphorylating addition of phosphate group (PO_4)

phosphorylation addition of the phosphate group PO_4^{3-}

phylogenetic related to the evolutionary development of a species

phylogeneticists scientists who study the evolutionary development of a species

phylogeny the evolutionary development of a species

plasma membrane outer membrane of the cell

plasmid a small ring of DNA found in many bacteria

plastid plant cell organelle, including the chloroplast

pleiotropy genetic phenomenon in which alteration of one gene leads to many phenotypic effects

point mutation gain, loss, or change of one to several nucleotides in DNA

polar partially charged, and usually soluble in water

pollen male plant sexual organ

polymer molecule composed of many similar parts

polymerase enzyme complex that synthesizes DNA or RNA from individual nucleotides

polymerization linking together of similar parts to form a polymer

polymerize to link together similar parts to form a polymer

polymers molecules composed of many similar parts

polymorphic occurring in several forms

polymorphism DNA sequence variant

polypeptide chain of amino acids

polyploidy presence of multiple copies of the normal chromosome set

population studies collection and analysis of data from large numbers of people in a population, possibly including related individuals

positional cloning the use of polymorphic genetic markers ever closer to the unknown gene to track its inheritance in CF families

posterior rear

prebiotic before the origin of life

precursor a substance from which another is made

prevalence frequency of a disease or condition in a population

primary sequence the sequence of amino acids in a protein; also called primary structure

primate the animal order including humans, apes, and monkeys

primer short nucleotide sequence that helps begin DNA replication

primordial soup hypothesized prebiotic environment rich in life's building blocks

probe molecule used to locate another molecule

procarcinogen substance that can be converted into a carcinogen, or cancer-causing substance

procreation reproduction

progeny offspring

prokaryote a single-celled organism without a nucleus

promoter DNA sequence to which RNA polymerase binds to begin transcription

promutagen substance that, when altered, can cause mutations

pronuclei egg and sperm nuclei before they fuse during fertilization

proprietary exclusively owned; private

proteomic derived from the study of the full range of proteins expressed by a living cell

proteomics the study of the full range of proteins expressed by a living cell

protists single-celled organisms with cell nuclei

protocol laboratory procedure

protonated possessing excess H^+ ions; acidic

pyrophosphate free phosphate group in solution

quiescent non-dividing

radiation high energy particles or waves capable of damaging DNA, including X rays and gamma rays

recessive requiring the presence of two alleles to control the phenotype

recombinant DNA DNA formed by combining segments of DNA, usually from different types of organisms

recombining exchanging genetic material

replication duplication of DNA

restriction enzyme an enzyme that cuts DNA at a particular sequence

retina light-sensitive layer at the rear of the eye

retroviruses RNA-containing viruses whose genomes are copied into DNA by the enzyme reverse transcriptase

reverse transcriptase enzyme that copies RNA into DNA

ribonuclease enzyme that cuts RNA

ribosome protein-RNA complex at which protein synthesis occurs

ribozyme RNA-based catalyst

RNA ribonucleic acid

RNA polymerase enzyme complex that creates RNA from DNA template

RNA triplets sets of three nucleotides

salinity of, or relating to, salt

sarcoma a type of malignant (cancerous) tumor

scanning electron microscope microscope that produces images with depth by bouncing electrons off the surface of the sample

sclerae the "whites" of the eye

scrapie prion disease of sheep and goats

segregation analysis statistical test to determine pattern of inheritance for a trait

senescence a state in a cell in which it will not divide again, even in the presence of growth factors

senile plaques disease

serum (pl. sera) fluid portion of the blood

sexual orientation attraction to one sex or the other

somatic nonreproductive; not an egg or sperm

Southern blot a technique for separating DNA fragments by electrophoresis and then identifying a target fragment with a DNA probe

Southern blotting separating DNA fragments by electrophoresis and then identifying a target fragment with a DNA probe

speciation the creation of new species

spindle football-shaped structure that separates chromosomes in mitosis

spindle fiber protein chains that separate chromosomes during mitosis

spliceosome RNA-protein complex that removes introns from RNA transcripts

spontaneous non-inherited

sporadic caused by new mutations

stem cell cell capable of differentiating into multiple other cell types

stigma female plant sexual organ

stop codon RNA triplet that halts protein synthesis

striatum part of the midbrain

subcutaneous under the skin

sugar glucose

supercoiling coiling of the helix

symbiont organism that has a close relationship (symbiosis) with another

symbiosis a close relationship between two species in which at least one benefits

symbiotic describes a close relationship between two species in which at least one benefits

synthesis creation

taxon/taxa level(s) of classification, such as kingdom or phylum

taxonomical derived from the science that identifies and classifies plants and animals

taxonomist a scientist who identifies and classifies organisms

telomere chromosome tip

template a master copy

tenets generally accepted beliefs

terabyte a trillion bytes of data

teratogenic causing birth defects

teratogens substances that cause birth defects

thermodynamics process of energy transfers during reactions, or the study of these processes

threatened likely to become an endangered species

topological describes spatial relations, or the study of these relations

topology spatial relations, or the study of these relations

toxicological related to poisons and their effects

transcript RNA copy of a gene

transcription messenger RNA formation from a DNA sequence

transcription factor protein that increases the rate of transcription of a gene

transduction conversion of a signal of one type into another type

transgene gene introduced into an organism

transgenics transfer of genes from one organism into another

translation synthesis of protein using mRNA code

translocation movement of chromosome segment from one chromosome to another

transposable genetic element DNA sequence that can be copied and moved in the genome

transposon genetic element that moves within the genome

trilaminar three-layer

triploid possessing three sets of chromosomes

trisomics mutants with one extra chromosome

trisomy presence of three, instead of two, copies of a particular chromosome

tumor mass of undifferentiated cells; may become cancerous

tumor suppressor genes cell growths

tumors masses of undifferentiated cells; may become cancerous

vaccine protective antibodies

vacuole cell structure used for storage or related functions

van der Waal's forces weak attraction between two different molecules

vector carrier

vesicle membrane-bound sac

virion virus particle

wet lab laboratory devoted to experiments using solutions, cell cultures, and other "wet" substances

wild-type most common form of a trait in a population

Wilm's tumor a cancerous cell mass of the kidney

X ray crystallography use of X rays to determine the structure of a molecule

xenobiotic foreign biological molecule, especially a harmful one

zygote fertilized egg

Topic Outline

APPLICATIONS TO OTHER FIELDS

Agricultural Biotechnology
Biopesticides
Bioremediation
Biotechnology
Conservation Biology: Genetic Approaches
DNA Profiling
Genetically Modified Foods
Molecular Anthropology
Pharmacogenetics and Pharmacogenomics
Plant Genetic Engineer
Public Health, Genetic Techniques in
Transgenic Animals
Transgenic Microorganisms
Transgenic Plants

BACTERIAL GENETICS

Escherichia coli (*E. coli* bacterium)
Ames Test
Antibiotic Resistance
Chromosome, Prokaryotic
Cloning Genes
Conjugation
Eubacteria
Microbiologist
Overlapping Genes
Plasmid
Transduction
Transformation
Transgenic Microorganisms
Transgenic Organisms: Ethical Issues
Viroids and Virusoids
Virus

BASIC CONCEPTS

Biotechnology
Crossing Over

Disease, Genetics of
DNA
DNA Structure and Function, History
Fertilization
Gene
Genetic Code
Genetics
Genome
Genotype and Phenotype
Homology
Human Genome Project
Inheritance Patterns
Meiosis
Mendelian Genetics
Mitosis
Mutation
Nucleotide
Plasmid
Population Genetics
Proteins
Recombinant DNA
Replication
RNA
Transcription
Translation

BIOTECHNOLOGY

Agricultural Biotechnology
Biopesticides
Bioremediation
Biotechnology
Biotechnology and Genetic Engineering, History
Biotechnology: Ethical Issues
Cloning Genes
Cloning Organisms
DNA Vaccines

Genetically Modified Foods
HPLC: High-Performance Liquid Chromatography
Pharmaceutical Scientist
Plant Genetic Engineer
Polymerase Chain Reaction
Recombinant DNA
Restriction Enzymes
Reverse Transcriptase
Transgenic Animals
Transgenic Microorganisms
Transgenic Organisms: Ethical Issues
Transgenic Plants

CAREERS

Attorney
Bioinformatics
Clinical Geneticist
College Professor
Computational Biologist
Conservation Geneticist
Educator
Epidemiologist
Genetic Counselor
Geneticist
Genomics Industry
Information Systems Manager
Laboratory Technician
Microbiologist
Molecular Biologist
Pharmaceutical Scientist
Physician Scientist
Plant Genetic Engineer
Science Writer
Statistical Geneticist
Technical Writer

CELL CYCLE

Apoptosis
Balanced Polymorphism
Cell Cycle
Cell, Eukaryotic
Centromere
Chromosome, Eukaryotic
Chromosome, Prokaryotic
Crossing Over
DNA Polymerases
DNA Repair
Embryonic Stem Cells
Eubacteria
Inheritance, Extranuclear

Linkage and Recombination
Meiosis
Mitosis
Oncogenes
Operon
Polyploidy
Replication
Signal Transduction
Telomere
Tumor Suppressor Genes

CLONED OR TRANSGENIC ORGANISMS

Agricultural Biotechnology
Biopesticides
Biotechnology
Biotechnology: Ethical Issues
Cloning Organisms
Cloning: Ethical Issues
Gene Targeting
Model Organisms
Patenting Genes
Reproductive Technology
Reproductive Technology: Ethical Issues
Rodent Models
Transgenic Animals
Transgenic Microorganisms
Transgenic Organisms: Ethical Issues
Transgenic Plants

DEVELOPMENT, LIFE CYCLE, AND NORMAL HUMAN VARIATION

Aging and Life Span
Behavior
Blood Type
Color Vision
Development, Genetic Control of
Eye Color
Fertilization
Genotype and Phenotype
Hormonal Regulation
Immune System Genetics
Individual Genetic Variation
Intelligence
Mosaicism
Sex Determination
Sexual Orientation
Twins
X Chromosome
Y Chromosome

DNA, GENE AND CHROMOSOME STRUCTURE

Antisense Nucleotides
Centromere
Chromosomal Banding
Chromosome, Eukaryotic
Chromosome, Prokaryotic
Chromosomes, Artificial
DNA
DNA Repair
DNA Structure and Function, History
Evolution of Genes
Gene
Genome
Homology
Methylation
Multiple Alleles
Mutation
Nature of the Gene, History
Nomenclature
Nucleotide
Overlapping Genes
Plasmid
Polymorphisms
Pseudogenes
Repetitive DNA Elements
Telomere
Transposable Genetic Elements
X Chromosome
Y Chromosome

DNA TECHNOLOGY

In situ Hybridization
Antisense Nucleotides
Automated Sequencer
Blotting
Chromosomal Banding
Chromosomes, Artificial
Cloning Genes
Cycle Sequencing
DNA Footprinting
DNA Libraries
DNA Microarrays
DNA Profiling
Gel Electrophoresis
Gene Targeting
HPLC: High-Performance Liquid Chromatography
Marker Systems
Mass Spectrometry
Mutagenesis

Nucleases
Polymerase Chain Reaction
Protein Sequencing
Purification of DNA
Restriction Enzymes
Ribozyme
Sequencing DNA

ETHICAL, LEGAL, AND SOCIAL ISSUES

Attorney
Biotechnology and Genetic Engineering, History
Biotechnology: Ethical Issues
Cloning: Ethical Issues
DNA Profiling
Eugenics
Gene Therapy: Ethical Issues
Genetic Discrimination
Genetic Testing: Ethical Issues
Legal Issues
Patenting Genes
Privacy
Reproductive Technology: Ethical Issues
Transgenic Organisms: Ethical Issues

GENE DISCOVERY

Ames Test
Bioinformatics
Complex Traits
Gene and Environment
Gene Discovery
Gene Families
Genomics
Human Disease Genes, Identification of
Human Genome Project
Mapping

GENE EXPRESSION AND REGULATION

Alternative Splicing
Antisense Nucleotides
Chaperones
DNA Footprinting
Gene
Gene Expression: Overview of Control
Genetic Code
Hormonal Regulation
Imprinting
Methylation
Mosaicism
Nucleus

Operon
Post-translational Control
Proteins
Reading Frame
RNA
RNA Interference
RNA Polymerases
RNA Processing
Signal Transduction
Transcription
Transcription Factors
Translation

GENETIC DISORDERS

Accelerated Aging: Progeria
Addiction
Alzheimer's Disease
Androgen Insensitivity Syndrome
Attention Deficit Hyperactivity Disorder
Birth Defects
Breast Cancer
Cancer
Carcinogens
Cardiovascular Disease
Chromosomal Aberrations
Colon Cancer
Cystic Fibrosis
Diabetes
Disease, Genetics of
Down Syndrome
Fragile X Syndrome
Growth Disorders
Hemoglobinopathies
Hemophilia
Human Disease Genes, Identification of
Metabolic Disease
Mitochondrial Diseases
Muscular Dystrophy
Mutagen
Nondisjunction
Oncogenes
Psychiatric Disorders
Severe Combined Immune Deficiency
Tay-Sachs Disease
Triplet Repeat Disease
Tumor Suppressor Genes

GENETIC MEDICINE: DIAGNOSIS, TESTING, AND TREATMENT

Clinical Geneticist
DNA Vaccines

Embryonic Stem Cells
Epidemiologist
Gene Discovery
Gene Therapy
Gene Therapy: Ethical Issues
Genetic Counseling
Genetic Counselor
Genetic Testing
Genetic Testing: Ethical Issues
Geneticist
Genomic Medicine
Human Disease Genes, Identification of
Pharmacogenetics and Pharmacogenomics
Population Screening
Prenatal Diagnosis
Public Health, Genetic Techniques in
Reproductive Technology
Reproductive Technology: Ethical Issues
RNA Interference
Statistical Geneticist
Statistics
Transplantation

GENOMES

Chromosome, Eukaryotic
Chromosome, Prokaryotic
Evolution of Genes
Genome
Genomic Medicine
Genomics
Genomics Industry
Human Genome Project
Mitochondrial Genome
Mutation Rate
Nucleus
Polymorphisms
Repetitive DNA Elements
Transposable Genetic Elements
X Chromosome
Y Chromosome

GENOMICS, PROTEOMICS, AND BIOINFORMATICS

Bioinformatics
Combinatorial Chemistry
Computational Biologist
DNA Libraries
DNA Microarrays
Gene Families
Genome
Genomic Medicine

Genomics
Genomics Industry
High-Throughput Screening
Human Genome Project
Information Systems Manager
Internet
Mass Spectrometry
Nucleus
Protein Sequencing
Proteins
Proteomics
Sequencing DNA

HISTORY

Biotechnology and Genetic Engineering, History
Chromosomal Theory of Inheritance, History
Crick, Francis
Delbrück, Max
DNA Structure and Function, History
Eugenics
Human Genome Project
McClintock, Barbara
McKusick, Victor
Mendel, Gregor
Morgan, Thomas Hunt
Muller, Hermann
Nature of the Gene, History
Ribosome
Sanger, Fred
Watson, James

INHERITANCE

Chromosomal Theory of Inheritance, History
Classical Hybrid Genetics
Complex Traits
Crossing Over
Disease, Genetics of
Epistasis
Fertilization
Gene and Environment
Genotype and Phenotype
Heterozygote Advantage
Imprinting
Inheritance Patterns
Inheritance, Extranuclear
Linkage and Recombination
Mapping
Meiosis
Mendel, Gregor
Mendelian Genetics

Mosaicism
Multiple Alleles
Nondisjunction
Pedigree
Pleiotropy
Polyploidy
Probability
Quantitative Traits
Sex Determination
Twins
X Chromosome
Y Chromosome

MODEL ORGANISMS

Arabidopsis thaliana
Escherichia coli (*E. coli* Bacterium)
Chromosomes, Artificial
Cloning Organisms
Embryonic Stem Cells
Fruit Fly: *Drosophila*
Gene Targeting
Maize
Model Organisms
RNA Interference
Rodent Models
Roundworm: *Caenorhabditis elegans*
Transgenic Animals
Yeast
Zebrafish

MUTATION

Chromosomal Aberrations
DNA Repair
Evolution of Genes
Genetic Code
Muller, Hermann
Mutagen
Mutagenesis
Mutation
Mutation Rate
Nondisjunction
Nucleases
Polymorphisms
Pseudogenes
Reading Frame
Repetitive DNA Elements
Transposable Genetic Elements

ORGANISMS, CELL TYPES, VIRUSES

Arabidopsis thaliana
Escherichia coli (*E. coli* bacterium)

Archaea
Cell, Eukaryotic
Eubacteria
Evolution, Molecular
Fruit Fly: *Drosophila*
HIV
Maize
Model Organisms
Nucleus
Prion
Retrovirus
Rodent Models
Roundworm: *Caenorhabditis elegans*
Signal Transduction
Viroids and Virusoids
Virus
Yeast
Zebrafish

POPULATION GENETICS AND EVOLUTION

Antibiotic Resistance
Balanced Polymorphism
Conservation Biologist
Conservation Biology: Genetic Approaches
Evolution of Genes
Evolution, Molecular
Founder Effect
Gene Flow
Genetic Drift
Hardy-Weinberg Equilibrium

Heterozygote Advantage
Inbreeding
Individual Genetic Variation
Molecular Anthropology
Population Bottleneck
Population Genetics
Population Screening
Selection
Speciation

RNA

Antisense Nucleotides
Blotting
DNA Libraries
Genetic Code
HIV
Nucleases
Nucleotide
Reading Frame
Retrovirus
Reverse Transcriptase
Ribosome
Ribozyme
RNA
RNA Interference
RNA Polymerases
RNA Processing
Transcription
Translation

Volume 3 Index

Page numbers in **boldface type** indicate article titles; those in *italic type* indicate illustrations. A cumulative index, which combines the terms in all volumes of **Genetics**, can be found in volume 4 of this series.

α, defined, 212

α-helices
conversion to β-sheets, 188–189, *189*, 190, *190*
protein structure role, 200–201, *203*

α-protobacteria, as mitochondrial ancestor, 57

β, defined, 212

β-sheets
conversion of α-helices, 188–189, *189*, 190, *190*
protein structure role, 200–201, *203*

2-D PAGE (two-dimensional polyacrylamide gel electrophoresis), 207–208

A

A3243G gene, diabetes and, 54

ABO blood group system, multiple alleles, 82

Acetylation
of DNA, gene repression role, 77
of proteins, post-translational, 203, 205

Acidity
defined, 200
impact on protein structure, 201

Actin proteins, pseudogenes for, 213

AD. *See* Alzheimer's disease

Addiction
alcoholism, 213
drug dependence, 214
smoking, 214

Adenine
and DNA structure, 94
mutagenic base analogs, 87
structure, 115, *116*, *118*, 119, *119*
See also Base pairs

Adenosine triphosphate (ATP)
aerobic *vs.* anaerobic respiration, 52
defined, 178, 204
metabolism and, 41–42, 51–52, 55–56, 117
post-translational phosphorylation by, 178, 204

Adenovirus, PML body mutations, 124

ADHD (attention deficit hyperactivity disorder)
genetic components, 213
Tourette's and OCD co-morbidity, 215

Adoption studies
psychiatric disorders, 213–215
See also Family studies/family histories

Adrenal glands, metabolic diseases, 42

Adult-onset diabetes (type 2), 125, 154

Aerobic respiration, advantages and disadvantages, 52

AFP (alpha fetoprotein) test, 185

Africa, as origin of early man, 66, 67, 167–170, 174

Africans (African-Americans), genotype frequencies, 172

Aggregate, defined, 188

Aging
and birth defects, 99, 182, 184
longevity as complex trait, 81
and mitochondrial genome mutations, 80
and nondisjunction, 112

reactive oxygen species contribution to, 52
See also Alzheimer's disease

Agricultural biotechnology. *See* Biotechnology, agricultural

Alanine-glyoxylate transaminase defect, 41

Alcoholism, genetic influence, 213

Algae
Chlamydomonas, as model organism, 61
Guillardia, overlapping genes, 136

Alkalinity
basic, defined, 200
impact on protein structure, 201

Alkaptonuria, symptoms and treatment, 37, 40, 43

Alkylation errors, and DNA damage, 88

Allele frequencies. *See* Hardy-Weinberg equilibrium

Alleles
defined, 5, 6, 11, 73, 94, 101, 144, 158, 171
fixed alleles, 171
inheritance patterns, 102
location of, 11
Mendel's concept of, 101
wild-type, 100
See also Crossing over (recombination); Dominant alleles; Recessive alleles

Alleles, multiple, **82–83**
ABO blood group system, 82
distinguished from polymorphisms, 82
gene mapping applications, 83
MHC proteins, 82
in noncoding DNA, 82–83
See also Polymorphisms, DNA

Allelic disorders, defined, 84

Allolactose
defined, 134
lac operons and, 134–135
Allopolyploid plants, 165–166
Alper, Tikvah, 188
Alpha fetoprotein (AFP) test, 185
Alpha particles, defined, 89
Alternative splicing. *See* Splicing, alternative
Alu sequences
copies in human genome, 211–212
evolutionary origin, 211–212
neurofibromatosis and, 153
Alzheimer's disease (AD)
"Disease Model" mouse strains, 62
genetic components, 213
genetic testing, 186
pharmacogenomics to predict, 146
protein conformation role, 190
Ambrose, Stanley, 168
American Society for Reproductive Medicine, 187
Ames test, applications, 87, 92
Amino acid metabolism, disorders, 40, 42–43
Amino acids
alphabet, 199–200, 205
C and N termini, 181, 197–198, 207
in diet, 41
extein role, 181
glycosylation, 178–179, *180*, 205
intein role, 181
phosphorylation, 20, 178, *180*, 204, 205
polypeptide synthesis, 103
and protein primary structure, 197, 200
sequences, mutations and, 94–95
structure, 197, 200
ubiquitination, 179, *180*
See also Protein sequencing; Proteins; *specific amino acids*
Amish populations
bipolar disorder, 214
founder effect, 174
Amniocentesis, 183, 184
Amphibians, polyploid, 165
Amplification
defined, 1, 51, 91, 154
of mtDNA, 51
PCR as tool, 68, **154–159**, 161
of proto-oncogenes, 128–129
Anaerobic respiration, defined, 52

Anaphase
I, meiosis, 26, 27, *28*
II, meiosis, 27, *28*
mitosis, 59
Anemia, Fanconi, 186
Aneuploidy
aging and, 112, 182
chromosomal mosaicism, 79
consequences, 164–165
monosomy, 109, 111
nondisjunction, **108–112**, 166
nonfatal human conditions, 110–111
in plants, 165–166
trisomy, 10, 79, 109, 111
See also Down syndrome
Annealing, defined, 157
Anthrax detection
gel electrophoresis, *71*
mass spectrometry, *19*
Anthropology, molecular, **62–70**
DNA comparisons, advantages, 63–64, *63*
DNA comparisons, caveats, 64–65
DNA comparisons, types, 65
genotype frequency tools, 174
mitochondrial DNA studies, 66–67
mitochondrial Eve, 63, 67–68
molecular clocks, 63, 67, *94*, 98, 100–101
mtDNA D-loop as tool, 53
population bottlenecks, **167–171**
Y chromosome analysis, 65–66, 168
Antibiotic resistance
public health concerns, 217
resistance plasmids, 151
of selectable markers, 16–17
Antibiotics, ganciclovir, 17
Antibodies
function, 199
H and L polypeptides, 202
quaternary structure, 202
Antigens, defined, 82, 202
Antioxidants, mutation rate and, 89, 100
Apoptosis
and cancer, 130
defined, 51, 125
mitochondrial role, 51
PML bodies and, 124
Arabidopsis thaliana (thale cress)
characteristics, 60
as model organism, 149
transposon mutagenesis, 97–98

Arabs, evolutionary origin, 66
Archaea, Taq DNA polymerase, 157
Arthritis, "Disease Model" mouse strains, 62
Asbestos, as mutagen, 89
Ashkenazic Jews, Tay-Sachs disease, 174
Asians (Asian-Americans)
evolutionary origin, 66
genotype frequencies, 172
Asparagine (Asn), glycosylation, 178–179
ATP. *See* Adenosine triphosphate (ATP)
"Attempts at Plant Hybridization" (G. Mendel), 31–32
Attention deficit hyperactivity disorder (ADHD)
genetic components, 213
Tourette's and OCD co-morbidity, 215
Attenuation
defined, 134
operon regulation by, 134
Attorneys, patent, 138
Aurbach and Robson, mustard gas studies, 90
Australasia, population bottlenecks, 168, 170
Autism, genetic components, 213
Autopolyploid plants, 165–166
Autosomal dominant disorders. *See* Dominant disorders, autosomal
Autosomal recessive disorders. *See* Recessive disorders, autosomal
Autosomes
aberrations, detecting, 183
defined, 76, 85, 164, 176, 183, 215
human, number, 64, 76, 106, 108–109, 111, 182
primate, number, 64
Avery, Oswald, 103, 105

B

Bacillus thuringiensis (BT), 149
Back mutations, 65, 100
Bacteria. *See* Eubacteria
Bacteriophages
defined, 105, 131
in Hershey-Chase experiment, *104*, 105
operons, 131
PhiX174, 135
Bands (banding patterns). *See* Chromosomal banding
Barbacid, Mariano, 127
Barr bodies. *See* Mosaicism

Base analogs, mutagenic, 87, *88*, 90–91

Base pairs
  as complementary, 94, 118–119, 155
  defined, 13, 100, 162
  and DNA structure, 115, *116*, 117
  structure, 115, *116*
  See also Codons; Genetic code; Nucleotides; *specific nucleotides*

Base substitution mutations, 95–96

Base-altering mutagens, 88

Basic, defined, 200

Basic Local Alignment Search Tool (BLAST), 196

Bateson, William, 36, 102

Bcr proto-oncogene, 129

Beadle, George, 76, 103, *103*

Becker, Peter, 84

Behavior
  ADHD, 213, 215
  alcoholism, 213
  animal model studies, 61
  drug dependence, 214
  smoking, 214
  See also Psychiatric disorders

Benzer, Seymour, 105

Benzo[*a*]pyrene, as mutagen, 89

Beta carotene, 149

Beta particles, defined, 89

Beta-hexosaminidase A enzymes, 45

BFB (breakage-fusion-bridge) cycle, 21, 22

B-glucoronidase gus A (uid A) enzymes, 18

Bioactivation, as mutagen source, 89

Biodiversity. *See* Genetic diversity

Bioethics. *See* Ethical issues

Biohazards, recombinant bacteria as, 152

Bioinformatics
  defined, 18, 106
  in mass spectrometry analyses, 18
  pharmaceutical scientist role, 143

Biologists
  cellular, 70
  computational, 196
  evolutionary, 70
  microbiologists, **50–51**
  molecular, **70–72**, *71*

Biology, conservation, 167

Biopesticides, insect-resistant crops, 149

Biopterin, 43

Biotechnology
  defined, 216
  ethical issues, 152

Biotechnology, agricultural
  genetically modified foods, 3, 149–150
  plant genetic engineer role, **149–150**
  polymorphisms as tools, 162

Bipolar disorder, 213–215

Bivalents, in crossing over, 26

Blakeslee, F. A., 165

BLAST (Basic Local Alignment Search Tool), 196

Blending theory, 30, 34, 102

Blindness, Tristan da Cunha islanders, 174

BLOCKS database, 209

Blood. *See* Globin proteins; Hemoglobin

Blood clotting factors, post-translational control, 181

Blood disorders
  "emia" suffix, 41
  hemoglobinopathies, 177
  hemophilia, 78, 97, 186

Blood samples, privacy of, *191*

Blood type
  ABO blood group system, 82
  codominance, 36
  ethnic differences, 172
  multiple alleles, 82

Bloom's syndrome
  DNA repair role, 100
  PML body mutations, 124

Blotting, DNA purification tools, 220–221

BMD (Becker muscular dystrophy), 84

Bombay, population of, *173*

Boveri, Theodor, 102, 127

Bovine spongiform encephalopathy (BSE), 188, 189

Brachydanio danio (zebrafish), 60, 61

Brackish, defined, 149

Brain development, DNA methylation role, 49

Brain disorders
  ADHD, 213, 215
  mitochondrial diseases, 54
  prion diseases, 188–190
  See also Alzheimer's disease; Mental retardation; Psychiatric disorders

BRCA1 and *BRCA2* genes, genetic testing, 218

Breakage-fusion-bridge (BFB) cycle, 22

Breast cancer. *See* Cancer, breast

Bridges, Calvin Blackman, 73, 75, 76, 80

5-bromo-deoxyuridine (5BU), as mutagen, 87, *88*

Brown, Robert, 119

BSE (bovine spongiform encephalopathy), 188, 189

BT (*Bacillus thuringiensis*), 149

Buffers, defined, 156, 221

C

C-Abl proto-oncogene, 129

Caenorhabditis elegans (roundworms)
  behavior studies, 61
  as model organism, 60, 61

Café-au-lait spots, 153, *154*

Cajal bodies, 120, 124

Calico (tortoiseshell) cats, mosaicism, 77, *78*

Cancer
  aberrant phosphorylation and, 178
  apoptosis and, 130
  carcinogens, 87, 89
  cervical, Pap smears, 175
  colon, oncogene activation, 131
  diffuse B-cell lymphoma, 146
  DNA methylation role, 49
  genetic predisposition to, 127
  mutations and, 98
  oncogenes, 98, **127–131**
  ovarian, 218
  prostate gland, prostatic antigen screening, 175
  retinoblastomas, 99, 124, 130, 131
  treatments, 146
  tumors, 153
  viral role, 127, 130

Cancer, breast
  BRCA genes, 218
  mammograms, 175
  pharmacogenomics to predict, 146

Cann, Rebecca, 67

Cannibalism, kuru disease and, 189

Carbohydrate metabolism
  disorders, 43–44
  normal function, 41

Carboxylase deficiency, symptoms and treatment, 40

Carboxypeptidase A and B, as protein sequencing tool, 198

Carcinogens
  defined, 89
  distinguished from mutagens, 87
  See also Mutagens

Cardiomyopathy, defined, 42

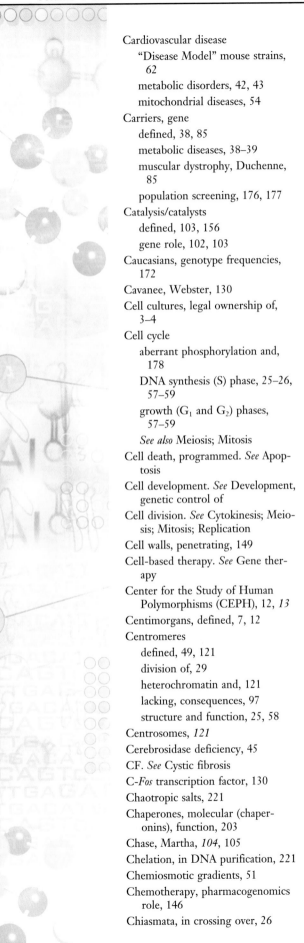

Cardiovascular disease
"Disease Model" mouse strains, 62
metabolic disorders, 42, 43
mitochondrial diseases, 54

Carriers, gene
defined, 38, 85
metabolic diseases, 38–39
muscular dystrophy, Duchenne, 85
population screening, 176, 177

Catalysis/catalysts
defined, 103, 156
gene role, 102, 103

Caucasians, genotype frequencies, 172

Cavanee, Webster, 130

Cell cultures, legal ownership of, 3–4

Cell cycle
aberrant phosphorylation and, 178
DNA synthesis (S) phase, 25–26, 57–59
growth (G$_1$ and G$_2$) phases, 57–59
See also Meiosis; Mitosis

Cell death, programmed. See Apoptosis

Cell development. See Development, genetic control of

Cell division. See Cytokinesis; Meiosis; Mitosis; Replication

Cell walls, penetrating, 149

Cell-based therapy. See Gene therapy

Center for the Study of Human Polymorphisms (CEPH), 12, 13

Centimorgans, defined, 7, 12

Centromeres
defined, 49, 121
division of, 29
heterochromatin and, 121
lacking, consequences, 97
structure and function, 25, 58

Centrosomes, 121

Cerebrosidase deficiency, 45

CF. See Cystic fibrosis

C-Fos transcription factor, 130

Chaotropic salts, 221

Chaperones, molecular (chaperonins), function, 203

Chase, Martha, 104, 105

Chelation, in DNA purification, 221

Chemiosmotic gradients, 51

Chemotherapy, pharmacogenomics role, 146

Chiasmata, in crossing over, 26

Children's Bureau (Maternal and Child Health Bureau), PKU screening, 219

Chimpanzees, genetic diversity, 167–168, 168

Chi-square test, 194–195

Chlamydomonas, as model organism, 61

Chlorophyll, albino plants, 10

Chloroplasts
defined, 149
genome, 55, 56–57

Chorionic villus sampling (CVS), 184, 187

Chorismate, as tryptophan metabolite, 132

Chromatids, defined, 106, 210

Chromatin
condensation, 25
methylation of, 77
in nucleus, 120, 120, 121

Chromatography
in mass spectrometry analysis, 19
as protein sequencing tool, 197

Chromosomal aberrations
assays for, 92–93, 182–185
birth defects, 182
breakage-fusion-bridge cycle, 21, 22
cancer and, 127, 130
centromeres lacking, 97
DNA methylation role, 49
duplications, 97
fragile X syndrome, 49
intersex organisms, 77, 164
inversions, 97
Klinefelter's syndrome, 111
mosaicism, 79
nomenclature, 107
nondisjunction, 79, **108–112**, 109, 166
polyploidy, 97, **163–167**
polyspermy, 163–164
prenatal diagnosis, 182–187
translocations, 96–97, 127, 128–129
Turner's syndrome, 111
See also Aneuploidy; Down syndrome

Chromosomal banding, 106, 107

Chromosomal territories, 121–122, 122

Chromosomal theory of inheritance, history, 102, **1:129–132**
McClintock's contributions, 10, 21–22, 75, 97, 105

See also Mendelian genetics; Morgan, Thomas Hunt; Muller, Hermann

Chromosome mapping. See Mapping genes

Chromosome painting (spectral karyotyping), 183

Chromosome terminology, 106–107

Chromosomes, eukaryotic
haploid complements, 24
karyotype, 108
"knobs," staining, 10
in nucleus, 119–126
overlapping genes, **135–136**
p and q arms, 106–107
ribonuclease role, 114–115
See also Autosomes; Eukaryotes; Meiosis; Mitosis

Chromosomes, sex, 76–77
aneuploidy, 109–112
as barrier to polyploidy, 164
human, number, 24, 76, 106, 108–109, 111, 182
lacking in plants, 164
monosomy, 110–111
trisomy, 110–111

Chrysanthemums, polyploid, 166

CJD (Creutzfeldt-Jakob disease), 188, 189–190

C-Jun transcription factor, 130

Classical hybrid genetics. See Mendelian genetics

ClB chromosomes, 91–92

Cleave/cleavage, defined, 113, 152

Climatic bottlenecks, 168–169, 169

Clinical geneticists, probability tools, 193–194

Clinical Laboratory Improvement Act (CLIA), 218

Cloning genes
DNA purification tools, 220–221
expression cloning, 152
as oncogene study tools, 127
PCR as tool, 158
plasmid vectors, 152

Cloning organisms
autopolyploid plants, 166
human, ethical issues, 3

Clotting factors. See Blood clotting factors

CML (chronic myelogenous leukemia), 127, 129

C-myc gene, mutations, 96

Codominance
AB blood types, 36
defined, 36

Codons
defined, 96, 200, 209

start, 135

stop, 96, 209

See also Genetic code

Cofactors, metabolic role, 37, 39, *39*, 43

Coiled-coil motif, 201

Col plasmids, 151

Colchicine, 165, 166

Collagen proteins, function, 199

Colonization bottlenecks. *See* Founder effect

Color blindness, inheritance patterns, 78

Complementary
base pairs as, 94, 118–119, 155
defined, 118

Complementary DNA. *See* DNA, complementary (cDNA)

Complex (polygenic) traits
intelligence as, 81
longevity as, 81
in Mendelian genetics, 102
psychiatric disorders as, 214–216
See also Gene-environment interactions

Computational biologists, 196

Computer technology
BLAST, 196
to draw and record pedigrees, 140
in linkage and recombination studies, 7
in mass spectrometry analysis, 18
as molecular anthropology tool, 67
as proteomics tool, 208–209

Confidentiality. *See* Privacy

Conformation
defined, 201
of proteins, 200–202, 204

Congenital adrenal hyperplasia, newborn screening, 176, 177

Conjugation, plasmid role, 151

Consanguinity. *See* Inbreeding

Continuous (quantitative) traits, in maize, 9

Cooper, Geoff, 127

Corn. *See* Maize

Correns, Carl, 102, 103

Cows. *See* Livestock

CpG dinucleotides, methylation of, 46–49, *47*

CpG islands, 47, *48*

Creighton, Harriet, 10, 21

Creutzfeldt-Jakob disease (CJD), 188, 189–190

Crick, Francis, DNA structure research, 105

Cristae, in mitochondrial membranes, 55, *56*

Crossing over (recombination)
chiasmata role, 26
ClB chromosome to prevent, 91–92
defined, 4–5
discovery, 10, 21, 75, 105
DNA comparison studies, 65
as gene mapping tool, 11–13, 26, 83
genetic diversity role, 24, *25*, 26, 29, 60
interference, 12–13
knock-out and knock-in genes, 91
lacking in Y chromosomes, 65
mechanisms, 24, *25*, 26, 29
pseudogenes, 210, *210*
recombination fraction, 6–7, 12–13
recombination rates, 26
unequal, 7
See also Linkage and recombination (linkage analysis); Meiosis

C-terminus, of amino acids/proteins, 181, 197–198, 207

CVS (chorionic villus sampling), 184, 187

Cyanobacteria, as chloroplast ancestor, 57

Cystic fibrosis (CF)
"Disease Model" mouse strains, 62
genetic testing, 177, 186
inheritance patterns, 142, 193–194
preimplantation genetic diagnosis, 186
Punnett squares to predict, 193–194, *194*

Cystine (Cys), chemical properties, 200

Cytochrome C protein, evolution of, *94*

Cytochrome P450, poor and ultrarapid metabolizers, 144–145

Cytogenetic analysis, defined, 10

Cytogeneticists, defined, 182

Cytokinesis
defined, 29
in meiosis, 29
in mitosis, 59

Cytologists, defined, 21

Cytomegalovirus, PML body mutations, 124

Cytoplasm, defined, 29, 57, 125, 201

Cytosine
and DNA structure, 94

methylation of, 46–49, *47*, 95, 100

mutagenic base analogs, 87

structure, 115, *116*, 118–119, *118*, *119*

See also Base pairs

D

Darwinism, 32

Databases
BLOCKS, 209
of chromosomal abnormalities, 182
of DNA samples, 192–193
of gene and protein sequences, 18, 198, 208–209
Human PSD, 209
Pfam, 209
PRINTS, 209
PROSITE, 209
SWISS-PROT, 209
Yeast Proteome, 209

DCMP (deoxycytosine monophosphate), 117

De Vries, Hugo, 102

Deamination
errors, DNA damage, 47, 88
of methyl cytosine, 47, *47*

Delano, Susana, *71*

Delbrück, Max, 103

Deleterious, defined, 89

Dementia
defined, 189
See also Alzheimer's disease; Brain disorders; Psychiatric disorders

Denaturation, in PCRs, 157

Deoxycytosine monophosphate (dCMP), 117

Deoxyribonuclease enzymes, function, 112, 113–114

Deoxyribonucleic acid. *See* DNA

Deoxyribose sugars, structure, 115, *116*, *117*

Detoxifying agents, to treat metabolic diseases, 39, 45

Development, genetic control of
DNA methyltransferase role, 49
early studies, 103
hedgehog protein role, 181
McClintock's hypothesis, 21–22
mitosis and, 57
polyploids, 164, 166
sexual development, *65*
zebrafish to study, 61

Diabetes
A3243G mutations, 54

Diabetes (continued)

"Disease Model" mouse strains, 62

insulin, 181, 199

as metabolic diseases, 42

type 2 (T2DM), 125, 154

Diagnostic and Statistical Manual (DSM-IV), 213

Diagnostic tests, *vs.* population screening, 175

Diamond v. *Chakrabarty*, 136–137

Dictyostelium discoideium (slime mold), as model organism, 60

Diet. *See* Nutrition

Diffuse B-cell lymphoma (DLBCL), chemotherapy, 146

Digestive enzymes, processing of, 204

Diploid

accumulation of recessive mutations, 93

defined, 163

somatic cells as, 24, 60

zygotes as, 24

Directive on Data Protection, 192

Disaccharides, defined, 132

"Disease Model" mouse strains, 62

Diseases, genetics of

DNA methylation role, 48, 49

inbreeding role, 174

linkage and recombination tools, **4–8**, 26

Mendelian inheritance, 36, 37

molecular biologist role, 70

mouse models, 62

mutation rates, 99–100

mutation role, 93

pedigrees as tools, 7, 36, **138–142**

pharmacogenomics research, 146–147

polymorphisms as tools, 162

population screening, **175–178**

public health role, 216–219

X chromosome inactivation, 77–80

See also Dominant disorders, autosomal; Metabolic diseases; Mitochondrial diseases; Recessive disorders, autosomal; *specific diseases*

Disulfide bridges, 200, 201

DLBCL (diffuse B-cell lymphoma), chemotherapy, 146

D-loop, of mtDNA, 53

DMD. *See* Muscular dystrophy, Duchenne

DMPK gene, 86

DNA, circular

eubacteria, 132

mitochondrial and chloroplast genomes, 55–56

See also Plasmids

DNA, complementary (cDNA)

as pseudogenes, 211

in reverse-transcription PCR, 159

DNA damage. *See* DNA repair; Mutagens; Mutations

DNA fingerprinting. *See* DNA profiling

DNA methylation. *See* Methylation, DNA

DNA methyltransferases (DNMTs), 46, 49

DNA microarrays, as pharmacogenomics tool, 146

DNA, mitochondrial (mtDNA)

characteristics, 55–56

D-loop, 53

in eggs and sperm, 51

elimination of genes, 57

as evolutionary study tool, 66–67

heavy and light strands, 56

mutations, 51, 53–54

sequence comparisons, 57

transcription of, 56

See also Mitochondrial genome

DNA polymerases

E. coli, 157

incorrect polymerization, 113–114

in PCR, 154–159

repair function, 113–114

Taq, 157

DNA (deoxyribonucleic acid), polymer formation, 117–118, *118*

DNA polymorphisms. *See* Polymorphisms, DNA

DNA profiling (fingerprinting)

DNA polymorphisms, 163

DNA purification tools, 220

legal concerns, 3, 4

mtDNA D-loop as tool, 53

PCR as tool, 154, 156

privacy concerns, 192–193

DNA purification, **220–221**

in molecular anthropology, 68

procedures, 220–221

uses, 220

DNA, recombinant

ethical issues, 152

as mapping tool, 7

molecular biologist role, 71

patents on, 137–138

for PCRs, 157

plasmid vectors, 152

in selectable markers, 17

DNA repair

cytochrome P450s, 89

deoxyribonuclease enzyme role, 113–114

DNA mismatch repair, 114

in mitochondria, 79

mutation rate role, 100

mutations, role in creating, 95, 100

DNA repetitive sequences

Alu, 153, 211–212

in kinase gene, 153–154

STRPs, 6, 14

VNTRs, 6, 14, 162

See also Short tandem repeats (microsatellites)

DNA replication

base substitution mutations, 95

cell cycle phases, 25–26, *27*, 57–59

directionality, 117–118, 155–156

DNA coiling in nucleus, 122

in mitochondria, 56

nuclear envelope breakdown and re-forming, 59, 120

by plasmids, 151–152

See also Crossing over; Meiosis; Mitosis; Polymerase chain reaction; Transcription

DNA sequencing

BLAST applications, 196

to detect SNPs, 162

DNA purification tools, 220–221

mass spectrometry tools, **18–20**

as molecular anthropology tool, 63–65

mouse models, 61–62

pairwise distributions, 167–168, *168*

patent issues, 137

procedures, 1

See also Human Genome Project; Mapping genes

DNA structure

3' and 5' ends, 115, *116*, 117–118, *118*, 155–156

coiled in nucleus, 121, 122

components, 115, *116*, 117

double-helix model, 88, 94, 105, 118–119

hydrogen bonding, *116*, 119

See also Base pairs; Deoxyribose sugars; Nucleotides

DNA structure and function, history

discovery of DNA, 103

DNA as transforming factor, 103, *104*, 105

Watson and Crick's model, 105

See also Gene nature, history

DNA thermal cyclers, 157

DNase, in DNA purification, 221

DNMTs (DNA methyltransferases), 46, 49

Dobzhansky, Theodosius, 76

Dominant alleles

defined, 194

incomplete dominance, 36

Mendel's hypotheses, 31, 33, 35–36

X-linked inheritance, 73–74, 91–92

See also Inheritance patterns

Dominant disorders, autosomal

defined, 153

gain-of-function mutations, 96

muscular dystrophies, 85–86, 153–154

neurofibromatosis, 153

psychiatric disorders, 214

Dosage compensation, defined, 81

Double-helix DNA model, 118–119

disrupted by mutagens, 88, 94

distortion, repairing, 114

of Watson and Crick, 105

Down syndrome

chromosomal basis, 29, 97, 110, 111, 164

clinical features, 110

trisomy 21, 29

Drosophila. See Fruit flies

Drug dependence, genetic components, 214

Drugs (medications)

chemotherapy, 146

developing new, 206

how they work, 206

for muscular dystrophies, 87

pharmaceutical scientist role, **142–144**

poor and ultrarapid metabolizers, 144–145

side effects, preventing, 144–146

See also Antibiotic resistance; Antibiotics; Pharmacogenetics and pharmacogenomics

DSM-IV (Diagnostic and Statistical Manual), 213

Duchenne de Boulogne, 84

Dunnigan-type lipodystrophy, 125

Duty of candor, for patents, 138

Dwarfism, prenatal genetic testing, 186

Dystrophia myotonica. *See* Muscular dystrophy, myotonic

Dystrophin gene, 84–85, 100

E

E. coli bacterium. *See Escherichia coli*

Edman degradation technique, 197

Edman, Pehr, 197

EDTA (ethylenediaminetetraacetic acid), 221

Eggs

developmental processes, 29

as haploid, 24, 60, 163

meiosis and, **24–29**, 60, 75

mtDNA in, 51

mutations, 99

Eigsti, O. J., 165

Electron transport chain (ETC)

coding, mitochondrial *vs.* nuclear genomes, 52–53, *52*

mitochondrial role, 41–42, 51–53, *52*

Electrophoresis. *See* Gel electrophoresis

Electrospray ionization (ESI), 20, 208

Ellis-van Creveld syndrome, 174

Embryonic development. *See* Development, genetic control of

Emerin proteins, 125

Emerson, R. A., 10

"Emia," as suffix, 41

Employment discrimination, genetic testing and, 4, 193

Endangered species, genetic diversity, 167

Endocrine system, metabolic diseases, 42

Endonuclease enzymes

defined, 47, 112

DNA repair, 114

function, 112, *113*

restriction, 13, 47, 114, 152, 161–162

structure preferences, 112–113

Endoplasmic reticulum (ER)

defined, 202

structure and function, *120, 121,* 125, 202

Endosymbionts

chloroplasts as, 55, 56–57

mitochondria as, 52, 55, 56–57

serial endosymbiotic theory, 56–57

Environment-gene interactions. *See* Gene-environment interactions

Enzyme replacement therapy

Gaucher disease, 45

metabolic diseases, 39

mucopolysaccharidosis, 45

Enzymes

defined, 2, 13, 17, 37, 100, 103, 105, 113, 144, 176, 198, 199, 220

genes as, 102–103

metabolism and, 37–38

mutations, and metabolic disorders, 38

one gene-one enzyme model, 76, 103

See also Proteins; *specific enzymes*

Epidemiology

defined, 217

public health role, 217–218

Epigenetic, DNA methylation as, 46

EPP (erythropoietic protoporphyria), *38*

ER. *See* Endoplasmic reticulum

Erythropoietic protoporphyria (EPP), *38*

Escherichia coli

binding protein tetramers, 202

conjugation, 151

DNA polymerases, 113–114, 157

endonuclease enzymes, 114

as model organism, 60

mutation rates, 99

operons, 131–135

plasmids, 151

uid A markers, 18

ESI (electrospray ionization), 20

Estriol, unconjugated, 185

ETC. *See* Electron transport chain

Ethical issues

human cloning, 3

population screening, 177–180

preimplantation genetic diagnosis, 186–187

recombinant DNA, 152

See also Genetic discrimination; Genetic testing, ethical issues

Ethidium bromide, as mutagen, 88

Ethylenediaminetetraacetic acid (EDTA), 221

Etiology, defined, 213, 214

Eubacteria

as chloroplast ancestor, 55, 56–57

chromosomes, 132

DNA methylation in, 47

as mitochondrial ancestor, 52, 55, 56–57

mutation rates, 99

operons, 105, **131–135**

recombinant, as biohazards, 152

Eubacteria (continued)
ribonuclease role, 114
transformation, 16–18, 152
transposable genetic elements, 22
See also Plasmids; Prokaryotes; *specific bacteria*
Euchromatin, structure and function, 121–122
Eugenics
defined, 81
germinal choice, 81
Muller's contributions, 81
population screening and, 177–178
positive, 81
prenatal genetic testing, 186–187
Eukaryotes
defined, 16, 135
See also Chromosomes, eukaryotic
Evolution, human. *See* Anthropology, molecular
Evolution, multiregional, 67
Evolution, of genes
Alu sequences, 211–212
chloroplasts, 55, 56–57
chromosomal aberration role, 97
cytochrome C, *94*
early studies, 32
globin proteins, 212–213, *212*
histone proteins, *94*, 96
mitochondria, 52, 55, 56–57
molecular clocks, 63, 67, 98
mouse *vs.* human chromosomes, *96*, 97
natural selection and, 93–94
overlapping genes, 136
pseudogenes, 211–213
Exons, function, 106, 181
Exonuclease enzymes
function, 112, 113–114, *113*
proofreading, 113–114
structure preferences, 112–113
Exportins, 126
Expression cloning, 152
Exteins, *179*, 181
Extinct species
DNA analysis, 68–69
population bottlenecks, 167
Extranuclear inheritance. *See* Inheritance, extranuclear
Extremophiles, Taq DNA polymerase, 157
Eye disorders
blindness, hereditary, 174
color blindness, 78
mitochondrial diseases, 54
retinoblastomas, 99, 124, 130, 131

F
Fabry disease, symptoms and treatment, 41
Family studies/family histories
CEPH, and gene mapping, 12, *13*
disease concordance studies, 7–8, 177
in molecular anthropology, 65
pedigrees, 7, 36, **138–142**, *141*
of psychiatric disorders, 213, 214, 214–215
of Thomas Jefferson, 65
Fatal familial insomnia, 190
Fats, breakdown in mitochondria, 41
Fatty acid metabolism
disorders, 40, 43
normal function, 41
Ferns, polyploidy, 164
Ferritin
function, 199
pseudogenes for, 213
Fertility factors (conjugative plasmids), 151
Fertilization
gamete fusion, 24
polyploidy, 163, 164, 166
See also Gametes
Fetal cell sorting, 184–185
Fixed alleles, 171
Fluorescence *in situ* hybridization (FISH), as prenatal diagnostic tool, 182–183
Fluorescence-activated cell sorters, 184
Fly lab, 73, 80
FMR-1 gene, 49
Forensics
defined, 220
DNA purification tools, 220–221
See also DNA profiling (fingerprinting)
Founder effect
Amish populations, 174
Ashkenazic Jews, 174
early human populations, 67, 168–169, *169*
genotype frequencies and, 174
mitochondrial Eve, 68
Tristan da Cunha islanders, 174
Fragile X syndrome, DNA methylation role, 49
Frameshift mutations, 88, *95*, 96, 153, 209
Free radicals, as mutagens, 88, 89
Frog, African clawed (*Xenopus laevis*), 60, *61*

Fruit flies (*Drosophila*), as model organism, 60, 73
Fruit flies, genomes
DNA methylation, 46
effects of radiation on, 81, 90–92, *90*, 166–167
Fly lab, 73, 80
mapping studies, 7, 12–15, *74*, 75, 80, 83
naming genes, 107
polyploidy, 166
transposon mutagenesis, 97–98
X-linked inheritance studies, 73–74, 80, 91–92, 102
Fungi, plasmids in, 150

G
Gain-of-function mutations
inheritance patterns, 96
in pseudogenes, 211
Gajdusek, D. Carleton, 189
Galactose operons, 134
Galactosemia, symptoms and treatment, 40, 44
Galactose-1-phosphate uridyl transferase, 44
β-galactosidase enzyme, 132, 134, 135
Gametes
defined, 24, 34, 109, 166
fertilization role, 24
haploid complements, 24, 60, 163
imprinting of, 48
independent assortment of, 34–35
mutations, consequences, 93
segregation of, 34
See also Eggs; Meiosis; Sperm
Gamma rays, as mutagens, 88
Ganciclovir, as nucleotide analog, 17
Gangliosides, 45
Garrod, Sir Archibald, 37, 43
Gastro-intestinal disorders
colon cancer, 131
mitochondrial diseases, 54
Gating process, of nuclear pores, 126
Gaucher disease, symptoms and treatment, 40, 45
Gel electrophoresis
2-D PAGE, 207–208
anthrax analysis, *71*
defined, 18, 206, 221
in DNA purification, 221
ethidium bromide, 88

to identify polymorphisms, 160–162, *161*

isoelectric focusing, 208

in PCRs, 158

in protein purification, 18–19

as proteomics tool, 206

Gems, subnuclear bodies, 124

Gender differences

recombination frequencies, 7

See also Sexual development; X chromosomes; Y chromosomes

Gene capture, 127

Gene cloning. *See* Cloning genes

Gene discovery

linkage and recombination tools, **4–8**

mapping tools, 11

naming genes, 107–108

polymorphisms as tools, 162

Gene dosage

nondisjunction and, 111

of polyploids, 164

Gene duplication

Bar eyes mutation, 81

consequences, 111

globin gene family, 212

pseudogenes, 209–210, *211*

Gene expression

in marker systems, 15–18

vs. mRNA abundance, 206

Gene expression, control overview

DNA coiling in nucleus, 122

early studies, 103–106

epigenetic effects, 46

imprinting, 48

intein and extein role in, 181

mRNA role, 70

operons, 105, **131–135**

See also Mosaicism; Pleiotropy; Transcription; Translation (protein synthesis)

Gene families

evolution of, 212–213, *212*

globin proteins, 211, 212–213, *212*

Gene knock-ins. *See* Knock-in mutants

Gene knock-outs. *See* Knock-out mutants

Gene machines (oligonucleotide synthesizers), 155

Gene mapping. *See* Mapping genes

Gene nature, history, **101–106**

DNA as heredity material, 103, *104,* 105

gene action and mutation, 102–103

genes as hereditary particles, 101–102

See also Chromosomal theory of inheritance; DNA structure and function, history; Mendelian genetics

Gene patents. *See* Patenting genes

Gene pools, 171–172

Gene silencing. *See* Imprinting; Methylation, DNA

Gene splicing. *See* Splicing, alternative

Gene targeting, homologous recombination method, 91

Gene therapy

metabolic diseases, 39

muscular dystrophy, 86–87

patent issues, 137

Gene-environment interactions

and divergence from common ancestry, 63

as eugenics rebuttal, 81

and mutation rates, 99–100

psychiatric disorders, 214

public health role, 217

See also Complex (polygenic) traits

Genes

as catalysts, 102, 103

as enzymes, 102–103

human, number, 5, 24, 106, 107, 205

modifier, 81

mutable, 21

naming, 107–108

one gene-one enzyme model, 76, 103

one gene-one polypeptide model, 103

one gene-one protein model, 103

origin of term, 73, 101

overlapping, **135–136**

paralogous, 210

physical *vs.* genetic linkage, 5–6, *5*

pseudogenes, **209–213**, *210, 212*

size *vs.* mutation rate, 99–100

structure and function, 70

syntenic, 5

See also Genomes

Genes, evolution. *See* Evolution, of genes

Genetic code

coding and noncoding regions, 82–83, 93, 99, 105, 160

homologies, comparing, 61

mutations, consequences, 94–95

overlapping genes, **135–136**

as universal, 60

See also Codons

Genetic counseling

pedigrees as tools, 139–140, 142

prenatal genetic testing, 185, 186

Genetic counselors, roles, 186

Genetic discrimination

Health Insurance Portability and Accountability Act, 191–192

legal issues, 4, 192, 193

pharmacogenomics and, 146

population screening and, 177–178

public health role, 216

See also Genetic testing

Genetic diversity

of chimpanzees, 167–168, *168*

crossing-over role, 24, *25, 26, 29,* 60

of humans, 167–170, *168,* 174, 177–178

individual genetic variation, 144–147, 167–168, *168,* 192–193

of maize, 9–10, *9*

population bottlenecks and, **167–171**

Genetic drift

and divergence from common ancestry, 63, 174

founder effect role, 174

Hardy-Weinberg equilibrium, 174

random events and, 174

Genetic engineering. *See* Biotechnology

Genetic ID, genetic testing, 2

Genetic load, calculated by Muller, 81

Genetic markers

polymorphic, 6, 7

RFLPs, 6, 13–14, 161–162, *161*

SNPs, 6, 14, 65, 162

STRPs, 6, 14

VNTRs, 6, 14, 162

See also Marker systems; Short tandem repeats (microsatellites)

Genetic medicine. *See* Clinical geneticists

Genetic predisposition

discrimination based on, 190, 192, 193

pharmacogenomics to predict, 144–147

population screening, **175–178**

public health role, 216–217

Genetic structure, in populations, 172

Genetic testing

to detect polyploidy, 166

Guthrie test, 176

hemoglobin electrophoresis, 176

Human Genome Project contributions, 176, 187, 215, 216

laboratory technician role, *2*

mutation detection procedures, 92–93

newborns, 42–43, 44

patent issues, 137

PCR contamination concerns, 158

population screening, **175–178**

prenatal diagnosis, **182–187**

public health role, 218

Genetic testing, ethical issues

CLIA standards, 218

eugenics concerns, 177–178

genetic discrimination, 4, 146, 177, 216

informed consent, 175

population screening, 177–178

privacy concerns, **190–193**

risks and benefits, 218

Genetic testing, specific diseases

congenital adrenal hyperplasia, 176, 177

cystic fibrosis, 177

hemoglobinopathies, 177

hypothyroidism, 176–177

muscular dystrophies, 85, 86

phenylketonuria, 42–43, 176

sickle-cell disease, 177

Tay-Sachs disease, 174, 177

Genetically modified (GM) foods

ethical issues, 3

plant genetic engineer role, **149–150**

Geneticists

clinical, 193–194

statistical, 193, 194–195

See also specific geneticists

Genetics

biochemical, 103

See also Mendelian genetics; Population genetics

Genograms, distinguished from pedigrees, 139

Genomes

comparisons, molecular anthropology, 63–70

defense, DNA methylation role, 46–49

defined, 22, 57, 119, 127, 144, 198

distinguished from proteomes, 205

McClintock's concept of, 22

in nucleus, 119–126

See also specific organisms and organelles

Genomics

defined, 143

high-throughput screening tools, 205

See also Human Genome Project

Genotype, phenotype and

adverse drug reactions, 144–145

CEPH families, 12

induced mutagenesis to study, 90–93

in maize, 9–10

in Mendelian genetics, 33–37, 102

silent mutations, 93, 96, 99

See also Alleles; Dominant alleles; Phenotypes; Polymorphisms; Recessive alleles

Genotypes

defined, 140, 144, 171

frequencies in gene pool, 171–172

represented in pedigrees, 140

Genotyping techniques, 160–161

Germ cells

defined, 93, 101

mutations, consequences, 93, 99

Germinal choice, 81

Gertsmann-Sträussler-Scheinker (GSS) syndrome, 189–190

GFP (green fluorescent protein), 16, 17–18

Globin proteins

alpha, beta, and gamma chains, 153, 212–213, *212*

evolution of, 211–213, *212*

gene family, 211, *212*

pseudogenes in, 211

Glutamic acid (Glu)

chemical properties, 200

substitution, sickle-cell disease, 200

Glyceraldehyde 3-phosphate dehydrogenase genes, 213

Glycine (Gln), chemical properties, 200

Glycogen, function, 41

Glycogen storage diseases, symptoms and treatment, 40, 44

Glycolysis

ATP production, 52

defined, 43, 52

metabolic disorders, 43–44

Glycoproteins

defined, 178, 188, 203

in prions, 188

Glycosylation

congenital disorders, 40

of proteins, 178–179, *180*, 205

GM foods. *See* Genetically modified foods

Goldschmidt, Richard, 102–103

Gout, and uric acid, 44

Green fluorescent protein (GFP), *16*, 17–18

Greenland, analysis of ice sheets, 168

Griffith, J. S., 188

Growth factor receptor mutations, and cancer, 129–130

Growth factors, oncogene impact on, *128*, 129–130

GTP (guanosine triphosphate), structure, 115, 117

Guanine

in CpG dinucleotides, 46–49, *47*

and DNA structure, 94

mutagenic base analogs, 87

structure, 115, *116*, 118, *118*, *119*

See also Base pairs

Guanosine triphosphate (GTP), structure, 115, 117

Guidelines for Research Involving Recombinant DNA Molecules, 152

Guillardia, nucleomorphs, 136

Guthrie test, for phenylketonuria, 176

H

Hagedoorn, A. L. and A. C., 102

Haldane, John Burdon Sanderson, 13

Haldane map functions, 13

Haploid

defined, 163

gametes as, 24–29, 60, 163

Haplotypes

defined, 140

represented in pedigrees, 140

Hardy-Weinberg equilibrium

allele frequency calculations, 172–173

assumptions, 173–174

Chi-square test application, 195

HCG (human chorionic gonadotropin), 185

Health Insurance Portability and Accountability Act (HIPAA), 191–192

Heart disease. *See* Cardiovascular disease

Hedgehog proteins, 181

Helix-loop-helix motif, 201

Heme proteins, defined, 202

Hemings, Sally, 65

Hemoglobin
amino acid alterations, 200
pseudogenes, 212, *212*
structure and function, 199, 202
See also Globin proteins

Hemoglobin electrophoresis, 176

Hemoglobinopathies
genetic testing for, 177
See also Sickle-cell disease

Hemophilia
inheritance patterns, 78
prenatal genetic testing, 186
transposons and, 97

Heritability
defined, 144
See also Inheritance patterns

Herpes simplex virus (HSV)
PML body mutations, 124
in selectable markers, 17

Hershey, Alfred, DNA as transforming factor, *104*, 105

Heterochromatin, structure and function, 121–122

Heteroplasmic organelles, mitochondrial disease role, 54

Heterozygote advantage, in maize, 10

Heterozygous
defined, 9, 171
in Mendelian genetics, 33–34

High-performance (high-pressure) liquid chromatography (HPLC), as protein sequencing tool, 197–198

High-throughput screening (HTS), proteomics role, 205–206

HIPAA (Health Insurance Portability and Accountability Act), 191–192

Histone proteins
defined, 77, 123
evolution of, *94*, 96
function, 199
methylation of, 77
movement within nucleus, 123

Homocystinuria, symptoms and treatment, 40

Homogentistic acid, 43

Homologous chromosomes (homologues)
defined, 5, 14, 62, 75, 82, 108, 109
of degenerate PCR primers, 158–159
of model organisms, 61

See also Crossing over (recombination); Meiosis; Mitosis; Nondisjunction

Homozygous
defined, 171
fixed alleles, 171
in Mendelian genetics, 33–34

Hormone receptors
movement within nucleus, 123
pleiotropic effects on, 154

Hormones
defined, 206, 216
processing, 204

Hox genes, 107

HPLC. *See* High-performance liquid chromatography

HPRT (hypoxanthine phosphoribosyltransferase), 44

Hprt gene, and somatic mutations, 93

HSV (herpes simplex virus), in selectable markers, 17

HTS (high-throughput screening), proteomics role, 205–206

HUGO (Human Genome Organization), 23

Human chorionic gonadotropin (hCG), 185

Human cloning, ethical issues, 3

Human disease genes, identification of
linkage and recombination tools, **4–8**, 26
mapping studies, 11
McKusick's contributions, 23
Mendelian inheritance ratios, 36
mouse models as tools, 62
pedigrees as tools, 7, 36, **138–142**
pharmacogenomics research, 146–147
polymorphisms as tools, 162
population screening, **175–178**
SNP tools, 146–147
See also Diseases, genetics of; Metabolic diseases; Mitochondrial diseases; Population genetics; Prenatal diagnosis; *specific diseases and disorders*

Human Gene Mapping Workshop, 23

Human genome
Alu sequences, 211–212
archaic, DNA analysis, 68–69
autosomes, 64, 76, 106, 108–109, 111, 182
chromosomal translocations, 96–97
as diploid, 60, 97

distinguished from mice, *96*, 97
distinguished from primates, 63, 64–65
DNA methyltransferases, 46, 49
endonuclease enzymes, 114
gene loci, number, 107
genes, number, 5, 24, 106, 205
genotype frequencies, 171–174
mapping, 7–8, 15
mtDNA, 53–54
mutation rates, 99, 177
naming genes, 107–108
noncoding genes, 65, 83, 99
nucleotide pairs, number, 155
overlapping genes, 135–136
in pharmacogenomics studies, 144–147
rRNA, number, 123
sex chromosomes, 24, 76, 106, 108–109, 111, 182
SNP loci, 162
transposable genetic elements, 64
triploids, 164–165

Human Genome Organization (HUGO), 23

Human Genome Project
genetic testing contributions, 176, 187, 216
McKusick's contributions, 23
mtDNA compared to nuclear DNA, 57
pharmacogenomics applications, 146–147
protein sequencing contributions, 198

Human PSD database, 209

Humans (*Homo* sp.)
archaic, 66–67
divergence from chimpanzees, 167–168, *168*
geographic origin, *64*, 66–67
H. erectus distinguished from Neandertals, 66–67, *66*, 69, 167
replacing Neandertals, 170
See also Anthropology, molecular

Hunter syndrome (iduronate sultatase deficiency), symptoms and treatment, 41, 45

Huntington's disease
Chi-square test to predict, 194–195
mutation rates, 100
pharmacogenomics to predict, 146
protein conformation role, 190
as triplet repeat disorder, 96

Hurler syndrome (α-iduronidase deficiency), symptoms and treatment, 41, 45

Hybridization (molecular). *See In situ* hybridization

Hybridization (plant and animal breeding)

of maize, **8–10**

plant genetic engineer role, **149–150**

See also Mendelian genetics

Hydrogen bonds

defined, 119, 200

and DNA structure, *116*, 119

and protein secondary structure, 200–201

Hydrolysis reactions, of nucleases, 113, *114*

Hydrophilic, defined, 200

Hydrophobic, defined, 200

Hydroxylation of proteins, post-translational, 203

Hyperammonemia, 45

Hypercholesterolemia, symptoms and treatment, 40, 43

Hyperhomocysteinemia, symptoms and treatment, 40

Hyperlipidemia, symptoms and treatment, 40, 43

Hypertrophic muscles, 84

Hypotheses, defined, 7, 18

Hypothyroidism, population screening, 176–177

Hypoxanthine phosphoribosyltransferase (HPRT), 44

I

Ice ages, population bottlenecks, 168

ICF (Immunodeficiency, Centromere Instability, and Facial Anomalies) syndrome, 49

Iduronate sultatase deficiency (Hunter syndrome), 41, 45

α-iduronidase deficiency (Hunter syndrome), 41, 45

IGF2 (insulin-like growth factor II), DNA methylation, 48

Immigration and migration

limited, founder effect, 168–169, *169*

patterns of early humans, 62, *64*, 66–67, 70, 168–170

Immune systems

antibodies, 199, 202

DNA methylation role, 47–48, 49

MHC proteins, 82

Immunodeficiency, Centromere Instability, and Facial Anomalies (ICF) syndrome, 49

Importins, 126

Imprinting, 48

In situ hybridization (molecular)

fluorescence (FISH), 182–183

oncogene detection role, *129*

PCR as tool, 158

as prenatal diagnostic tool, 182–183

In vitro, defined, 18

In vitro fertilization, preimplantation genetic diagnosis, 182, *184*, 185–187

Inbreeding

and disease risk, 174

founder effect, 174

polymorphisms to study, 163

Incomplete dominance, 36

Indels (insertion or deletion polymorphisms), 160

Indians (American), evolutionary origin, 66

Individual genetic variation

among early humans, 167–168, *168*

and DNA profiling, 192–193

pharmacogenomics applications, 144–147

Infertility, chromosomal translocations, 97

Informants, in pedigree drawing, 140–141

Informed consent

defined, 218

genetic testing, 175

Ingram, Vernon, 95

Inheritance, extranuclear

mitochondrial genome, 51, 67–68, *69*

plasmids, 151

significance, 57

Inheritance patterns

blending theory, 30, 34, 102

gain-of-function mutations, 96

imprinting, 48

in linkage and recombination studies, 7–8

loss-of-function genetic mutations, 96

multifactorial and polygenic, 214

pedigrees as tools, 7, 36, **138–142**

of plasmids, 151

Punnett squares, *33*, 35, 193–194

See also Dominant alleles; Mendelian genetics; Recessive alleles; X-linked disorders

Inheritance patterns, specific diseases

color vision defects, 78

cystic fibrosis, 142, 193–194

hemophilia, 78

metabolic diseases, 38–39, 45

muscular dystrophies, 78, 83–84, 85, 86

psychiatric disorders, 214, 215

sickle-cell disease, 142

Initiation sites, transcription. *See* Start codons

Insect-resistant crops, *Bacillus thuringiensis*, 149

Insulin

function, 199

post-translational control, 181

Insulin receptors, pleiotropic effects on, 154

Insulin-like growth factor II (IGF2), DNA methylation, 48

Insurance discrimination, genetic testing and, 4, 146, 192, 193

Inteins, post-translational control, *179*

Intelligence

as complex trait, 81

See also Mental retardation

Interbreeding, 171, 173–174

Intercalating agents, 88

Interchromatin compartment, 120, *121*, 122–123, *122*

Interchromatin granules (speckles), 120, 125

Interference, crossover, 12–13

Intermediate filaments, 120, *121*

International Congress of Eugenics, 81

Internet

bioinformatics tools, 18

genetic maps, 15

Mendelian Inheritance in Man, 23

Interphase

DNA synthesis (S) phase, 25–26, 57–59

growth (G_1 and G_2) phases, 57–59

Intervention strategies, public health role, 218–219

Intestinal disorders. *See* Gastrointestinal disorders

Introns

defined, 56

discovery of DNA, 105–106

function, 181

lacking in mitochondria and chloroplasts, 56

lacking in mRNA, 124, 211

Inversions, and chromosomal aberrations, 97

Isoelectric focusing, 208

Isoforms, protein, 160

Isohaemagglutinin alleles, 82

Isoleucine-valine operon (*ilv*), 134

J

Jacob, Francois, 105, 131

Jefferson, Thomas, Y chromosome analysis, 65

Jellyfish genes, green fluorescent protein, 17

Jews

    Ashkenazic, Tay-Sachs disease, 174

    evolutionary origin, 66

Johansson, Wilhelm, 101

Jumping genes. *See* Transposable genetic elements

K

Karyotype analysis, *108*

    chromosomal banding, 106, *107*

    spectral (chromosome painting), 183

Kearns-Sayre syndrome, *53*

Kidney disorders

    metabolic diseases, 42

    mitochondrial diseases, 54

Kilobases, defined, 7

Kinases

    defined, 153, 178

    post-translational phosphorylation by, 178

    repetitive DNA sequences, 153–154

Kinetochores, structure and function, 25, 26, 27, 28, 29, 58–59

Klinefelter's syndrome, cause and symptoms, 111

Knock-in mutants, gene targeting, 91

Knock-out mutants

    "Disease Model" mouse strains, 62

    gene targeting, 91

    PrP genes, 188

Knudson, Alfred Jr., 131

Kosambi, Damodar, 13

Kosambi map functions, 13

Krebs cycle, defects, 42, 43, 44

Kremer bodies. *See* PML bodies

Kuru disease, cause, 189

L

Laboratory technicians, **1–3**, *2*

    automated sequencing by, 1

    microbiologists as, **50–51**

    molecular biologists as, **70–72**

    plant genetic engineers as, **149–150**

Lac operons

    discovery, 131

    function, *132*

    gene expression regulation, 132, *133*, 134–135

LacA gene, 132

Lactose, *lac* operons, 131–135, *132*, *133*

Lactose permease enzyme, 135

LacY gene, 132

LacZ gene, 132, 152

Lahr, Marta, 168

Lamin proteins, 125

Law of independent assortment

    genes on homologous chromosomes, 36, 75

    incomplete dominance, 36

    linked genes, 5–6, 14–15

    meiosis and, 26, 29, 60, 75

    Mendel's hypotheses, 31, 34–35

Law of segregation, Mendel's hypotheses, 31, 34

Lawyers. *See* Attorneys

Leber's hereditary optic neuropathy, cause, 53

Leewenhoek, Antoni van, 119

Legal issues, **3–4**

    DNA profiling, 3, 4

    genetic discrimination, 3–4

    genetically modified foods, 3

    patenting genes, 4, **136–138**

    right to privacy, 4, 191–193

    tissue ownership, 3–4

Leigh, Dennis, 54

Leigh's disease, *52*, 54

Lesch-Nyhan syndrome, symptoms and treatment, 40, 44

Leucine (Leu), chemical properties, 200

Leukemia

    acute myeloid, 96

    chronic myelogenous (CML), 127, 129

    human acute promyelocytic, 124

LGMD (limb-girdle muscular dystrophy), 86

Li Fraumeni syndrome, 98

Ligation, defined, 158

Light and high-energy particles, as mutagens, 88

Linkage and recombination (linkage analysis), **4–8**, *74*

    bipolor disorder, 215

    calculating linkage and map distance, 6

    CEPH families, 12, *13*

    genetic marker tools, 6, 11

    independent assortment role, 5–6, 75

    lod score linkage, 8

    of maize, 10

    Morgan's studies, 7, 12, 14–15, *74*, 75

    physical *vs.* genetic linkage, 5–6, *5*

    psychiatric disorders, 214–215

    Punnett and Bateson's contributions, 36

    statistical tools, 7–8, 12–13, 15

    See also Crossing over (recombination)

Linkage, defined, 4

Lipoprotein cholesterol, and cardiovascular disease, 43

Liquid chromatography. *See* High-performance liquid chromatography

Liver disorders, Gaucher disease, 40, 45

Liver function, poison detoxification, 89

Livestock, prion diseases, 188–190

Loci, gene

    defined, 5, 11, 82, 160

    human, number, 107

    marker, 11, 13–14

    physical *vs.* genetic linkage, 5–6, *5*

    See also Mapping genes

Lod score linkage analysis, 8

Los Alamos National Laboratory, *71*

Loss-of-function mutations

    hemophilia, 78, 97, 186

    inheritance patterns, 96

    metabolic diseases, 38–39

Lucas, William, *38*

Lumen (endoplasmic reticulum), defined, 202

Lyase reactions, of nucleases, 113

Lymphomas, diffuse B-cell, 146

Lyon Hypothesis, 76–77

Lyon, Mary, 76–77

Lysine (Lys), ubiquitination, 179

Lysosomes, storage disorders, 40–41, 44–45

M

MacLeod, Colin, 103, 105

Macromolecules, defined, 44, 132

Macrophages, defined, 45

Mad cow disease. *See* Bovine spongiform encephalopathy (BSE)

Maize (corn), **8–11**
  crossing-over studies, 21, 75
  genetic diversity, 9–10, *9*
  genetically engineered, 149
  hybrid vigor, 9
  mapping studies, 10
  as model organism, 8–9, 60, 149
  origin, 8
  production statistics, 8
  transposable genetic elements, 10, 21–22

Major histocompatibility complex (human leukocyte antigen) proteins, 82

MALDI (matrix-assisted laser desorption ionization), 20, 208

Mammals
  DNA methylation, 46, 47
  marker systems for, 16–17
  mutation rates, 101
  nuclear pores, 125–126

Mammograms, and breast cancer, 175

Map functions
  defined, 12
  Haldane, 13
  Kosambi, 13

Maple syrup urine disease, symptoms and treatment, 40

Mapping genes, **11–15**
  CEPH families, *12, 13*
  crossing-over frequency, 11–13, 26, 83
  disease identification role, 11
  fruit flies, 7, 12–13, 14–15, *74*, 75
  gene discovery role, 11
  human applications, 15
  Human Gene Mapping Workshop, 23
  interference, 12–13
  linkage and recombination tools, **4–8**, 26
  maize, 10
  physical maps, 7
  physical *vs.* genetic maps, 15
  polymorphic markers, 6, 7
  SNP tools, 162
  See also DNA sequencing; Human Genome Project

Marker loci, 11, 13–14

Marker systems, **15–18**
  how they work, 15–16
  screenable markers, 16, 17–18, 91

selectable markers, 16–17, 152, 162
  See also Genetic markers

Maroteaux-Lamy syndrome, symptoms and treatment, 41

Mass spectrometry, **18–21**
  applications, 20, 198
  procedure, 18–20, *19*
  as proteomics tool, *206*, 208

Maternal and Child Health Bureau (Children's Bureau), PKU screening, 219

Maternal serum marker screening, 185

Matrix-assisted laser desorption ionization (MALDI), 20, 208

McCarty, Maclyn, 103, 105

McClintock, Barbara, **21–22**, *21*
  breakage-fusion-bridge cycle, 21, 22
  crossing-over studies, 10, 75, 105
  transposable genetic elements, 10, 21–22, 97

McKusick, Victor, 10, **22–24**, *23*

MD. *See* Muscular dystrophy

The Mechanism of Mendelian Heredity, 76

MeCP2 protein, and Rett syndrome, 48, 49, 78

Media (for bacterial cultures), defined, 50

Medical genetics. *See* Clinical geneticists

Medical records, privacy of, 191–192

Medium-chain acyl-CoA dehydrogenase deficiency, newborn screening, 176

Meiosis, **24–30**
  centromere role, 25, 29
  chiasmata role, 26
  chromosomal aberrations, 29
  cytokinesis, 29
  defined, 5, 65
  distinguished from mitosis, 24, 26, 29, 59, 60
  gamete formation, 24–25, 75
  independent assortment of chromosomes, 26, 29
  kinetochore role, 25, 26, 27, *28*, 29
  microtubule role, 26, 27, 29
  nondisjunction, 79, **108–112**, *109*, 166
  purpose, 24
  spindle role, 26, 111–112
  See also Chromosomes, eukaryotic; DNA replication

Meiosis I, 24–26
  anaphase, 26, 27, *28*

gamete dormancy, 29
  metaphase, 26, 27, *28*
  prophase, 25–26
  telophase, 26

Meiosis II, 25, 29
  gamete dormancy, 29
  metaphase, 27, *28*
  telophase, 27

MELAS (mitochondrial encephalopathy, lactic acidosis, and stroke), 54

Membrane proteins, structure, 201

Mendel, Gregor, **30–32**, *30*

Mendelian genetics, **32–37**
  character pairs, 101–102
  dihybrid crosses, 35
  exceptions to, 35–36, 75
  law of independent assortment, 5–6, 14–15, 26, 29, 31, 34–35, 60, 75
  law of segregation, 31, 34
  The Mechanism of Mendelian Heredity, 76
  monohybrid crosses, 33
  procedures, 30–31, 32–34
  Punnett squares, *33*, 35
  trihybrid crosses, 35
  See also Chromosomal theory of inheritance, history; Gene nature, history

Mendelian Inheritance in Man (V. McKusick), 23

Mendelism. *See* Mendelian genetics

Menstruation, ovulation and, 29

Mental retardation
  genetic components, 213
  intelligence as inherited trait, 81
  newborn screening to prevent, 42–43
  phenylketonuria, 42–43, 176
  See also Down syndrome; Phenylketonuria

Metabolic diseases, **37–46**
  amino acid metabolism disorders, 40
  cofactor or vitamin enhancement, 39
  detoxifying agents, 39, 45
  diet modifications to treat, 39, 43, 44, 45
  enzyme defects as cause, 38–39, *39*
  enzyme replacement therapy, 39, 45
  fatty acid metabolism disorders, 40
  gene therapy, 39
  inheritance patterns, 38–39

organ transplants, 39, 45

organic acid metabolism disorders, 40

peroxisomal metabolism disorders, 41

prenatal genetic testing, 183–184, 186

purine and pyrimidine metabolism disorders, 40, 40–41

urea formation disorders, 41

Metabolism

ATP and, 51–52, 55–56, 117

defined, 37, 131

enzymatically controlled reactions, 37–38

inborn errors of, 37

mitochondrial role, 41, 55–56

normal *vs.* impaired sequences, *39*

Metabolites, defined, 142, 218

Metals (heavy), plasmid resistance to, 151

Metaphase

I, meiosis, 26, *27*, *28*

II, meiosis, *27*, *28*

mitosis, 59

prometaphase, mitosis, 59

Metaphase plate, 59

Methylation, DNA, **46–49**

CpG islands, 47, *48*

gene imprinting, 48

gene repression role, 48, 77, 100

host defense function, 46, 47–48

and human disease, 49

mutations and, 95, 100

process, 46–47, *47*

Methylation, of proteins, 203, 205

Methyl-binding proteins, gene repression role, 48

Methylmalonic acidemia, symptoms and treatment, 40, 43

Methylmalonyl-CoA, 43

MHC (major histocompatibility complex) proteins, 82

Mice. *See* Rodent models

Microbiologists, **50–51**

Microliters, defined, 156

Micronucleus test, 92–93

Microsatellites. *See* Short tandem repeats

Microtubules, structure and function, 26, *27*, 29, 59, *121*

Migration. *See* Immigration and migration

Minisatellites. *See* Variable number of tandem repeats

Mismatch distributions (pairwise differences), 167–168, *168*

Missense mutations, 153

Mitochondria

defined, 135, 149

electron transport chain, 41–42, 51–53, *52*

plasmids in, 150

structure and function, 41–42, 51, 55–56, *56*

Mitochondrial diseases, 41, **51–55**

aging and, 80

genetic components, 52–53

Kearns-Sayre syndrome, 53, *53*

Leber's hereditary optic neuropathy, 53

Leigh's disease, *52*, 54

MELAS, 54

mosaicism, 79–80

multisystem, list of, 54

oxidative metabolism disorders, 41–42

Mitochondrial DNA. *See* DNA, mitochondrial (mtDNA)

Mitochondrial encephalopathy, lactic acidosis, and stroke (MELAS), 54

Mitochondrial Eve, 63, 67–68

Mitochondrial genome, **55–57**

base pairs, number, 79

coding for ETC subunits, 52–53, *52*

evolutionary origin, 52, 55, 56–57

as evolutionary study tool, 66–67

heteroplasmic and homoplasmic, 54

inheritance patterns, 51, 67–68, *69*

mtDNA, 51–54, 55–56, 57, 66–67

mutations, 42, 79–80

overlapping genes, **135–136**

random replication, 56

repair mechanisms lacking, 79

Mitosis, **57–60**, *58*

anaphase, 59

cell cycle phases, 57–59

centromere role, 58

cytokinesis, 59

defined, 106

distinguished from meiosis, 24, 26, 29, 59, 60

kinetochore role, 58–59

metaphase, 59

prophase, 59

spontaneous tetraploids, 166

telophase, 59

See also Chromosomes, eukaryotic; DNA replication

Mob genes, 152

Model organisms, **60–62**

Arabidopsis thaliana, 60, 97–98, 149

examples, 60–61

roundworms, 60, 61

useful traits and attributes, 60–61, 73

yeasts, 46, 60, 181

zebrafish, 60, 61

See also Escherichia coli; Fruit flies; Maize (corn); Rodent models

Modified Gomori trichome stains, 53, *53*

Modifier genes, Muller's concept, 81

Molecular anthropology. *See* Anthropology, molecular

Molecular biologist. *See* Biologists: molecular

Molecular clocks

divergence from common ancestry, 63, *94*, 98, 100–101, 167–168

precision, 67

Molecular cloning. *See* Cloning genes

Monod, Jacques, 131

Monohybrid crosses, 33

Monomers, defined, 178

Monosomy, impact of, 109, 111

Moore v. *Regents of the University of California*, cell line ownership, 3–4

Morgan, Thomas Hunt, **72–76**, *73*

fruit fly genome mapping, 7, 12, 14–15, *74*, 75

genes described by, 32, 73

X-linked inheritance studies, 73–74, 102

Morgan, unit of measurement, 75

Morquio syndrome, symptoms and treatment, 41

Mosaicism, **76–80**

calico cat example, 77, *78*

chromosomal, 79

mitochondrial, 79–80

nonrandom inactivation, 78

X chromosome inactivation, 77–78

MRNA. *See* RNA, messenger

Mucopolysaccharidosis, symptoms and treatment, 45

Muller, Hermann, **80–81**, *80*

The Mechanism of Mendelian Heredity, 76

mutation rate measurements, 80

on polyploidy, 164

radiation-induced mutations, 81, 88, 103

X-linked inheritance studies, 73, 91–92

Mullis, Kary, 91, 154
Multimers, defined, 202
Multiple alleles. *See* Alleles, multiple
Multiple genetic hit theory, 131
Muscle cells, healthy tissue, *84*
Muscle disorders
 Kearns-Sayre syndrome, 53, *53*
 mitochondrial diseases, 54
 spinal muscular atrophy, 124
Muscular dystrophy (MD), **83–87**
 Becker (BMD), 84
 characteristics, 83
 drug therapies, 87
 Emery-Dreifuss, 125
 gene therapy, 86–87
 limb-girdle (LGMD), 86
 prenatal genetic testing, 186
Muscular dystrophy, Duchenne (DMD)
 causative gene, 84–85, 100
 characteristics, 84, *85*
 genetic testing, 85, 184
 inheritance patterns, 78, 83–84
 mutation rates, 100
Muscular dystrophy, myotonic
 causative gene, 86
 characteristics, 85–86
 genetic testing, 86
 inheritance patterns, 85
 pleiotropic effects, 153–154
 as triplet repeat disease, 86
Mustard gas, as mutagen, 90
Mutable genes, 21
Mutagenesis, **89–93**
 induced, research role, 89, 90–92, 97–98
 site-directed, 91
Mutagens, **87–89**
 Ames test, 87, 92
 base analogs, 87, *88*, 90–91
 base-altering, 88
 carcinogens, 89
 defined, 89
 distinguished from carcinogens, 87
 and increased mutation rates, 99–100
 intercalating agents, 88
Mutagens, specific
 asbestos, 89
 benzo[*a*]pyrene, 89
 5-bromo-deoxyuridine, 87
 ethidium bromide, 88
 free radicals, 88, 89
 light and high-energy particles, 88

mustard gas, 90
 nitrite preservatives, 88
 radiation, ultraviolet, 88–89, 128
 smoke, 88, 89
 transposons, 91, 97
 See also Radiation, ionizing
Mutation analysis, pedigrees as tools, 140
Mutation rates, **98–101**
 amino acids, *94*
 Duchenne muscular dystrophy, 100
 factors that influence, 99–100
 human genome, 99, 177
 Huntington's disease, 100
 mammals, 101
 measuring, 80, 99
 molecular clocks, 63, 67, 98, 100–101
 reduced by antioxidants, 89, 100
Mutations, **93–98**
 back, 65, 100
 cancer and, 98
 defined, 63, 140
 and divergence from common ancestry, 63, 65, 167–168
 enzyme defects, 38
 frameshift, 88, 95, 96, 153
 gain-of-function, 96, 211
 in germ-line cells, 93, 99
 knock-in, 91
 knock-out, 62, 91, 188
 loss-of-function, 38–39, 78, 96, 97, 186
 of maize, 9–10, *9*
 missense, 153
 in mitochondrial DNA, 42, 51, 53–54
 mosaicism, **76–80**, *78*
 Muller's hypotheses, 80–81
 neutral, overlapping genes and, 136
 in noncoding regions, 65, 82–83, 93, 99
 nonsense, 153
 pleiotropy and, 85, **153–154**
 point, 54, 95, *95*, 128–129
 in pseudogenes, 209–212
 rarity of, 90
 as recessive, 90
 selective advantages, 94
 silent, 93, 96, 99
 spontaneous, 90, 100
 transition, 95
 transversion, 95
 See also Chromosomal aberrations; DNA repair; Mosaicism

Mutations, detecting
 Ames test, 92
 chromosomal aberrations, 92–93
 and mutation rates, 100
 somatic mutations, 93
Mutations, specific genes
 A3243G, 54
 c-myc, 96
 DNMT3B, 49
 dystrophin, 85, 100
 FMR-1, 49
 hprt, 93
 NF1, 130, 153
 p53, 130
 proto-oncogenes, 127–130
 PrP, 188–189, 190
 RB, 130
 v-*erbB*, 130
 v-*fms*, 130
Mycobacteria, inteins, 181
Myoglobin proteins
 evolutionary origin, 212, *212*
 tertiary structure, 201
Myosin, function, 199

N

Na$^+$/K$^+$ ATPase pump, 204
Nash family, preimplantation genetic diagnosis, 186
National Institutes of Health (NIH), population screening, 177
Natural selection
 and divergence from common ancestry, 63
 mutations, consequences, 93–94
Nature of the gene. *See* Gene nature
ND10. *See* PML bodies
Neandertals
 distinguished from modern humans, 66–67, *66*, 69, 167
 DNA analysis, 68–69
 replaced by modern humans, 170
Negative selection marker systems, 16
Neural tube defects (NTDs), prenatal genetic testing, 184, 185
Neurofibromatosis, pleiotropic effects, 153, *154*
Neurological crises, metabolic diseases, 42
Neurological, defined, 42
Neurological disorders, mitochondrial diseases, 54
Neurons, defined, 45
Neurospora, biochemical genetics studies, 103
New Synthesis, 32

Newborn screening, specific diseases
congenital adrenal hyperplasia, 176, 177
cystic fibrosis, 177
galactosemia, 44
hemoglobinopathies, 177
hypothyroidism, 176–177
medium-chain acyl-CoA dehydrogenase deficiency, 176
phenylketonuria, 42–43, 176, 218–219
Newborn screening tests
DNA- *vs.* non-DNA based, 176–177
Guthrie, 176
hemoglobin electrophoresis, 176
NF1 gene, 130, 153
Nic sites, on DNA, 151, *151*
Nicotine addiction, inheritance of, 214
Nitrite preservatives, as mutagens, 88
Nobel Prizes
chromosomal theory of inheritance, 76, 80
Kuru transmission, 189
operon discovery, 131
PCR invention, 91, 154
prion hypothesis, 187
radiation-induced mutations, 81
site-directed mutagenesis, 91
transposable genetic elements, 10, 22
Nomenclature, **106–108**
Nondisjunction, **108–112**
aging and, 112
aneuploidy, 110–111, 166
and chromosomal mosaicism, 79
fatal *vs.* non-fatal conditions, 111
mechanism, 108–109, *109*
spindle checkpoint errors, 111–112
Non-insulin-dependent diabetes mellitus (diabetes type 2), 125, 154
Nonpolar, defined, 200
Nonprocessed pseudogenes, 210–211, *210*
Nonsense mutations, 153
NTDs (neural tube defects), prenatal genetic testing, 184, 185
N-terminus, of amino acids/proteins, 181, 197–198, 207
Nuclear lamina, 120, 121, *121*, 125
Nuclear localization signals, 126
Nuclear magnetic resonance, 71
Nuclear membranes (envelopes)

breakdown and reformation, 59, 120
structure and function, 120, *120*, 122, *123*, 125
Nuclear pores, structure and function, 120, *121*, 122–123, *123*, 125–126
Nuclease enzymes, **112–115**
deoxyribonucleases, 112, 113–114
DNase, 221
endonucleases, 112–113, *113*, 114–115
exonucleases, 112–113, 113–115, *113*
function, 112
hydrolysis reactions, 113, *114*
lyase reactions, 113
ribonucleases, 112, 114–115
structure preferences, 112–113
See also Restriction endonuclease enzymes
Nucleic acid
defined, 187
early studies, 103, *104*, 105
Nucleolar organizer, 123–124
Nucleolus, structure and function, 120, *121*, 123–124
Nucleomorphs, overlapping genes, 136
Nucleoplasm
defined, 121
structure and function, 121, 122–123
Nucleoporins, 125–126
Nucleosides, structure, 115, 117
Nucleotide analogs, ganciclovir as, 17
Nucleotides, **115–119**
defined, 82, 112, 146, 160
and DNA structure, 94, *116*
function, 115
structure, 115, *116*, 117
See also Base pairs; DNA sequencing; Genetic code; RNA sequencing
Nucleus, **119–126**, *121*
Cajal bodies, 120, 124
centrosomes, *121*
chromatin, 120, *120*, *121*
chromosomal territories, 121–122, *122*
discovery, 119
endoplasmic reticulum, 120, *120*, *121*, 125
gems, 124
interchromatin compartment, 120, *121*, 122–123, *122*
intermediate filaments, 120, *121*

microtubules, *121*
nuclear lamina, 120, 121, *121*, 125
nuclear pores, *121*
nucleolus, 120, *121*, 123–124
nucleoplasm, 121, 122–123
origin of term, 119
PML bodies, 124–125
ribosomes, 70–71, 119, 125
speckles, 120, 125
structure, 119–121, *120*, *121*
techniques for studying, 121, 122, 124, 125
Nutrition (diet)
amino acid requirements, 41
antioxidants in, 89
of early humans, 63, 68–69
phenylketonuria, 176
to treat metabolic diseases, 39, 43, 44, 45

O

Obesity, "Disease Model" mouse strains, 62
O'Brien, Chloe, 186
Obsessive-compulsive disorder (OCD), 213, 215
Oligonucleotide synthesizers, 155
Oncogenes, **127–131**
as dominant, 130
early research, 127
growth-signaling pathways and, *128*, 129–130
multiple genetic hits, 131
mutations and, 98
proto-oncogene activation, amplification, 128–129
proto-oncogene activation, retroviruses, 127
Ras, 128, 130
in situ hybridization tools, *129*
transcription factors as, *128*, 130
v-*erbB*, 130
v-*fms*, 130
v-*sis*, 130
Oocytes
defined, 24
primary, 24, 29
Open reading frames (ORFs), 209
Operator regions, of operons, 132, 134
Operons, **131–135**
discovery, 131
function, 105
functional relationships, 132
gene clustering, 132
lac, 131–135, *132*, *133*

Operons (continued)
  operators, 132, 134
  promoter regions, 132, 134
  regulation mechanisms, 134–135
  transcription of, 134–135
Organelles, defined, 79, 149
Organic acid metabolism, disorders, 40
Organic, defined, 88
Origin of replication (ori) sequences, 151, *151*, 152
Ornithine transcarbamylase deficiency, 45
Oryza sativa (rice), genetically engineered, 149
Outcrossing, of maize, 9
Ovalbumin, function, 199
Ovaries
  cancer, 218
  function, 24, 60
Overlapping genes, **135–136**
Ovulation, and meiosis, 29
Oxidative metabolism, 41–42
Oxidative phosphorylation
  ATP production, 52
  by mitochondria, 51–52, 55–56, 57
  reactive oxygen species, 52
Oxygen
  as electron acceptor, 41–42
  reactive species, as mutagens, 52, 100
Oxytocin, post-translational control, 181

P

P values, 194–196
P53 gene, mutations, 98, 130
Pairwise differences, in DNA sequences, 167–168, *168*
Pancreatic disorders, mitochondrial diseases, 54
Panic disorder, genetic components, 213
Pap smears, 175
Paralogous genes and pseudogenes, 210
Parkinson's disease, protein conformation role, 190
Patent attorneys, 138
Patent depositories, 138
Patenting genes, **136–138**
  history, 136–137
  international protection, 138
  legal issues, 3–4
  procedure, 138
  requirements, 137–138

Patents, defined, 136
Pathogenic/pathogenicity
  defined, 151, 187, 217
  virulence plasmids, 151
Pathogens
  prions, **187–190**
  See also Eubacteria; *specific pathogens*; Viruses
Pauling, Linus, 95, 200
PCR. *See* Polymerase chain reaction
Pearson, Karl, 102
Peas (garden), Mendelian genetics, 30–31, 32–36
Pedigrees, **138–142**
  confidentiality of, 142
  defined, 7
  degrees of relationship, *141*
  distinguished from genograms, 139
  drawing and recording, 140–141
  ethnic background, 141–142
  in Mendelian genetics, 36
  sample, *140*
  symbols, *139*, 140, 141
  terminology, 140
  uses, 139–140
Penetrance/nonpenetrance, reduced, 214
Peptide bonds, 197, 200
Peptide mass fingerprinting, 208
Peptide sequencing. *See* Protein sequencing
Peptides, defined, 20, 207, 208
Peroxisome metabolism, disorders, 41, 45–46
Pfam database, 209
PGD (preimplantation genetic diagnosis)
  ethical issues, 186–187
  uses, 185–186
Pharmaceutical scientists, **142–144**, *145*
Pharmacogenetics and pharmacogenomics, **144–147**
  DNA microarray tools, 146
  Human Genome Project contributions, 146–147
  pharmaceutical scientist role, **142–144**
  predisposition predictions, 146, 148
  SNP comparisons, 146–147
  See also Drugs (medications)
Phenotypes
  defined, 9, 16, 29, 54, 91, 93, 99, 102, 190
  See also Genotype, phenotype and

Phenylalanine (Phe), and PKU, 42–43
Phenylalanine hydroxylase, 42–43, 176
Phenylketonuria (PKU)
  Guthrie test, 176
  newborn screening, 42–43, 176, 218–219
  symptoms and treatment, 40, 42–43, 176
PhiX174 bacteriophage, overlapping genes, 135
Phosphatase enzymes, post-translational phosphorylation by, 178
Phosphate groups, and DNA structure, 115, *116*, 117
Phosphodiester bonds
  defined, 112, 113, 117
  and DNA structure, 117–118, *118*
  hydrolysis of, 113, *114*
  lyase reactions, 113
Phospholipids, in cell membranes, 55
Phosphorylation
  of antibiotics, 17
  defined, 51, 129
  of proteins, post-translational, 20, 178, *180*, 203–204, 205
  of resistance genes, 16–17
Physician, defined, 147
Physician scientists, **147–148**
Pilus, sex, 151
PKU. *See* Phenylketonuria
Placenta, chromosomal mosaicism, 79
Plant genetic engineers, **149–150**
Plasma membranes
  defined, 202
  structure and function, 202
Plasmids, **150–153**
  Col, 151
  conjugative (fertility factors), 151
  defined, 16
  degradative and catabolic, 151
  inheritance of, 151
  mobilization, 152
  in recombinant organisms, 152
  relaxed, stringent, and incompatible, 151
  replication of, 151–152
  resistance (R), 16–17, 151
  structure and function, 150–151
  as vectors, 91, 152
  virulence, 151
Plastids, defined, 149
Pleiotropy, **153–154**
  impact on hormone regulation, 154

myotonic muscular dystrophy, 85

sickle-cell disease, 153

signaling pathway defects, 153–154

Ploidy

diploid, 24, 60, 93, 163

haploid, 24–29, 60, 163

monosomy, 109

polyploid, 97, **163–167**

tetraploid, 97, 165, 166

triploid, 164–166, *164*

trisomy, 10, 79, 109, 111

PML bodies (PODs), 120, 124–125

Point mutations

base substitutions, 95

frameshift mutations, *95*

mitochondrial diseases, 54

and proto-oncogene activation, 128–129

Polar bodies, defined, 29

Polar, defined, 200

Pollination

outcrossing, 9

selfing, 9, 33–34

Polyacrylamide gel electrophoresis, two-dimensional (2-D PAGE), 207–208

Polycistronic mRNAs, 135

Polygenic traits. *See* Complex traits

Polylinkers, 152

Polymerase chain reaction (PCR), **154–159**

contamination concerns, 158

degenerate primers, 158–159

DNA purification tools, 220–221

as genotyping tool, 161

as molecular anthropology tool, 68

primers, designing, 155–156

procedures, 154–155, *155,* 156–158, *156*

reverse-transcription, 159

Polymerases

defined, 221

RNA, *133,* 134–135

Polymerization, defined, 43

Polymers

defined, 112, 198

DNA and RNA, formation, 117–118, *118*

Polymorphic markers, 6, 7

Polymorphisms, DNA, **159–163**

as agricultural biotechnology tool, 162

cause of, 160

defined, 1, 82, 146, 205

as disease study tool, 162

distinguished from multiple alleles, 82

in DNA profiling, 163

gel electrophoresis tools, 160–161, *161*

genotyping, 160–161

origin of term, 159

PCR tools, 161

public health role, 216

RFLPs, 6, 13–14, 161–162, *161*

selectively neutral, 160

STRPs, 6, 14

as taxonomy tool, 163

VNTRs, 6, 14, 162

See also Alleles, multiple; Short tandem repeats (microsatellites)

Polymorphisms, single-nucleotide (SNPs)

in DNA comparison studies, 65

in human genome, 162

in linkage analyses, 6, 14

as pharmacogenomics tool, 146–147

Polypeptide backbone, post-translational modifications, 179, *179,* 181

Polypeptides

defined, 105, 188

heavy and light, in antibodies, 202

one gene-one polypeptide model, 103

Polyploidy, **163–167,** *164*

allopolyploid, 165–166

in amphibians, 165

autopolyploid, 165–166

causes, 163–164

genetic analysis of, 166

in humans, 164–165

mutations, consequences, 97

in plants, 165–166, *165*

and sex determination, 164

See also Aneuploidy

Polyspermy, consequences, 163–164

Population

of Bombay, *173*

defined, 171, *172*

Population bottlenecks, **167–171**

ancient population sizes, 167–168

colonization and climatic, *169*

and expansions, human evolution, 168–169

technological and social influences, 169–170

Population genetics, **171–174**

disease prevalence studies, 194–195

gene pools, 171–172

genetic drift, 63, 174

genetic structure, 172

Hardy-Weinberg equilibrium, 171–174, 195

mutation role, 93

Population screening, **175–178**

for bipolor disorder, 215

for carriers, 176, 177

criteria for, 175

vs. diagnostic tests, 175

ethical issues, 177–180

for hypothyroidism, 176–177

for inherited disorders, 174, 175–176

public health role, 217–219

for sickle-cell disease, 177, 219

See also Newborn screening

Population studies. *See* Family studies/family histories; Population genetics

Positive selection marker systems, 16–17

Post-translational control, **178–182,** 203–204

amino acid alterations, 178–179, *180,* 205–206

analysis of protein structures, 71

detection tools, 20

exteins, *179,* 181

inteins, *179,* 180

polypeptide backbone alterations, 179, *179,* 181

PPD skin test, for tuberculosis, 175

Predisposition (susceptibility) to disease. *See* Genetic predisposition

Prednisone, for muscular dystrophies, 87

Pregnancy

miscarriage, and prenatal diagnoses, 182, 184

spontaneous abortions, 165

terminating, 187

in vitro fertilization, 182, *184,* 185–187

See also Prenatal diagnosis

Preimplantation genetic diagnosis (PGD)

ethical issues, 186–187

uses, *184,* 185–186

Prenatal diagnosis, **182–187**

alpha fetoprotein test, 185

amniocentesis, 183, 184

chorionic villus sampling, 184

chromosome painting, 183

cytogenetics analysis, 182–185

ethical issues, 186–187

fetal cell sorting, 184–185

genetic counseling, 185, 186

Prenatal diagnosis (continued)
    maternal serum marker screening, 185
    preimplantation genetic diagnosis, 182, *184*, 185–186
    in situ hybridization tools, 182
    ultrasound scans, 182, *183*, 185
Prenatal diagnosis, specific diseases
    Alzheimer's disease, 186
    cystic fibrosis, 186
    Down syndrome, 185
    dwarfism, 186
    Fanconi anemia, 186
    hemophilia, 186
    metabolic diseases, 183–184, 186
    muscular dystrophies, 85, 86, 184, 186
    neural tube defects, 184, 185
    Tay-Sachs disease, 184
    X-linked disorders, 186, 187
Primates
    chimpanzees, genetic diversity, 167–168, *168*
    DNA, compared to human, 63, 64–65
Primer nucleotides, *155*
    degenerate, 158–159
    in PCRs, 155–157
PRINTS database, 209
Prion protein (PrP), conformation conversions, 188–189, *189*, 190, *190*
Prions, **187–190**
    bovine spongiform encephalopathy, 188, 189
    Creutzfeldt-Jakob disease, 188, 189–190
    fatal familial insomnia, 190
    Gertsmann-Sträussler-Scheinker syndrome, 189–190
    kuru, 189
    origin of term, 188
    protein conformation conversions, 188–189, *189*, 190, *190*
    scrapie, 188, 189
    structure, 188
    transmissible spongiform encephalopathies, 188–190
Privacy, **190–193**
    Directive on Data Protection, 192
    DNA samples, justice system, 192–193
    genetic discrimination, 4, 146, 177–178, 191–192, 193, 216
    genetic testing and, 190
    of medical records, 191–192
    of pedigrees, 142
    pharmacogenomics and, 146

Probability, **193–196**
    BLAST, 196
    Chi-square test, 194–195
    Hardy-Weinberg equilibrium, 172–173, 173–174, 195
    in linkage analyses, 6–8
    in Mendelian genetics, 31, 34–35
    p values, 194–196
    Punnett squares, *33*, 35, 193–194
Probands
    defined, 140
    symbol for, *139*
Processed pseudogenes (retropseudogenes), 211–212
Professions
    cellular biologist, 70
    clinical geneticist, 193–194
    computational biologist, 196
    cytologist, 21
    evolutionary biologist, 70
    genetic counselor, 186
    laboratory technician, **1–3**
    microbiologist, **50–51**
    molecular biologist, **70–72**
    patent attorney, 138
    pharmaceutical scientist, **142–144**, *145*
    physician scientist, **147–148**
    plant genetic engineer, **149–150**
    statistical geneticist, 193, 194–195
Progeny, defined, 127
Programmed cell death. *See* Apoptosis
Prokaryotes
    defined, 16, 135
    operons, 105, **131–135**
    overlapping genes, **135–136**
    See also Eubacteria; Plasmids
Proline (Pro), chemical properties, 200
Prometaphase, 59
Promoter DNA sequences
    defined, 47, 100, 134
    DNA methylation role, 47–48, *48*, 49
    location of, 118
    mutation rates, 100
    of operons, 132, 134
Prophase
    meiosis, 25–26
    mitosis, 59
Propionic acidemia, symptoms and treatment, 40, 43
Propionyl-CoA, 43
Pro-proteins (zymogens), polypeptide cleavage by, *179*, 181

Prosecution application procedure, for patents, 138
PROSITE database, 209
Prostate gland, cancer, 175
Prostatic antigen screening (PSA), 175
Proteases
    N-terminal signal sequence activation, 181
    prion disease role, 188–189, 190
    role in ubiquitination, 179
Protein domains
    noncovalent interactions, 201
    and tertiary structure, 201, *203*
Protein evolution. *See* Evolution, of genes
Protein folding (secondary structure)
    α-helices, 188–189, 200–201, *203*
    β-sheets, 200–201, *203*
    by chaperones, 203
    coiled-coil motif, 201
    helix-loop-helix motif, 201
    hydrogen bonding, 200–201
    prions and, **187–190**
Protein sequencing, **196–198**
    BLAST applications, 196
    carboxypeptidase tools, 198
    databases, 18, 198, 208–209
    defined, 197
    Edman degradation technique, 197
    HPLC tools, 197–198
    mass spectrometry tools, **18–20**, 198
Protein splicing. *See* Splicing, proteins
Protein structure, *197*
    C and N termini, 181, 197–198, 207
    conformational changes, 204
    disulfide bridges, 200, 201
    peptide bonds, 200
    primary, 197, 200
    quaternary (multimeric), 202, *203*
    techniques for studying, 2
    tertiary, 201, *203*
    See also Protein folding (secondary structure)
Protein synthesis. *See* Translation
Proteinaceous infectious particles. *See* Prions
Proteinase K, 221
Proteins, **198–204**
    apparent mass, 207
    degradation of, 197–198
    digestion roles, 197
    DNA repair role, 197

drugs that target, 206–207

as enzymes, 103

immune system role, 197

induced mutagenesis to study, 89, 90–92

isoelectric points, 207

isoforms, 160

mass, 207

methyl-binding, 48

one gene-one protein model, 103

signal transmission role, 103, 197

structural roles, 103, 197, 199

See also Post-translational control; Proteomics; *specific proteins and enzymes*

Proteomes

databases, 209

defined, 205

distinguished from genomes, 205

Proteomics, 204, **205–209**

2-D PAGE tools, 207–208

bioinformatics tools, 205

challenges, 205–206

defined, 124, 198

gel electrophoresis tools, 206

human nucleoli analysis, 124

mass spectrometry tools, **18–20**, *206*, 208

yeast two-hybrid system, 209

Proteosomes, ubiquitination, 179, *180*

Protista, genome characteristics, 57

Proto-oncogenes

activation by retroviruses, 127

activation without retroviruses, 128–129

Bcr, 129

c-Abl, 129

transcription factors as, *128*, 130

Protozoans, as model organisms, 61

Prusiner, Stanley, 187–188, *188*

PSA (prostatic antigen screening), for prostate cancer, 175

Pseudogenes, **209–213**

distinguished from functional genes, 209–210

evolution of, 211–213

globin gene family, 211, 212–213, *212*

mutations, 209–212

nonprocessed, 210–211, *210*

paralogous, 210

processed, *210*, 211–212

Pseudohypertrophic muscles, 84

Psychiatric disorders, **213–216**

ADHD, 213, 215

alcoholism, 213

autism, 213

bipolar disorder, 213, 214, 215

genetic components, 213

obsessive-compulsive disorder, 213

panic disorder, 213

schizophrenia, 213, 214–215

social phobia, 214

Tourette's syndrome, 214, 215–216

See also Alzheimer's disease; Huntington's disease; Mental retardation

Public health, **216–220**

antibiotic resistance concerns, 217

approaches in genetics, 216–217

communication and information role, 219

epidemiology studies, 217–218

genetic testing role, 218

Human Genome Project and, 215

population intervention role, 218–219

population screening, **175–178**

surveillance activities, 217

Puffer fish, as model organism, 61

Punnett, Reginald, *33, 35,* 36

Punnett squares, *33, 35,* 193–194

Pure-breeding lines, Mendel's experiments, 30

Purification of DNA. *See* DNA purification

Purines

and DNA structure, 44, 118–119

metabolism disorders, 40, 44

structure, 115

transition mutations, 95

See also Adenine; Guanine

Pyrimidines

dimers, as mutagens, 88–89

and DNA structure, 44, 118–119

metabolism disorders, 40, 44

structure, 115

transition mutations, 95

See also Cytosine; Thymine; Uracil

Pyrococcus, inteins, 181

Pyruvate, 44

Pyruvate dehydrogenase deficiency, 54

Q

Quantitative (continuous) traits, in maize, 9

R

Racial and ethnic differences

blood types, 172

genotype frequencies, 172

molecular anthropology tools, 66, 67

in pedigrees, 141–142

See also Eugenics

Radiation, defined, 81

Radiation genetics, 81

Radiation, ionizing

and cancer, 128

and chromosomal aberrations, 99

and DNA damage, 89

dose-mutability relationships, 103

Muller's research, 81, 88, 90, 91–92, 103

Radiation, ultraviolet

and cancer, 128

and DNA damage, 88–89

Ragged red muscle fibers, 53, *53*

Ras genes and proteins, as oncogenes, 128, 130

RB gene, and retinoblastomas, 130

In re Application of Bergy, 137

In re Kratz, 137

Reading frames

frameshift mutations, 88, *95*, 96

overlapping genes, **135–136**

Recessive alleles

defined, 73, 90, 193

maize, 9

Mendel's hypotheses, 31, 33, 35–36

mutations as, 90, 93

in polyploids, 166

X-linked inheritance, 73–74, 91–92

Recessive disorders, autosomal

carrier testing for, 177

Leigh's disease, 54

loss-of-function diseases, 96

metabolic diseases, 38–39, 45

muscular dystrophy, limb-girdle, 86

stop codon mutations, 96

See also Cystic fibrosis; Phenylketonuria; Sickle-cell disease

Recombinant DNA. *See* DNA, recombinant

Recombinase enzymes, 17

Recombination. *See* Crossing over

Recombination fraction, 6–7, 12–13

Refsum disease, 41

Relaxosomes, *151*

Repetitive DNA sequences. *See* DNA repetitive sequences

Replication
cytokinesis, 29, 59
defined, 46, 60
fission, bacterial, 152
of prions, 187–190
random, of mitochondrial and chloroplast genomes, 56
See also DNA replication; Meiosis; Mitosis

Replicons, 151

Reporter genes. *See* Screenable markers

Reproductive technology, *in vitro* fertilization, 182, *184,* 185–187

Resistance to antibiotics. *See* Antibiotic resistance

Restriction endonuclease enzymes
as biotechnology tool, 114
defined, 152
DNA methylation role, 47
function, 114
RFLPs detected by, 13, 161–162

Restriction fragment length polymorphisms (RFLPs), 6, 13–14, 161–162, *161*

Retinoblastomas
mutation rates, 99
PML body mutations, 124
RB gene mutations, 130
two-hit theory, 131

Retroposons, 211–212

Retropseudogenes (processed pseudogenes), *210,* 211–212

Retroviruses (RNA viruses)
proto-oncogene activation, 127
simian sarcoma (SSV), 130

Rett syndrome, MeCP2 protein, 48, 49, 78

Reverse transcriptase. *See* Transcriptase, reverse

Ribonuclease enzymes
function, 112, 114–115
tertiary structure, 201

Ribonucleic acid. *See* RNA

Ribose sugars, structure and function, 115

Ribosomes
in nucleus, 119, 125
structure and function, 70–71
See also RNA, ribosomal

Ribulose 1,5-biphosphate carboxylase, 199

Rice (*Oryza sativa*), genetically engineered, 149

RNA (ribonucleic acid)
polymer formation, 117–118, *118*

structure, 115
susceptibility to cleavage, 115

RNA, 7SL, 211–212

RNA, messenger (mRNA)
coding for, 105
correlation to protein levels, 206
function, 70, 114
impaired, and metabolic disease, *39*
introns lacking in, 124, 211
maturation and degradation, 114–115
polycistronic, 135
processed pseudogenes and, 211

RNA polymerases
defined, 134
operon role, *133,* 134–135

RNA, ribosomal (rRNA)
coding for, 105
human, amount, 123
in mitochondria, 53, 55–56
processing, 119–120
as pseudogenes, 211
synthesis, 119
transcription of, 123–124

RNA sequencing, mass spectrometry tools, **18–20**

RNA, small nuclear (snRNA), 211

RNA, transfer (tRNA)
in mitochondria, 53, 55–56
as pseudogenes, 211

Robson and Aurbach, mustard gas as mutagen, 90

Rodent models
genome characteristics, 61–62
as model organisms, 60–61
mouse *vs.* human chromosomes, *96, 97*
mutation rates, 99
PrP gene knock outs, 188
transgenic mice, 62

Roundworms (*Caenorhabditis elegans*)
behavior studies, 61
as model organism, 60, 61

Rous, Peyton, 127

Rowley, Janet, 127

S

Saccharomyces sp. *See* Yeasts

Salmonella typhimurium, Ames test, 92

Salt tolerance, transgenic plants, 149

Sanfilippo syndrome, symptoms and treatment, 41

Sarcolemma, muscular dystrophy and, 84–85

Satellite sequences. *See* Short tandem repeats

Schizophrenia, 213–215

Scientist, defined, 147

Scrapie, 188, 189

Screenable markers (reporter genes), 16, 17–18, 91

SDS (sodium dodecyl sulfate), 208, 220

Sea slugs and urchins, as model organisms, 60

Selectable markers
gene cloning tools, 152
negative selection, 16
polymorphisms as, 162
positive selection, 16–17

Selfing (self-pollination)
of maize, 9
in Mendel's experiments, 33–34

Senility. *See* Alzheimer's disease

Sequencing. *See* DNA sequencing; Protein sequencing; RNA sequencing

Serial endosymbiotic theory, 56–57

Serine (Ser)
chemical properties, 200
glycosylation, 179
phosphorylation, 178

Serum albumin, post-translational control, 181

Sex chromosomes. *See* Chromosomes, sex

Sex determination
intersex organisms, 77, 164
by prenatal testing, 187
X chromosomes and, 76, 77, 106
Y chromosomes and, 65, 76, 106

Sex-linked inheritance
X-linked, 73–74
Y-linked, 65–66
See also X-linked disorders

Sexual development, normal human, 65

Sharp, Lester, 21

Short tandem repeats (microsatellites), *160*
gene evolution role, 65
as molecular markers, 13–14
multiple alleles, 82–83
as polymorphisms, 162
pseudogene creation, 210
VNTRs, 6, 14, 162

Shull, George Harrison, 10

Sickle-cell disease
asymptomatic carriers, 177
cause, 153, 200
inheritance patterns, 142

population screening, 177, 219

symptoms, 153

as transversion mutation, 95

Signal transduction

cancer role, *128*, 129–130

N-terminal signal sequences, 181

pleiotropic effects on, 153–154

in protein synthesis, 202–203

Silent mutations, 93, 96, 99

Simian sarcoma retrovirus (SSV), and cancer, 130

Simple tandem repeat polymorphisms (STRPs), 6, 14

Single-nucleotide polymorphisms (SNPs). *See* Polymorphisms, single-nucleotide

Slime mold (*Dictyostelium discoideium*), as model organism, 60

Small nuclear ribonucleoproteins (snRNP), 125

Smith, Michael, 91

SMN (survival of motor neurons) protein, 124

Smoking (tobacco)

addiction, inheritance patterns, 214

mutagens in smoke, 88, 89

SNPs. *See* Polymorphisms, single-nucleotide

Social phobia, genetic components, 214

Sodium dodecyl sulfate (SDS), 208, 220

Somatic cells

diploid number, 24, 60, 163

mutations, consequences, 93

mutations, detecting, 93

replication of, 58–59

Somatic, defined, 24, 58, 93, 163

Southern blotting, defined, 221

Speciation

chimpanzees, 167–168, *168*

H. erectus distinguished from Neandertals, 66–67, *66*, 69

Speckles (interchromatin granules), 120, 125

Sperm

developmental processes, 29

as haploid, 24, 60, 163

meiosis and, **24–29**, 60, 75

mutations, 99

polyspermy, 163–164

SRY gene and, 76

Spermatocytes, primary, 24

Spinal muscular atrophy, 124

Spindle apparatus

colchicine's impact on, 166

structure and function, 26, 59

Spindle checkpoint errors, nondisjunction, 111–112

Spleen disorders, Gaucher disease, 40, 45

Splicing, alternative

amount in human genome, 205

function, 106

mechanisms, 181

Splicing, proteins, post-translational control, 181

Splicosomes, 124

Spontaneous mutations, 90, 100

SRY gene, 76

SSV (simian sarcoma retrovirus), and cancer, 130

Start codons (transcription), overlapping genes, 135

Statistical geneticists, 193, 194–195

Statistics

in linkage and recombination studies, 7–8, 12–13, 15

See also Probability

Stevens, Nettie, 74

Stone ages, population bottlenecks, 169–170

Stoneking, Mark, 67

Stop codons (transcription)

lacking in pseudogenes, 209

mutations, consequences, 96

overlapping genes, 135

STRPs (short tandem repeat polymorphisms), 6, 14

STRs. *See* Short tandem repeats

Sturtevant, Alfred Henry, 7, 32, 73, 75, 76, 80

Substance abuse. *See* Addiction; Alcoholism; Smoking (tobacco)

Substrates, enzyme binding to, 37

Succinylcholine, adverse reactions to, 144

Succinyl-CoA, 43

Sugars

deoxyribose, 115, *116*, *117*

galactose, 134

glycosylation of amino acids, 178–179

lactose, *lac* operons and, 131–135

ribose, 115

Sungene Technologies Laboratory, *150*

Survival of motor neurons (SMN) protein, 124

Sutton, Walter, 73, 74, 102

SWISS-PROT database, 209

Symbionts

defined, 52, 55

endosymbionts, 52, 55–57

Syntenic genes, 5

Synthesis, defined, 132

Systematics. *See* Taxonomy

T

Tandem repeats. *See* Short tandem repeats

Taq DNA polymerase, 157

Tatum, Edward, *102*, 103

Taxonomy (systematics)

molecular anthropology role, 63–70

polymorphisms as tools, 163

Tay-Sachs disease

in Ashkenazic Jews, 174

genetic testing, 174, 177, 184

symptoms and treatment, 40, 45

Telomeres

defined, 121

structure and function, 121

Telophase

I, meiosis, 26

II, meiosis, 27

mitosis, 59

Templates

defined, 155

for PCR reactions, 155–158

Tenets, defined, 177

Termination sites. *See* Stop codons (transcription)

Testes, function, 24, 60

Tetrahymena, as model organism, 61

Tetraploidy, consequences, 97

Thale cress (*Arabidopsis thaliana*)

characteristics, 60

as model organism, 149

transposon mutagenesis, 97–98

Thermus aquaticus, 157

Thiogalactoside acetyltransferase enzyme, 132, 135

Thioguanine, 93

Threonine (Thr)

glycosylation, 179

phosphorylation, 178

Thymidine kinase (TK), in selectable markers, 17

Thymine

and DNA structure, 94

mutagenic base analogs, 87

structure, 115, *116*, *118*, 119, *119*

See also Base pairs

Time-of-flight (TOF) tubes, 20

Timofeeff-Ressovsky, Nikolay, 103

Tissue donors, legal ownership concerns, 3–4

TK (thymidine kinase), in selectable markers, 17

Toba volcano, 168

Topological, defined, 114

Tourette's syndrome
genetic components, 214
OCD and ADHD co-morbidity, 215
symptoms, 215

Traits. *See* Phenotypes

Transcriptase, reverse, defined, 159

Transcription
chromatin condensation/decondensation, 25
defined, 60, 70, 152
directionality, 118, 155–156
DNA methylation and, 48
in mitochondria, 56
of operon genes, *133*, 134–135
procedure, 70
RNA polymerases, *133*, 134–135
of rRNA genes, 123–124
start codons, 135
stop codons, 96, 135, 209

Transcription factors
c-*Fos*, 130
c-*Jun*, 130
defined, 48, 203
function, 199
as oncogenes, *128*, 130

Transduction. *See* Signal transduction

Transfer (*tra*) genes, of plasmids, 151, *151*

Transformation
in eubacteria, 152
marker systems, **15–18**

Transgenes
defined, 17
in marker systems, 15–18

Transgenic animals. *See* Knock-in mutants; Knock-out mutants

Transgenic technology, patent issues, 136–138

Transition mutations, 95

Translation (protein synthesis)
defined, 60, 71, 135, 152
procedure, 70–71
signal sequences, 202–203
See also Post-translational control; Proteins

Translocation
and chromosomal aberrations, 96–97, 127
defined, 81
and proto-oncogene activation, 128–129

in pseudogenes, 211

signal sequencing role, 202

See also Crossing over (recombination)

Transmissible spongiform encephalopathies (TSE), 188–190

Transplantation
organ, T cell-MHC interactions, 82
organ, to treat metabolic diseases, 39, 45
tissue donors, legal issues, 3–4

Transposable genetic elements (transposons)
defined, 47, 64
DNA methylation role, 47–48
gene evolution role, 97
gene expression role, 211–212
and hemophilia, 97
in human genome, 64, 97
McClintock's hypothesis, 10, 21–22
as mutagens, 91, 97
in primate genome, 64
as vectors, 91, 97–98

Transversion mutations, 95

Triplet code. *See* Codons; Genetic code

Triploidy, *164*
in animals, 164–165
defined, 164

Trisomy
chromosomal mosaicism, 79
consequences, 109, 111
defined, 10, 79
See also Down syndrome

Tristan da Cunha islanders, inbreeding, 174

TRNA. *See* RNA, transfer

Tryptophan (Trp), operons, 132, 134

Tschermak, Erich, 102

TSE (transmissible spongiform encephalopathies), 188–190

Tuberculosis, PPD skin test, 175

Tumor suppressor genes
discovery, 130
mutations of, 98
NF1 gene, 130, 153
p53 gene, 98, 130
as recessive trait, 130
sequential activation of, 131

Tumors
unregulated cell growth, 153
See also Cancer; Oncogenes

Turner's syndrome, cause and symptoms, 111

Twins, fraternal
bipolor disorder, 215
schizophrenia, 215

Twins, identical
bipolor disorder, 215
mosaicism, 79
schizophrenia, 215

Two-dimensional polyacrylamide gel electrophoresis (2-D PAGE), 207–208

Tyrosine (Tyr)
breakdown, alkaptonuria, 43
phosphorylation, 178

Tyrosine kinase, 129

Tyrosinemia, symptoms and treatment, 40

U

Ubiquitin, 179

Ubiquitination, of proteins, 179, *180*

Uid A (B-glucoronidase gus A) enzymes, 18

Ultrasound scans, 182, *183*, 185

Uracil
structure, 115, *116*, *119*
See also Base pairs

Urea, formation disorders, 41, 45

Uric acid, and gout, 44

V

Vaccines, developing, 71

Valine (Val)
chemical properties, 200
substitution, sickle-cell disease, 200

Variable number of tandem repeats (VNTR) analysis, 6, 14, 162

Vasopressin, post-translational control, 181

Vectors
defined, 91, 152
plasmids as, 91, 152
for site-directed mutagenesis, 91
transposons as, 91, 97–98

V-*erbB* oncogene, 130

V-*fms* oncogene, 130

Viruses
adenovirus, 124
bacteriophages, *104*, 105, 131
and cancer, 127, 130
cytomegalovirus, 124
DNA methylation to combat, 47–48
infection through nuclear pores, 126
overlapping genes, **135–136**
RNA (retroviruses), 127, 130

Vision, color, 78

Vitamins
A, transgenic rice, 149
metabolic role, 37, 39

VNTR (variable number of tandem repeats) analysis, 6, 14, 162

Vogelstein, Bert, 131

Volcanic winter, and population bottlenecks, 168

V-*sis* oncogene, 130

W

Watermelons, seedless, 165, *165*

Watson, James, DNA structure research, 105

Web sites. *See* Internet

Weinberg, Robert, 127

Weissmann, Charles, 188

Weldon, Walter F. R., 102

Wheat, polyploid, 166

White, Raymond, 130

Wigler, Michael, 127

Wild-type alleles/traits, defined, 100

Wilson, Allan, 67

Wilson, Edwin, 73

Wright, Sewall, 103

X

X chromosomes
aberrations, detecting, 183
fruit fly, 7, 73, 164
human, 106, 164
monosomy, 111

random inactivation of, 77–78, 111
trisomy, 111
See also Mosaicism

X inactive specific transcripts (*XIST*) gene, 77

Xenopus laevis (African clawed frog), as model organism, 60, *61*

X-linked disorders
assays for, 93
Klinefelter's syndrome, 111
Leigh's disease, 54
in males, 78
muscular dystrophy, Duchenne, 78, 83–84
ornithine transcarbamylase deficiency, 45
preimplantation genetic diagnosis, 186, 187
Rett syndrome, 48, 49, 78
Turner's syndrome, 111
See also Hemophilia; Mosaicism

X-linked inheritance, fruit fly studies, 73–74, 91–92, 102

X-ray diffraction, 71

X-rays
dose-mutability relationships, 103
as mutagens, 81, 88, 90–92, 97, 103, 166–167

Y

Y chromosome Adam, 67–68

Y chromosomes

aberrations, detecting, 183
fruit fly, 164
genome characteristics, 76
human, 106, 164
as indicators of human evolution, 66, 168
lack of recombination, 65
monosomy, 111
trisomy, 111

Yangtze River dolphin, 167

Yeast Proteome Database, 209

Yeast two-hybrid system, 209

Yeasts (*Saccharomyces* sp.)
DNA methylation lacking, 46
inteins, 181
as model organism, 60

Z

Zea mays. *See* Maize (corn)

Zebrafish (*Brachydanio danio*), as model organism, 60, 61

Zimmer, Karl, 103

Zygotes
defined, 109
as diploid, 24–29
monosomic, 109, 111
mtDNA in, 51
trisomic, 109, 111

Zymogens (pro-proteins)
polypeptide cleavage by, *179*, 181